THE DEEP SKY

An Introduction

Other *Sky & Telescope* Observer's Guides

Astrophotography: An Introduction by H. J. P. Arnold
City Astronomy by Robin Scagell
Meteors by Neil Bone

Also by Philip S. Harrington

Astronomy for All Ages (with Edward Pascuzzi)
Eclipse!
Short Bike Rides In and Around New York City
Star Ware: The Amateur Astronomer's Ultimate Guide to Choosing, Buying, and Using Telescopes and Accessories
Touring the Universe Through Binoculars

THE DEEP SKY

An Introduction

PHILIP S. HARRINGTON

Series Editor Leif J. Robinson

Sky Publishing Corporation
Cambridge, Massachusetts

Dedication

For my parents, Frank and Dorothy

For the many years of love and support that you have given me, and for buying me my first good telescope.

In memoriam

Leland S. Copeland (1886–1973)
Walter Scott Houston (1912–1993)
This century's deepest deep-sky observers

Published by Sky Publishing Corporation
49 Bay State Road, Cambridge, MA 02138

ISBN 0-933346-80-8

Library of Congress Cataloging-in-Publication Data

Harrington, Philip S.
The deep sky: an introduction / Philip S. Harrington
p. cm. — (Sky & Telescope observer's guides).
Includes bibliographical references and index.
ISBN: 0-933346-80-8 (pbk. : alk. paper)
1. Astronomy — Observer's manuals. I. Title. II. Series.
QB63.H318 1997 95-22380
523.8 — dc20 CIP

Edited by John Woodruff
Layout and design by Efrat Metser
Produced by the Maple Vail Book Manufacturing Group
Printed and bound in the United States of America

Contents

Foreword

"There it was. This was easier than I thought!" That's the 40-year-old entry in my observing log; I had just spotted M83, the last target needed to sweep Messier's list. I remember feeling fulfilled, though with a simultaneous tinge of emptiness. As a 15-year-old, I had finished what I had set out to do — see all the Messier objects within a year of getting my first real telescope. Its wooden-slat tube, cast-iron head, and sturdy tripod had been given to me by the Los Angeles Astronomical Society only after I had promised to use it. My parents provided the 6-inch f/8 mirror from Cave Optical.

But now what? What could I do next? I really didn't know, hence the empty feeling. In the mid-1950s there weren't many user-friendly guides to the deep sky. William Olcott's *Field Book of the Skies* had already celebrated its silver anniversary, and it would be several years before Dover reprinted Thomas Webb's classic *Celestial Objects for Common Telescopes*. At least among my circle, William Smyth's parallel classic *Bedford Catalogue* was virtually unknown; it would not be reprinted until 1986. I guess we amateurs of that era weren't very inventive, for we could have compiled our own lists of challenges beyond Messier. We had one ready source: the *New General Catalogue*, which I recall could be purchased quite cheaply. Yet few beyond the legendary Walter Scott Houston systematically sampled its treasures.

How lucky we are today to have expert guides like Phil Harrington to lead us across the skies to wonders unexpected. Reading his manuscript, I was impressed with his sense of what's interesting, what's worth ferreting out. Harrington, literally, has seen it all! Beyond his selection of captivating nebulae, clusters, galaxies, and multiple stars are his STAR objects, those in his Small Telescope Asterism Roster. Harrington compiled this offbeat list from his own accidental encounters as well as those of others. I've checked out a few, and I'll guarantee they will spice up any evening under the stars.

Perhaps I've been too hard on the amateurs of my youth. I think the reason we didn't dig further into the deep sky was that everyone was a telescope maker of one stripe or another. At that time commercial telescopes were very limited, mainly small refractors from Unitron and medium-size reflectors from Cave. (Six- and 8-inchers were commonplace; 10s and 12s were monsters.) So the homespun tools became the focus of our attention rather than the challenges provided by the sky itself. We went back again and again to the same handful of sights, using them simply to test one scope's performance against another's.

In that context we can perhaps be excused for nearsightedness. And there was another stumbling block to accessing the deep sky — the lack of good charts. This came home to me instantly one night. A supernova had appeared in some wisp of a galaxy in the Virgo Cluster. I tried to find this new critter with my friend Claude Carpenter's 18-inch. I carefully calibrated the setting circles, pointed the telescope, and looked. The field was awash with faint galaxies, as was any adjacent field I turned to! Unfortunately for me, over 90 percent of those ghostly island universes weren't plotted on the best atlas of the day.

That was the original *Skalnate Pleso Atlas of the Heavens*. Until that frustrating evening I had thought it was the cat's pajamas. In vivid black and white it showed stars to 7¾ magnitude and a couple of thousand deep-sky objects as well. That was the epitome then. Please don't laugh too loudly; you'll wake up the neighbors! Only in the 1960s would a true 9th-magnitude star atlas become available to amateurs, but it didn't have any deep-sky objects! You had to plot each one yourself (after precessing the NGC's epoch 1860 coordinates — by hand, of course, since the pocket calculator hadn't been invented). Is it any wonder that profound encounters with the deep sky were largely accidental?

Wasn't astronomy "fun" back then? In fact, it was! We all learned a lot — we *had* to, or we couldn't play the game. But I believe, in many respects, that we were better off than amateurs today who depend on software to find their way around the sky. We knew the basics of "practical astronomy," a discipline now sadly obsolete on campus and at home. Solid-state telescope accessories, nebula filters, and desktop computers lay a generation in the future.

Of course, the future is now. Most of us live in urban, light-polluted environments, and technology may provide our only means for enjoying the night sky. A computer-guided scope with pinpoint accuracy may be the only means for finding objects visually — the subtlest object is much easier to see if you *believe* it is actually in the field of view. No matter, I'm sure that a decade from now we will find CCDs used as often as today's "supereyepieces." The neat thing about Harrington's book is that it has something for everyone's equipment arsenal, from binoculars to big Dobs. Read his text. Then go out and blast away, armed with whatever's at hand or whatever it takes.

Leif J. Robinson
Editor in Chief, *Sky & Telescope*

P.S. So, what did I end up doing after I had exhausted the Messier catalog? First, I had a love affair with all the nifty double stars I could learn about. This was spectacular fun, whether splitting jobs at the limit of the telescope's capability or looking at wide ones through binoculars. Then came the planets, especially during the fabulously close Mars opposition of 1956 and the appearance of Saturn's great white spot in 1960. Finally, I was seduced by the infinite challenges provided by variable stars; they have remained my favorites ever since.

Preface

Many of us find that to leave bright room and cozy chair for the dark world outside is contrary to nature, like a moth flying from the light. As a bather plunges into cold water, so the sky hunter must immerse himself in darkness before he will find it comfortable in the night. . . . So rich is this nocturnal wonderland that even for the smallest telescopes numerous objects await observation. . . . A larger lens or mirror is not an assured benefit. Devotion and patience are as important as light grasp.

Then outward turn an optic tube
From some high, lonely hill,
That we may glance at cosmic nooks
And marvels rich, until
The morning glow conceals those realms
Where precious things distill,
Far-forth beyond the utmost reach
Of human hope and will.

So wrote Leland S. Copeland in his article "All Night with the Stars" that appeared in the November 1949 issue of *Sky & Telescope*. In this article, he conveyed the wonder felt by amateur astronomers when observing the splendors of the night sky. Copeland was one of his day's most prolific deep-sky observers. Indeed, he originated and penned the Deep-Sky Wonders column in *Sky & Telescope* from 1942 to 1946, when it was passed on to Walter Scott Houston. "Scotty" continued what Copeland began, until his death in December 1993; his last column appeared in the July 1994 issue. These two extraordinary observers and writers are responsible for introducing an uncountable number of amateur astronomers — including me — to the joys of deep-sky observing. Together, they have left us a legacy of more than 50 years of insight and inspiration. It is my hope that, through this book, I can pass on to you some of the enthusiasm and love of deep-sky observing that they gave me over the years.

This book is much more than "just another observing handbook." In fact, a good subtitle for it might be *Deep Sky 101,* for it is my hope that it will serve not only to inspire you, but also to train you to be a deep-sky observer. I can recall observing with a friend in the winter of 1969. I had just gotten my first telescope, a 100-mm (4-inch) reflector, and was anxious to view some of the wondrous objects I had heard about in books and magazines. I quickly found, however, that it takes more than a telescope and youthful enthusiasm to find even the brightest deep-sky treasures. I envied my friend and his 150-mm (6-inch) reflector, for he was able to spy deep-sky objects at what I thought was harrowing speed. “Wouldn’t it be wonderful if I could find things as quickly as he does?” I thought to myself.

After years of persistence, I slowly became oriented to the heavens above. I hope this book helps make your learning progress a little faster. Think of it as a laboratory guide to the deep sky. The first five chapters introduce you to the many types of deep-sky objects that populate our universe, the history and origin of the numerous catalogs that have been compiled over the years, what equipment is needed to see these objects, and how to go about finding them. Chapters 6 through 9 are expanded adaptations of articles I’ve written for *Sky & Telescope* over the years entitled “An Observer’s Guide to . . .” Ordered by season, these chapters describe more than 300 of the finest deep-sky objects down to approximately declination 60°S. Advice is offered on how to find each of them, as well as what they look like through different instruments.

If this is your first experience with deep-sky observing, enjoy it — but have patience. There is a whole universe of amazing sights awaiting your scrutiny. Be sure to savor each new object that you add to your list of conquests. If you are an old hand at deep-sky observing, then I hope you and I can share some sights together. Nothing is more enjoyable than observing with a friend. That was always the message of both Copeland and Houston, and it is one that is timeless.

Clear skies.
Phil Harrington

Acknowledgments

This book would never have come to fruition were it not for the help, advice, and general support offered by numerous people. First, I must thank my proofreaders Susan and Alan French, Dave Kratz, Jack Megas, and Richard Sanderson. Each was a tireless source of help in developing the manuscript as well as in checking the many facts and figures.

Many thanks to Dean Williams, creator of NGP, the New General Program of Nonstellar Astronomical Objects. As mentioned at the end of chapter 5, the NGP is an excellent compilation of observations culled from many amateurs' notes and letters. If you own an IBM-compatible computer, you ought to have a copy of the NGP.

All of the photographs were taken by amateurs like you and me. Their excellent quality serves as testimony to the genius of the amateur astronomer and astrophotographer. My thanks to each of those diligent souls whose work adorns these pages.

I would also like to take this opportunity to thank all at Sky Publishing who have helped make this book a success. I would especially like to single out Sally MacGillivray, who was pivotal in resolving all the fine points that inevitably crop up in such a data-intensive book, and Roger W. Sinnott, creator of the many fine star charts found throughout these pages.

Finally, to my wonderful wife, Wendy. This is our fifth book together, and through it all she has been a continuous source of love, enthusiasm, and encouragement. I always leave it to her careful eye to do the final checking, to help identify my linguistic flaws and save me embarrassment. Both she and our daughter, Helen, understand (and tolerate!) my inborn need to go out in the dead of night, while the rest of the world around us sleeps, and commune with the universe. Their love, above all else, has made this book possible.

CHAPTER 1

Galaxies and clusters and stars — oh my!

The night sky opens up to us beautiful vistas without equal here on Earth. Our wondrous universe is populated with an astonishing variety of celestial objects. Pick up just about any astronomy book and you will find dazzling photographs of these objects in alien spacescapes that will surely stir deep emotions.

Broadly speaking, the universe can be divided into two very general categories: the *shallow sky* and the *deep sky*. The shallow sky is an unofficial term that may be applied to anything that belongs to our solar system, such as the Moon, Sun, and all nine planets and their families of satellites, as well as comets, meteoroids, and asteroids. Members of the deep sky may be said to include any object that lies beyond the solar system, except single stars. These *deep-sky objects* include double and multiple stars, variable stars, star clusters, nebulae, galaxies, and quasars. Let's take a look at each of them in turn.

Double stars

Contrary to appearances, many of the stars we see in the night sky are not single points of light, but are instead two or more closely set stars, such as the example shown in Figure 1.1. These

Figure 1.1. *Albireo, in the summer constellation Cygnus, is one of the northern sky's finest double stars. The brighter star appears yellow, its fainter companion blue. Photo by Edward Pascuzzi.* *(See pages 104-105 for more about Albireo.)*

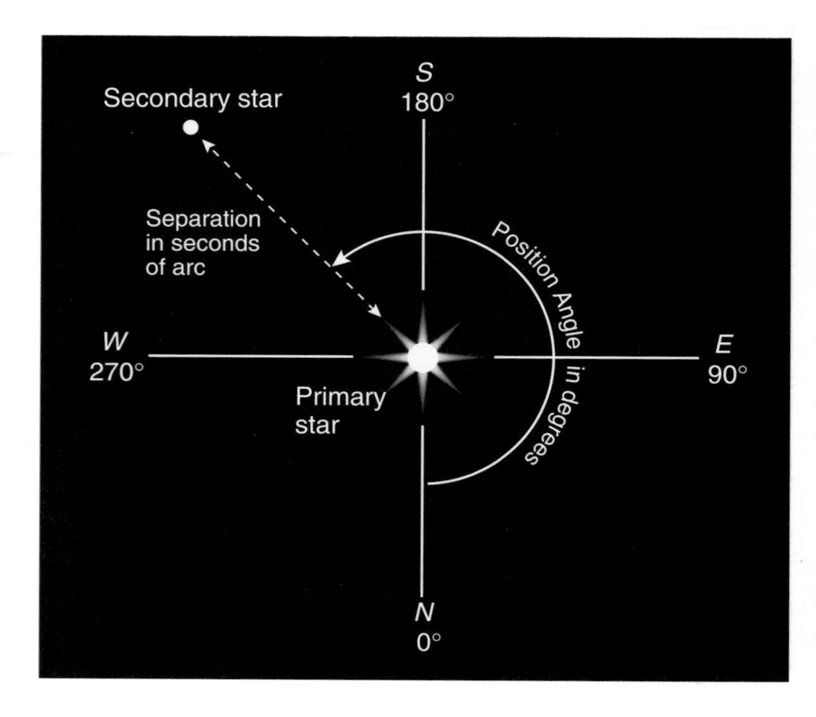

Figure 1.2. *The position angle (P.A.) of the companion in a double-star system is its direction relative to the brighter star, measured counterclockwise from north (0°). Their apparent separation is measured in arcseconds. Here the P.A. is 225°.*

tight stellar families are called either *double stars* or *binary stars,* or, if there are more than two suns, *multiple stars.*

The brightest star in a double-star system is always referred to as the *primary,* or A, star, and the fainter companion is always the *secondary,* or B, star. If there are more than two stars, the others are assigned letters, C, D, and so on (in order of discovery, which is usually, but not always, the order of brightness). For example, our North Star, Polaris, is a double star. Polaris A shines at magnitude 2.0, while its companion, Polaris B, is a dimmer magnitude 9.0. (Actually, Polaris is a quadruple star. Polaris C shines at magnitude 13.0, while Polaris D shines at magnitude 12.0.)

Distances in the sky are normally expressed as angles. There are 90° from horizon to zenith, or from, say, north to east along the horizon. Each of those degrees may be subdivided into 60 equal parts called *arcminutes* (symbol '), each of which may in turn be subdivided into 60 *arcseconds* (symbol "). (Arcminutes and arcseconds are angular measurements and are not related to time.) The gap between two members of a double-star system, called the *separation,* is customarily measured in arcseconds. Of course, in reality the stars may be millions or even billions of kilometers apart.

Astronomers specify the direction of a secondary star relative to

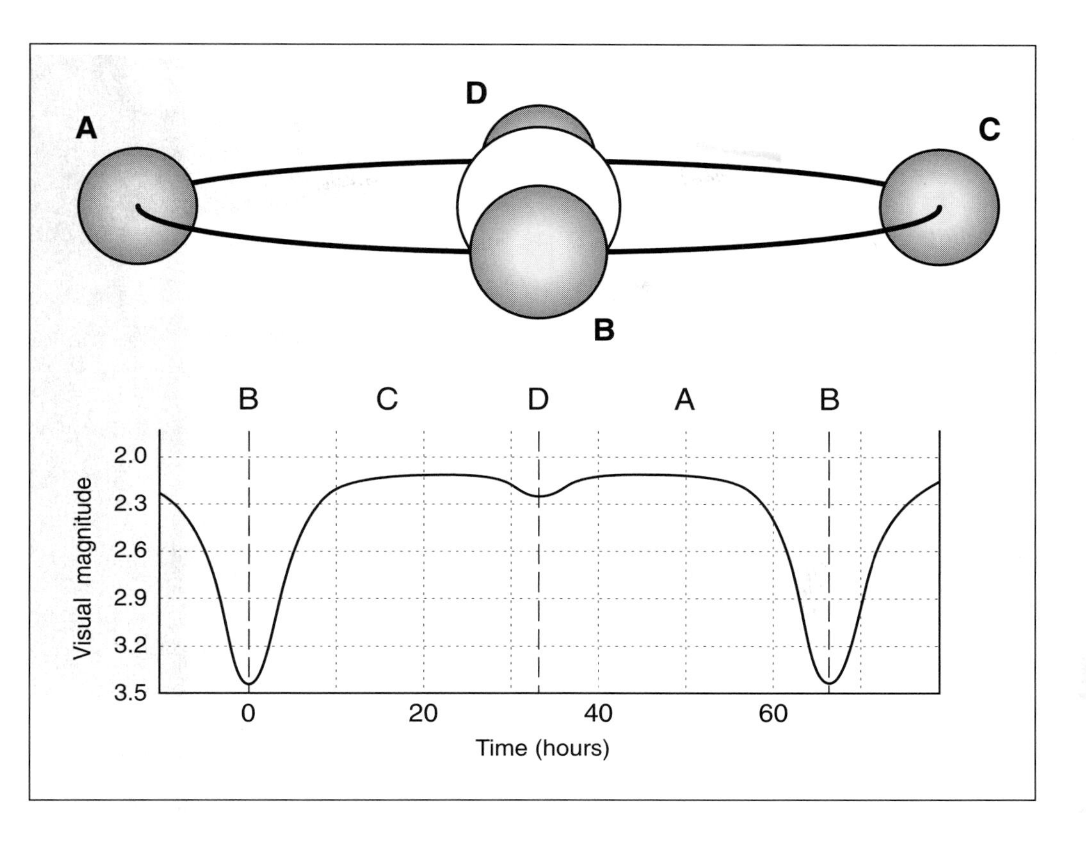

Figure 1.3. *Eclipsing binary stars fluctuate in brightness as the secondary star alternately passes in front of and behind the primary. In this example, brightness diminishes most when the secondary star is in front of the primary (point B).*

the primary star according to its *position angle* (abbreviated P.A. — see Figure 1.2). If the secondary lies due north of the primary, then its position angle is 0°. If it is due east of the primary, it is at P.A. 90°; due south, 180°; and due west, 270°. Intermediate orientations are specified accordingly.

In all double-star systems the two stars orbit around their common center of gravity. If they are sufficiently far apart for them to be resolvable (distinguished in a telescope), the system is called a visual binary. Some doubles are not resolvable through even the largest telescopes, yet we know them to be authentic binaries from analysis of their light with a spectroscope. Even if a star's image appears to be single, astronomers can deduce that a second star is present if the lines in the star's spectrum oscillate in wavelength. Such stars are referred to as *spectroscopic binaries*.

Still other binaries can be detected only by their peculiar behavior. If we happen to be seeing the star system edge on, as with the example shown in Figure 1.3, then the system's total light output will dip

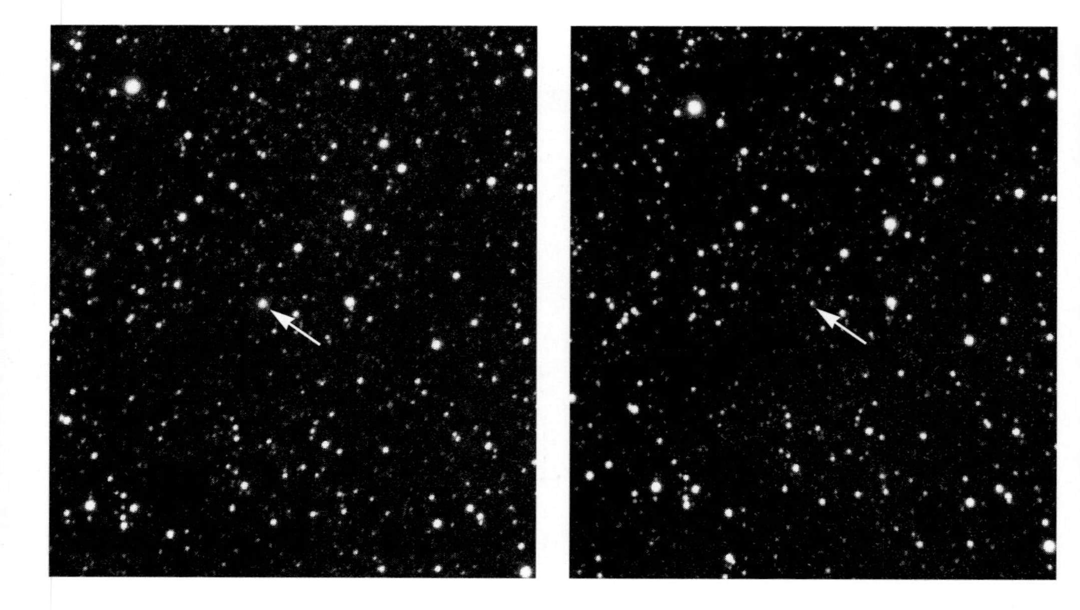

Figure 1.4. *Chi (χ) Cygni in central Cygnus is a Mira-type long-period variable star. John Chumack of Dayton, Ohio, captured it at magnitude 7.1 on June 16, 1993* (left) *and at magnitude 9.2 on August 6, 1994* (right). *North is up.*

every time one star passes in front of or behind the other. These are called *eclipsing binaries*. When both stars are visible side by side, their collective brightness is greatest. As the fainter companion star passes in front of the primary and eclipses it, the system's combined magnitude drops off dramatically. The light output returns to normal when the eclipse ends, only to decrease a second time, but less markedly, when the secondary passes behind the primary.

Then, there are those impostors that appear close to each other but are actually nowhere near each other in space, being merely aligned along our line of sight from here on Earth. These celestial charlatans are called *optical doubles*. The only way to tell a fraud from the genuine article is to study the motions of the two stars for weeks, months, or even years. If orbital motion is observed, or if the stars are found to share a common motion and direction of travel in our galaxy, then the odds are that they form a true pair. If not, they might well be impostors.

The double and multiple stars highlighted in this book are just a small sampling of what each season's sky has to offer. Many display striking contrasts in magnitude; others are nearly equal in brightness. Some seem to shine pure white, while others display tints of blue, yellow, red, and orange — sometimes conspicuously so — and a few pairs present a striking color contrast. Best of all for observers, double stars do not suffer as badly as many nebulae and galaxies do from the detrimental effects of light pollution, making them great targets on nights when the Moon shines bright or for city-based astronomers.

Variable stars

While we normally think of stars as unwavering in nature, many change in brightness over periods that range from, in extreme cases, fractions of a second to several years. Some do so with great regularity, and others behave erratically. Collectively, they are known as variable stars (Figure 1.4). Nearly all variable stars fall into one of three categories, depending on their behavior. The first, eclipsing binaries, has already been discussed under "Double Stars."

The most common type is the *pulsating variable* — a star that actually expands and contracts in size. Long-period pulsating variables typically take several months to complete their cycles, going from maximum brightness to minimum light output and then back to maximum again with fair regularity. The most famous long-period variable is Mira (also known as Omicron [o] Ceti) in the constellation Cetus. At maximum, Mira can attain 3rd or even 2nd magnitude, while at minimum it dips below 10th magnitude. Some pulsating variables are semiregular or irregular, with less of a regular period or none at all.

Other types have comparatively short periods. For instance, RR Lyrae variables, named for the variable RR Lyrae in the constellation Lyra, range from maximum to minimum with a very well defined period of less than a day. Cepheid variables, named for the star Delta (δ) Cephei, also vary rhythmically, with regular periods of one day to several weeks. Both Cepheid and RR Lyrae variables have been instrumental in helping astronomers establish distances within our galaxy, and Cepheids have provided the key to gauging distances to other, nearby galaxies as well.

Cataclysmic variables are stars that spend most of their time shining at a fixed brightness, then suddenly rise by several magnitudes. For instance, U Geminorum stars (named for the prototype star in Gemini) are usually very faint but flare up unexpectedly by many magnitudes in a matter of hours. These stars remain at peak brightness for several days and then slowly fade back to their pre-outburst magnitudes, where they remain for a period that varies unpredictably from days to years. More extreme examples of erupting variables include the explosive novae and supernovae.

A good example of a "backward" erupting variable is the star R Coronae Borealis. It normally shines at around 6th magnitude but fades quickly to about 11th magnitude when it periodically gives off an opaque cloud of soot that masks it from view. Once the cloud has dissipated, the star returns to its normal brightness.

For their research, professional astronomers rely partly on amateurs for observational data of variable stars. To help collect and process these data and disseminate them to professionals, the American Association of Variable Star Observers was established. The AAVSO is an international band of amateurs devoted to the study of variable stars. If you yearn to make a valuable contribution to the science of astronomy, consider joining their legions (the address, together with those of other organizations, is given at the end of the book).

Figure 1.5. *Open clusters come in all shapes and varieties, as you can see from these two photographs. The photo on the left, taken by George Viscome, shows the densely packed open cluster M11 in the summer constellation Scutum. The photo on the right, taken by Chuck Vaughn, is a view of the comparatively loosely packed Pleiades in the winter constellation Taurus. North is up in both photos.*

Star clusters

As well as double and multiple systems, stars exist in larger groupings called clusters. Scattered mostly along or near the Milky Way are stellar congregations called *open clusters,* randomly shaped families of mostly young, hot stars all gravitationally linked to one another. Over a thousand are known, each containing anything from 20 or so to a few thousand stars. All of the stars in a given cluster are about the same age, having formed from the same nebula. Although the stars in clusters we observe at present are traveling through space together, they will eventually be scattered by gravitational interactions with one another and with other bodies in our galaxy.

In this book we'll explore some of the finest open clusters visible from the Northern Hemisphere. For some, a moderately large telescope (one with an aperture of 200 to 254 mm, or 8 to 10 inches) is needed to show them at their best, while others are better observed with binoculars or even the naked eye. As shown in Figure 1.5, some, such as M11 in Scutum, look like dozens, even hundreds, of stellar fireflies swarming tightly together; others, like the Pleiades in Taurus, appear as loose collections of celestial jewels spilled across black velvet. Some are so weakly concentrated that they are barely discernible from background stars. Because of this great diversity, I can promise that you will never be bored!

Surrounding our galactic nucleus, much as moths gather around the flicker of a flame, are gigantic, spherical agglomerations of stars called *globular clusters* (Figure 1.6). Each globular cluster contains from hundreds of thousands to several million stars. Astronomers believe that the stars in most globular clusters are among the oldest ever discovered, perhaps as old as 10 billion years.

Figure 1.6. *The globular cluster M13 in the summer constellation Hercules. Through binoculars and small telescopes its 100,000 stars merge to form what looks like a small ball of celestial cotton. In this image, south is to the top left. Photo by Preston Scott Justis.*

The story of globular clusters begins in 1665. While attempting to observe Saturn, German astronomer Abraham Ihle (n.d.) stumbled across a smudge of light near the star Kaus Borealis (Lambda [λ] Sagittarii). Although he didn't realize it at the time, Ihle had discovered the first of the more than a hundred globular clusters that surround our galaxy. Between 1765 and 1781, when Charles Messier (1730–1817) published his list of nebulae and clusters, over thirty more were discovered. William Herschel (1738–1822) was the first to use the name "globular cluster" to identify some of the entries in his catalog of 1786.

Although globular clusters possess many more stars than open clusters, the stars are much harder to resolve because of the clusters' greater stellar densities and distances from Earth. The densities of globular clusters are measured on a 12-point scale of classes. Class I globulars are very compressed, while Class XII clusters are quite loosely structured. Typically, at least a 150-mm (6-inch) telescope is necessary to begin to resolve even the loosest globulars, though as noted throughout this book, some reveal a few individual suns through smaller instruments.

Asterisms

Although technically not deep-sky objects in the truest sense of the word, asterisms are nonetheless interesting to look for and at through binoculars or a telescope. Unlike open clusters, whose stars are close enough to exert a gravitational effect upon one another, most asterisms are simply chance alignments, as seen from Earth, of widely separated stars.

Most of us are familiar with many large, naked-eye asterisms such as the Big Dipper or the Summer Triangle. But how many others are you familiar with? There are many small asterisms scattered throughout the sky that are visible through telescopes and binoculars and worthy of attention. More than a dozen are mentioned in the chapters to come.

Nebulae

"Nebula" (plural nebulae) is the Latin word for cloud. Once, any fuzzy deep-sky object that was not a star was called a nebula. The advent of the telescope allowed some types of objects to be differentiated, though it was not until the advent of photography that many star clusters and galaxies were correctly identified. When

Figure 1.7. *The North America Nebula (NGC 7000), a bright nebula in the summer constellation Cygnus. Although easy to photograph, its deep red color makes it difficult to detect visually. North is up. Photo by John Gleason.*

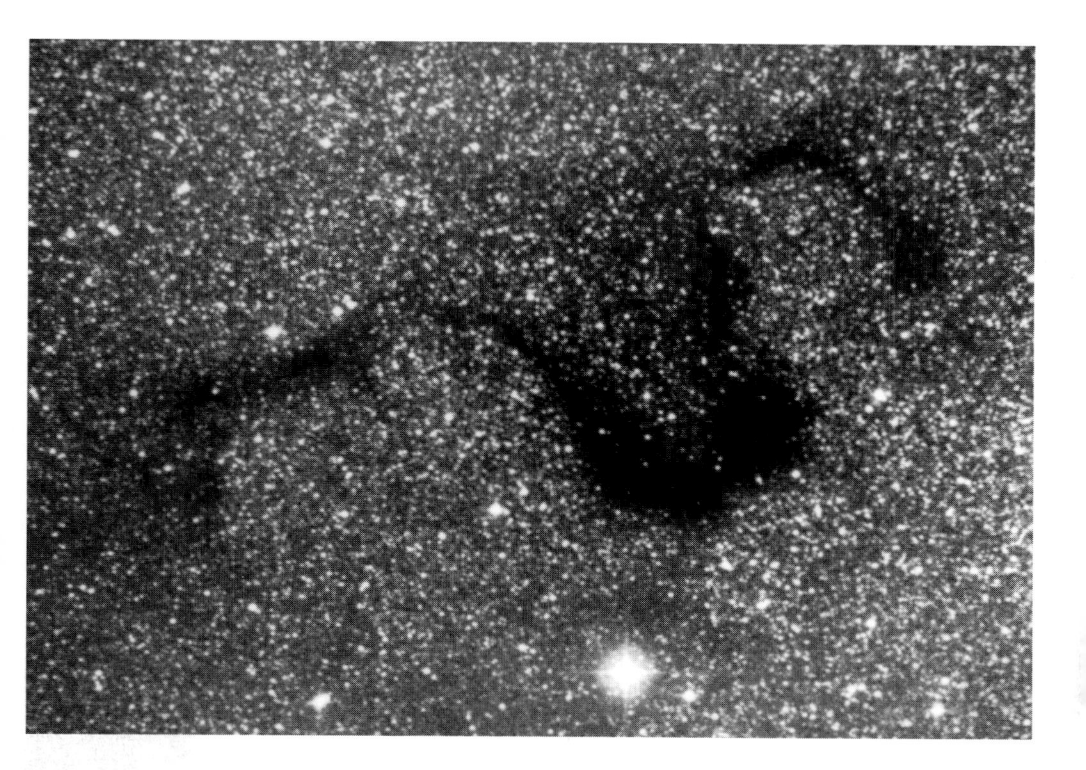

Figure 1.8. *The Snake Nebula, a dark nebula in Ophiuchus. Dark nebulae are considered by many to be the most difficult type of deep-sky object to spot because of their low contrast against the surrounding sky. Unless viewed under dark skies, most will evade detection. North is toward upper right. Photo by Michael Stecker.*

present-day astronomers refer to a nebula, they are talking about a cloud of gas, primarily hydrogen, and dust. Nebulae may be broken down into several categories according to their characteristics.

Bright nebulae (Figure 1.7), sometimes called *diffuse nebulae,* are among the most beautiful objects in the night sky. Some shine by reflecting the light from stars embedded in them; these are known as *reflection nebulae.* A good example of a reflection nebula is the blue cloud surrounding the stars of the Pleiades cluster. Other bright nebulae are excited into fluorescence by the energy from very hot, nearby stars; these are *emission nebulae.* The Great Nebula in Orion is classified as an emission nebula since it glows with the light of ionized hydrogen. As such, emission nebulae are frequently referred to as H II regions.

For all their magnificence bright nebulae are among the most challenging deep-sky objects to observe visually. The difficulty arises from a combination of conditions. Most reflection nebulae are hard to see simply because the overwhelming brightness of the stars illuminating them renders the nebulae themselves undetectable. Emission nebulae are no less difficult because — unlike stars, clusters, and galaxies (all of which shine fairly consistently across the visible spectrum) — they emit light only at certain discrete wavelengths that, unfortunately, are poorly received by the human eye.

The situation is further complicated by the large angular size of many bright nebulae. Instead of being concentrated into a small area, many are spread across wide tracts of sky. This can result in low surface brightness, making it difficult to tell where the sky ends and the objects begin. An observer can easily pass right over a faint nebula without ever suspecting its presence, causing many to think of the name "bright nebula" as an astronomical oxymoron!

In recent years many previously "unobservable" emission nebulae have become common sights through backyard telescopes, thanks primarily to the availability of large-aperture instruments, high-quality eyepieces, and contrast-enhancing filters. These form a powerful nebula-busting team from which only the wispiest clouds escape unnoticed. Smaller instruments can also take advantage of these advanced eyepiece/filter combinations. Nowadays, 100- to 150-mm (4- to 6-inch) instruments can spy objects that were once considered impossible for telescopes two to three times the size.

The second major class of nebulae are *dark nebulae,* sometimes called absorption nebulae (Figure 1.8). These abysmally black clouds are far from any source of illumination and are actually invisible! They are seen only in silhouette, as ghostly wisps of inky blackness in an otherwise brilliant universe. Were they not situated in front of a brighter backdrop, they would remain completely undetectable.

Their elusiveness meant that dark nebulae were largely overlooked by early telescopic observers. William Herschel spotted and cataloged several examples, but he was at a loss to explain exactly what he was seeing. It was not until the advent of photography a century later that the study of dark nebulae really got under way. Most of our early knowledge of these black clouds came about through the diligence of the American astronomer Edward Emerson Barnard (1857–1923).

Trying to see dark nebulae will take some getting used to. For instance, the first thing you would normally consider when deciding to search for a new object is its apparent magnitude. Clearly, this cannot be the case with dark nebulae, for it's what you don't see that counts. So, rather than magnitude, dark nebulae are ranked according to their opacity, on a scale of 1 to 6. The darkest nebulae rate a 6, and the least opaque clouds a 1. However, whether a dark nebula is visible to an observer depends more on its contrast with its surroundings than on its inherent opacity. For instance, a Class 6 dark nebula seen against a sparse star field will always prove more difficult to see than a Class 3 dark nebula seen in silhouette against a rich Milky Way starcloud.

Unlike many bright nebulae (which may be spotted through moderate light pollution by using a narrowband filter, discussed in chapter 3), dark nebulae are usually seen only in transparent, rural

Figure 1.9. *The Helix Nebula (NGC 7293), a planetary nebula in the autumn constellation Aquarius. Through small telescopes, most planetary nebulae look like tiny bluish or greenish disks of light. These objects were dubbed "planetary" by 18th-century astronomers because they resemble Uranus and Neptune. South is up. Photo by Chuck Vaughn.*

skies. The successful "dark nebulist" will plan his or her attack to coincide with the new Moon, waiting for those special evenings when the air is dry and exceptionally clear. Fortunately, large optics are not needed for the finest dark nebulae; indeed, some can be seen through binoculars or even with the unaided eye. The key to success is a slow, methodical low-power scan of the nebula's region. But even on the best evenings, the gentle nature of dark nebulae challenges our skills as observers.

Many bright and dark nebulae contain pockets of condensing gas — evidence of stellar birth. By contrast, there are two varieties of nebula associated with stellar death. As a star of similar mass to our Sun goes through its death throes, its interior will begin to contract under gravity. During this process its outer layers expand and cool as it passes into the red-giant stage of its evolution. Eventually, this outer shell separates from the inner core and begins to expand into space. To astronomers of the 18th and 19th centuries, these expanding shells of gas looked like small disks similar to the then recently discovered planets Uranus and Neptune, and though they bear no relation to the solar system they were named *planetary nebulae* (Figure 1.9), or planetaries for short. Most planetaries last less than 50,000 years (a blink of the eye on the cosmic time scale) before their shells dissipate.

Many planetaries appear bluish or greenish in telescopes, testifying to the fact that they contain ionized oxygen, which is known to glow with a turquoise hue under certain conditions. Unlike bright and dark nebulae, which can span a degree or more of sky, most planetary nebulae are notoriously small objects, requiring high magnification and superior skies to be seen.

When very massive stars die, they may erupt in violent explosions called *supernovae*. While the exact triggering mechanism is a complicated mix of forces internal and external to the victim star, all

Figure 1.10. *The Crab Nebula (M1), the sky's best-known supernova remnant, is found in the winter constellation Taurus. The explosion that gave birth to the Crab was recorded by Chinese astronomers nearly 1,000 years ago. North is up. Photo by Chuck Vaughn.*

supernovae are detonated by either thermonuclear ignition or the collapse of their cores. (Figure 1.10)

Nebulae are among the most difficult but rewarding celestial objects for amateurs to locate. Some are stunning in all optical instruments, while others appear as little more than faint smudges of light, even in large backyard telescopes. A few reveal to the eye minute detail that rivals what can be seen in the best photographs, but most show only a soft, ill-defined glow when first viewed. On each subsequent visit, however, the observer's eye becomes more attuned to their subtle shadings. Before long a visual observer may actually perceive fine detail that is frequently lost in long-exposure photographs.

Galaxies

All the objects so far discussed belong to our galaxy. This immense assembly of more than 200 billion stars, and thousands of star clusters and nebulae, is structured like a gigantic pinwheel, with several arms spiraling away from a central galactic core. Most nebulae and open clusters lie within the spiral arms, as do the Sun and its planets.

A survey of the universe beyond the confines of our galaxy reveals a multitude of similar pinwheel-shaped stellar gatherings, some smaller than the Milky Way, some larger. Astronomers have named them *spiral galaxies* (Figure 1.11a) from their appearance. Spiral galaxies are classified according to how tightly wound their arms appear. Those with very tight arms are designated Sa. Others, less tightly wound, are designated Sb, while the loose spirals are designated Sc (with Sd recently introduced for the very loosest). Rather than curling directly away from the core, the arms of some spirals extend from the ends of barlike features that protrude from their cores. These are called *barred spirals,* and are designated in the same way as for ordinary spirals, but with a B inserted: SBa, SBb, and so on. Some evidence suggests that our galaxy is a barred spiral.

Figure 1.11a. *Andromeda, spiral galaxy M31, is one of autumn's deep-sky showpieces. The two dark dust lanes near the galaxy's core are visible in telescopes with apertures as small as 75 mm. Photo by Jim Riffle, Astrospace Works.*

Other galaxies bear no resemblance to the spirals at all. Instead, they appear only as gigantic oval and circular spheres of stars. These are the *elliptical galaxies* (Figure 1.11b), and they outnumber the spirals about two to one. Like spirals, ellipticals are also classified

Figure 1.11b. *The elliptical galaxy M110 (NGC 205), one of the companion galaxies to M31. M110 may also be seen in Figure 1.11a to the lower left of M31. South is up. Photo by Preston Scott Justis.*

Figure 1.11c. *The Large Magellanic Cloud, one of the Milky Way's companion galaxies, shows the asymmetric appearance that is common among irregular galaxies. North is up. Photo by Dennis di Cicco.*

according to appearance and structure, but on a seven-point scale. An E0 galaxy appears as a perfectly round sphere, and an E7 galaxy is flattened strongly at either end, while intermediate points on the scale indicate varying degrees of ellipticity. Strangely, elliptical galaxies do not contain any bright or dark nebulosity.

Finally, galaxies that fit into neither the spiral nor the elliptical families are termed *irregular galaxies* (Figure 1.11c). As the name implies, irregular galaxies share no common shape. Instead, most appear as amorphous glows.

Few things in this universe are as fascinating to view through amateur telescopes as galaxies. Some appear as huge, elongated masses that reveal subtle detail when studied with care. Others are comparatively small and faint, glowing indistinctly behind the nearby stars of our galaxy. Still more barely reveal themselves, being merely vague glows that play a game of galactic hide-and-seek with the observer's eye.

Figure 1.12. *3C 273, the sky's brightest quasar, is in the spring constellation Virgo. Even though it is two billion light years from us, 3C 273 is bright enough to be seen in 150-mm telescopes. South is up. Photo courtesy Harvard College Observatory. (See figure 6.14 for directions on locating 3C 273.)*

Quasars

We come finally to quasars (Figure 1.12). Quasars, short for quasi-stellar objects (so named for their starlike appearance), are the most distant objects yet detected. The distances to even the closest quasars are measured in billions of light-years, in contrast to the millions of light-years that separate us from the galaxies visible in modest telescopes. For them to appear as bright as they do (keep in mind that, in this context, "bright" is a relative term; the brightest quasar shines at 13th magnitude), each quasar must be as luminous as a hundred or more galaxies. Yet it is believed that all of this power comes from an object no larger than the volume of our solar system!

Since quasars appear so small and faint, spotting them through amateur instruments is usually close to impossible. Yet one quasar — 3C 273 in Virgo (described in chapter 6) — is bright enough to be seen through telescopes as small as 150 mm (6 inches) on clear, dark nights.

CHAPTER 2

Number, please

Not long after humans began looking at the night sky, they began the daunting task of recording what they saw. Pictures were drawn among the stars to illustrate stories and legends but also to serve as celestial markers for important seasonal events. These pictures have come down to us as constellations. Many of their names, such as Hercules, Perseus, and Orion, are recognizable to nonastronomers as heroic figures from times long ago. Others, such as Ophiuchus, Lyra, and Sagittarius, may not be as familiar, but for many of them we can nonetheless trace their origins to before the birth of Christ. Many bright stars also acquired strange-sounding proper names like Betelgeuse, Vega, and Zubenelgenubi. But what about the myriads of other stars and almost innumerable deep-sky objects? Let's let our fingers do the walking through the universe's directory.

Stars, double stars, and variable stars

In 1603 astronomer Johann Bayer (1572–1625) compiled that era's most extensive atlas and catalog of naked-eye stars, the *Uranometria.* In it Bayer assigned lowercase Greek letters to the brightest stars in each constellation. In most cases the brightest star in each constellation was labeled alpha (α), followed by beta (β), gamma (γ), delta (δ), and so on (a complete list of Greek letters is given at the end of the book). Bayer's contribution has had a lasting effect, as astronomers continue to use these *Bayer letters* to this day.

Another way astronomers refer to specific stars is by their *Flamsteed number.* In a star atlas published posthumously in 1725, John Flamsteed (1646–1719) numbered many stars in each constellation. In his system the numbers increase from west to east within each constellation. For instance, the brightest star in the summer constellation Lyra may be called by its proper name, Vega, or by its Bayer designation, Alpha (α) Lyrae, or by its Flamsteed number, 3 Lyrae. In each case it is the genitive form of the constellation (see Appendixes) — in this instance Lyrae — that is appended to the designation.

When specifying bright double stars, it is not unusual for astronomers to rely on either their proper names, or Bayer's or Flamsteed's designations. For instance, one of the best-known double stars is Albireo, also known as Beta (β) Cygni and 6 Cygni. However, if you look up Albireo in a catalog such as volume 2 of *Sky Catalogue 2000.0,* edited by Alan Hirshfeld and Roger W. Sinnott, you will also find it

Figure 2.1. *Charles Messier, the 18th-century comet-hunter who compiled the famous Messier catalog of deep-sky objects. Courtesy San Diego State University.*

cross-referenced as Σ I 43 and ADS 12540. Σ I 43 is its designation in the first supplement to the extensive double-star listing compiled by F. G. W. Struve (1793–1864). This is the most comprehensive inventory of double stars assembled in the 19th century. His catalog entries are signified by the prefix Σ (sigma), while those added by his son Otto (1819–1905) are prefixed by either OΣ or OΣΣ. The ADS designation above refers to the *New General Catalogue of Double Stars,* compiled in 1932 by Robert Aitken (1864–1951). Aitken's roster includes most of the double stars resolvable through amateurs' telescopes.

Like double stars, many bright variable stars were assigned proper names and Greek designations, such as Algol (Beta [β] Persei) and Mira (Omicron [ο] Ceti). Most variables, however, seem to speak a language all their own. Traditionally, variable stars are cataloged within their home constellations using single and double letters of the alphabet; examples are R Orionis, Z Ursae Majoris, and RT Cygni. This system can account for a total of 334 variables per constellation. Although usually adequate, some constellations along the Milky Way contain more variables than can be accommodated in this scheme. In those cases, the 335th variable is labeled V335 (e.g., V335 Cygni), and so on. Each new variable is given the next available number, in order of discovery. Sagittarius contains more than 4,000 known or suspected variable stars — more than any other constellation.

The Messier catalog

Literally dozens of catalogs of star clusters, nebulae, and galaxies have been compiled since the invention of the telescope four centuries ago. While many have faded into oblivion, some have remained popular to this day. Of these, the most famous is that compiled by Charles Messier, an 18th-century French comet-hunter (Figure 2.1). The Messier catalog was created not in an effort to record the locations of deep-sky objects, but rather to record the locations of annoying cometlike objects that hindered the comet-hunting efforts of Messier and his contemporaries. It reads like a "who's who" of deep-sky objects. The first entry, M1, is the famous Crab Nebula in Taurus, the Bull. Other outstanding members are M8 (the Lagoon Nebula), M13 (the Great Cluster in Hercules), M27 (the Dumbbell Nebula), M42 (the Orion Nebula), and M45 (the Pleiades), to name but a few. A complete list of Messier objects is given at the back of the book.

Finding all of the Messier objects is a great way to begin your career as a deep-sky observer. The finest are sprinkled through chapters 6 through 9. Begin with any object you wish, and slowly work your way through the list. There is no need to rush, so take your time and enjoy the view. Most new observers take three years or more to observe the entire Messier catalog. They begin the project knowing little about the sky, and earn their "eyes" along the way. By the time they spot their last M-object, they are seasoned veterans fluent in the language of both the sky and their telescopes.

Best of all, a large, expensive instrument is not needed to see all the Messier objects. Messier himself used no telescope greater than 150 mm (6 inches) in aperture to catalog them in the first place. Today, some amateurs complete the project with telescopes half that size. In fact, all of the Messier objects can be seen in 11 × 80 binoculars if used in a rural setting, while more than 80 of them can be seen through 7× binoculars from suburbia (depending, of course, on your visual acuity).

The Astronomical League, a confederation of astronomy clubs across the United States, gives its members a certificate for finding 70 of the Messier objects. To qualify for the award, you must belong to the League through an affiliated club (or be a member-at-large) and be able to produce a logbook noting which objects you have found. A similar award is offered to members of the Royal Astronomical Society of Canada. More information from these organizations may be obtained by writing to them at the addresses given at the back of the book. A listing of North American and European astronomy clubs is available on the World Wide Web at http://www.skypub.com/astrodir/astrodir.html.

The Herschel list

In addition to discovering Uranus and being appointed Astronomer Royal to King George III, William Herschel (Figure 2.2) compiled an exhaustive inventory of more than 2,500 star clusters and nebulae at the end of the 18th century. In 1834 his son John Herschel (1792–1871) traveled to the Cape of Good Hope to study the southern sky. The younger Herschel collated both his own and his father's observations and published the *General Catalogue of Nebulae* in 1864.

The *General Catalogue of Nebulae* organizes its objects into eight categories according to their visual appearance: Class I (bright nebulae), Class II (faint nebulae), Class III (very faint nebulae), Class IV (planetary nebulae), Class V (very large nebulae), Class VI (very compressed star clusters), Class VII (compressed star clusters), and

Figure 2.2. *William Herschel, arguably the finest deep-sky observer of all time. From* Great Astronomers *by Sir Robert Ball, 1912.*

Class VIII (scattered star clusters). These classifications are confusing at best and are frequently incorrect by our modern reckoning of the universe. (For one thing, they make no mention of galaxies, which in the Herschels' day were lumped with the nebulae.)

The Astronomical League also offers a certificate of merit for observers who have successfully found 400 selected objects from the Herschels' *General Catalogue*. Award criteria are similar to those of the League's Messier certificate.

The New General Catalogue and *Index Catalogues*

The traditional comprehensive anthology of deep-sky objects is the *New General Catalogue of Nebulae and Clusters,* known by its abbreviation NGC, which was compiled by John L. E. Dreyer (1852–1926) in 1888. The NGC lists 7,840 star clusters, nebulae, and galaxies covering the entire celestial sphere. All entries are ordered by increasing right ascension, with the first entry, NGC 1, specifying a faint galaxy in Pegasus located at R.A. $00^h\ 07^m.3$, Dec. +27° 43' (in coordinates for the year 2000, referred to as 2000.0 coordinates). With few exceptions, all of the Messier objects are also included in the NGC. The Orion Nebula, M42, for instance, is also known as NGC 1976.

Dreyer assembled a pair of supplementary *Index Catalogues* (IC), which contains new objects discovered after the NGC was published. The first of these includes objects found between 1889 and 1894, while the second covers discoveries from 1895 to 1907.

Other celestial inventories

Many other deep-sky inventories have been compiled since the NGC and IC listings. While the Messier catalog, the NGC, and the IC are general compilations, most 20th-century catalogs confine themselves to a single type of object. Dark nebulae, for example, are

Figure 2.3. *The American astronomer Edward Emerson Barnard used photography to discover and catalog dozens of dark nebulae. Courtesy Lick Observatory.*

usually specified by their "Barnard" number, from the *Catalogue of 349 Dark Objects in the Sky* by Edward Emerson Barnard (Figure 2.3). For instance, B33 identifies the famous (and infamously faint) Horsehead Nebula in Orion. The most extensive list of dark nebulae was assembled by Beverly T. Lynds (usually abbreviated LDN) in 1962 and subsequently published as the *Catalogue of Dark Nebulae* in the *Astrophysical Journal Supplement Series,* volume 7.

Several 20th-century astronomers have compiled lists of open clusters not already documented in the Messier, NGC, and IC collections. These include the Collinder (abbreviated Cr), Harvard (H), Melotte (Mel), and Stock (St) catalogs. While most of these unsung clusters cannot compete with their better-known Messier and NGC counterparts, some prove quite striking.

Finally, we come to an informal listing of stellar asterisms, which continues to grow to this day: the Small Telescope Asterism Roster, or STAR. In *Touring the Universe Through Binoculars,* I began compiling a list of star patterns and asterisms that I called "Harrington objects." In response to a request I made in *Sky & Telescope* in 1994, many amateurs submitted new asterisms that they had come across while viewing the heavens. Because of the great response, it seemed only fair to drop the "Harrington" designation in favor of a more general name, Small Telescope Asterism Roster. A few, up to STAR 12, are carryovers from my original binocular list, while STAR 13 and above appear here for the first time. I hope you enjoy looking at each of these asterisms as much as I have.

CHAPTER 3

Tools of the astronomer

Many people just starting out in the hobby of astronomy, and especially deep-sky observing, are often confused by all the telescopes, binoculars, and accessories that flood today's astronomical marketplace. What kind of telescope is best? Yes, a simple question, but it has a complicated answer. The problem is that there is no such thing as an ideal telescope. No one telescope is perfect for every amateur to view every type of celestial object from every observing site. Additional questions arise about eyepieces, finderscopes, and filters, to name a few more sources of confusion, but the list goes on and on. In an effort to help put things into perspective, this chapter presents a quick overview of today's equipment market.

Let's begin with the least expensive and work our way up. The cheapest optical instrument for viewing the sky is the human eye. Under dark skies, our eyes alone can detect stars to below 6th magnitude as well as dozens of deep-sky objects. In fact, many observers enjoy the challenge of seeing just how many deep-sky objects they

Figure 3.1. *The author's daughter, Helen, demonstrates the usefulness of binoculars for observing.*

can spot without additional optical aid. Try it — you might just be surprised.

Binoculars

Every observer should own a pair of binoculars (Figure 3.1). For portability and ease of operation, they cannot be beaten. While many telescopes require 15 minutes or more just for setup, binoculars are ready in an instant. Their wide fields of view make it possible to take in vast cosmic panoramas that are impossible to grasp with the narrow fields of most telescopes. And then there is the inevitable question of economics. While a good telescope will cost hundreds, even thousands of dollars, a good pair of binoculars can be had for less than $100 — if you know what to look for.

Pick up just about any pair of binoculars and you will see somewhere on them a pair of numbers, such as 7 × 35 (read "seven by thirty-five") or 10 × 70. The first number in the pair defines the binoculars' *magnification* (or power); the second defines the *aperture* (or lens diameter).

Magnification is an important consideration when selecting binoculars. Before making a choice, ask yourself how the binoculars will be held. If they will be held by hand, then the best choice is either 7× or 8× binoculars. Higher magnifications prove difficult to support steadily by hand for even short periods of time. If, on the other hand, the binoculars will be mounted on a tripod or other support, then higher magnifications may prove beneficial.

The second number specifies the diameter, or aperture, of the binoculars' front lenses, or objectives, in millimeters (mm). For example, 7 × 35 binoculars have front lenses that are 35 mm across, while 10 × 70s have 70-mm objectives. (Converting these values into inches — there are 25.4 mm in an inch — indicates that the objective lenses measure nearly 1½ inches and 3 inches, respectively.)

Selecting the right aperture is just as critical as choosing the right magnification. Ideally, for binoculars to deliver optimum results, they should fit both the observer and the observing site. The diameter of the beam of light exiting the binoculars' eyepieces (called the *exit pupil*) should equal the diameter of the observer's dark-adapted pupils (see chapter 4). While this value varies from one person to the next, it is generally 7 mm for young adults, but only 4 mm for seniors.

The exit pupil for any pair of binoculars is easily found by dividing the aperture by the magnification. A pair of 7 × 35 binoculars has a 5-mm exit pupil (35 ÷ 7 = 5), while a pair of 7 × 50s has approximately a 7-mm exit pupil (50 ÷ 7 = 7.1). If you are relatively young (up to the age of 30) and plan on doing most of your skywatching from a dark, rural setting, then you would do best with binoculars that yield a 7-mm exit pupil, such as 7 × 50s or 10 × 70s. For us over-the-hill observers (say, 40 years old and up), our eyes' pupils may not dilate beyond 5 mm, regardless of observing conditions. If you fall into this category, or if you are limited by the light-pollution of a city or suburb, then a better choice would be binoculars yielding smaller exit pupils, such as 7 × 35s or 10 × 50s.

Figure 3.2. *Binoculars with BaK-4 prisms* (left) *produce brighter, crisper images than those equipped with BK-7 prisms* (right). *BaK-4 prisms are uniformly round and bright; BK-7 prisms are diamond-shaped, with grayish areas at the edges.*

Just about all modern binoculars feature optics coated with a thin layer of magnesium fluoride both to reduce flare and "ghosting" as well as increasing image contrast. Preferred models feature "fully coated" or, better still, "fully multicoated" optics. Superior binoculars now use internal prisms made from a type of glass called BaK-4, while less expensive models have prisms made of BK-7 glass. To tell them apart take a look at the eyepieces' exit pupils (Figure 3.2). What do they look like? If they are uniformly round and bright, the binoculars have BaK-4 prisms. If the exit pupils are diamond-shaped with grayish areas squaring off the edges, the prisms are made from BK-7 glass and will yield dimmer images.

I have long been a proponent of the idea that binoculars are not just a beginner's substitute for a telescope but instead are useful observational tools in their own right. And, unlike many telescopes, binoculars are comparatively inexpensive and readily available.

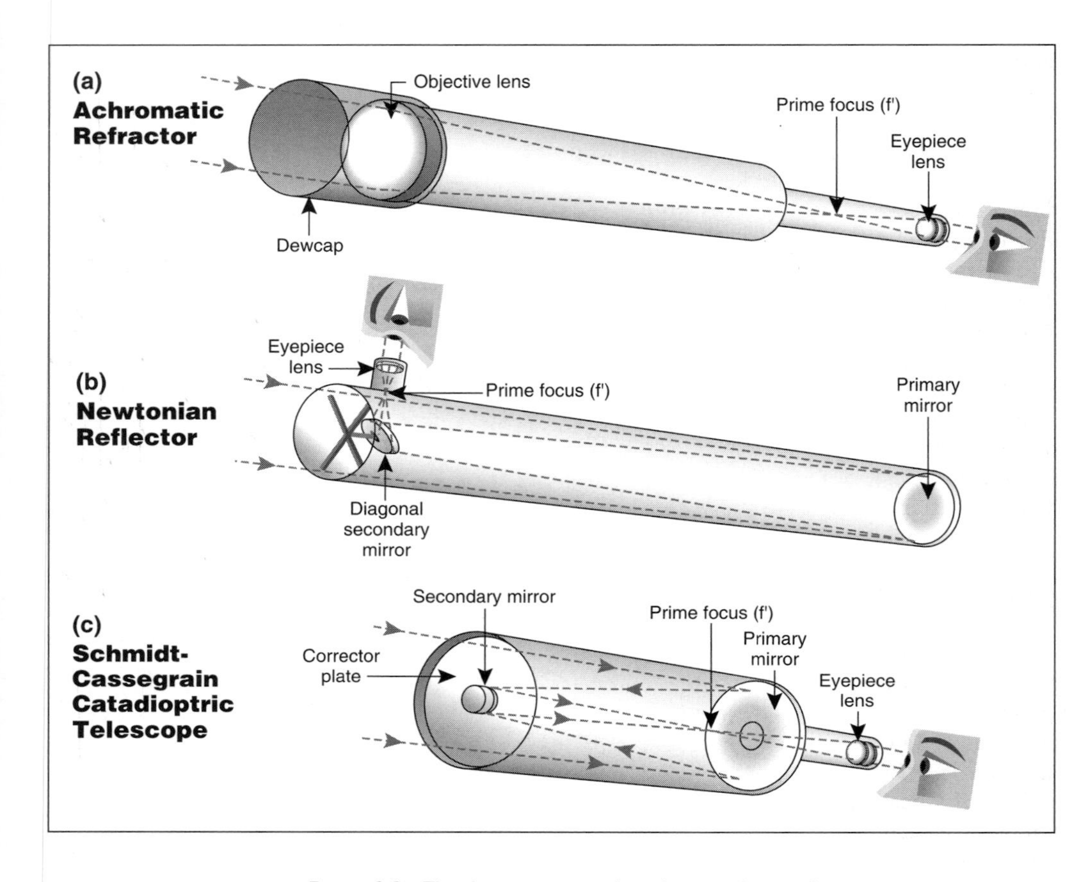

Figure 3.3. *The three most popular telescope designs for amateur astronomers: (a) the refractor, in this case an achromatic design; (b) the Newtonian reflector; and (c) the Schmidt-Cassegrain catadioptric.*

Telescopes

Although one could spend years touring the universe through binoculars, most skywatchers eventually succumb to the desire to own a telescope. Like binoculars, a telescope's main purpose is to augment the performance of the human eye. The opening, or aperture, of the human eye is regulated by its pupil. A telescope increases our own innate ability to see the sky by gathering more light with its larger aperture.

A telescope's aperture specifies the diameter of its main optical element, be it a lens or a mirror. This is vital when it comes to the instrument's usefulness for deep-sky observing: the larger the aperture, the more light the telescope will gather. A telescope with an aperture of 100 mm (4 inches) will gather more than 200 times as much light as the human eye. A 150-mm (6-inch) telescope has more than twice the light-gathering area of a 100-mm, while a 200-mm (8-inch) will gather nearly twice as much light as a 150-mm.

If deep-sky observing is your passion, then aperture should be your main, but *not* your only, concern. A deep-sky telescope must address other needs as well. The optical and mechanical quality of a telescope is just as important as its aperture, for what good is a large telescope if it cannot produce clear, sharp images? I can recall looking through a very large homemade telescope not long ago. The images it produced were inferior to my own 200-mm telescope, even though its aperture was over four times as large! The reason? Poor-quality optics.

The moral to all this is *never* sacrifice optical quality for aperture. Always purchase a telescope from a reputable company. Ask a company representative to send you literature on their entire range of telescopes. If possible, have them supply you with a list of references. Read as much as you can before making a decision. *Sky & Telescope* magazine features regular test reports on telescopes and accessories, while my book *Star Ware* describes over 90 binoculars and 170 telescopes. Perhaps most importantly, join a local astronomy club. Attend club star parties, look through as many telescopes as you can, and talk to their owners.

Many less-than-reputable telescope manufacturers play on people's ignorance by advertising their product's magnification. Magnification is *not* an important criterion for judging the usefulness of a telescope. High power does *not* mean high quality. In fact, it is completely inconsequential. Any telescope can be made to work at any power simply by changing eyepieces, but too much magnification will only yield dim, fuzzy images. For further discussion on this subject, see the "Eyepieces" section later in this chapter.

Broadly speaking, all telescopes may be classified into one of three different categories depending on their optical design, as shown in Figure 3.3. *Refracting telescopes* use large front lenses (called *objective lenses*) to gather starlight, while *reflecting telescopes* employ large mirrors (called *primary mirrors*) to do the job. The third type of telescope, the *catadioptric telescope*, is a hybrid design that combines a *corrector lens* and a primary mirror.

Refractors

Refractors, the instruments of choice during the 17th, 18th, and 19th centuries, are enjoying a revival in popularity among amateur astronomers. Well-made refractors are second to none when it comes to sharp, high-contrast views of the universe.

Most refractors sold today are either *achromatic* (Figure 3.3a) or *apochromatic*. The achromatic type uses a two-element objective lens to overcome the optical aberrations that result from using a single-element objective. Even better at suppressing optical aberrations is the costly apochromatic refractor, whose objective is made from a special low-dispersion optical material, such as fluorite, and often consists of three elements. Apochromatic refractors are wonderful for viewing bright deep-sky objects, such as most of those described in this book.

The biggest advantage of refractors is that their apertures are unobstructed. To redirect the incoming light toward their eyepieces, reflectors and catadioptric telescopes require secondary mirrors —

mirrors that are mounted inside the telescope tube and block a small portion of the telescope's aperture. Anything that interferes with the transmission of light through the instrument is bound to cause some loss of contrast and image degradation.

Refractors are not without their drawbacks, however. Their biggest disadvantage is their small apertures. Most commercial refractors range in aperture from 50 to 180 mm (2 to 7 inches), yet they can cost more than reflecting telescopes of twice the aperture. Their small light-gathering areas mean that faint objects such as nebulae and galaxies will appear dimmer than they do through larger (and cheaper) reflectors. In addition, the common practice of using a 90° star diagonal with refractors results in mirror-images of star fields, making it difficult to compare the view with star charts.

Reflectors

For those who want the largest aperture for the least money, the standard *Newtonian* type of reflecting telescope (Figure 3.3b) is the way to go. The Newtonian's simplicity of design means that, in terms of dollars per unit aperture, you can buy a larger instrument than you could with any other type of telescope.

Unlike refractors and catadioptrics, Newtonian reflectors do not require the use of a star diagonal for comfortable viewing. You therefore see a "correct" (albeit upside-down) image that is easy to compare with charts and other references. For those new to telescopic astronomy (and yes, even for many veterans too), this can save a lot of frustration.

Another plus for Newtonians is that, unlike other designs, their optics are less apt to fog over. For those of us who live in a temperate climate, it is not unusual for the relative humidity to rise noticeably in the evening. When that happens, the air becomes saturated, causing dew to form on cold surfaces. The front lens of a refractor or catadioptric is just such a cold surface, and is especially prone to dewing, while the long tube of a Newtonian shields the primary mirror and prevents it from fogging over.

Of course, as mentioned above, reflectors lose some light and image contrast through obstruction by their secondary mirrors. Just how much light is blocked depends on the size of the secondary, which in turn depends on the focal length of the primary. In general, the shorter the primary's focal length, the larger the secondary required to bounce the light toward the eyepiece. That's one reason why, when comparing, say, a 200-mm f/7 reflector (7 × 200 mm = 1.4-meter focal length) with a 200-mm f/4.5 (4.5 × 200 mm = 0.9-meter focal length), the f/7 instrument will probably show better contrast between the target and the surrounding sky.

The optics in reflectors tend to go out of alignment, or *collimation*, more readily than refractors and catadioptric instruments. To ensure that everything is as it should be, most owners make a quick check of their telescope's collimation at the start of each observing session. Figure 3.4 illustrates a four-step procedure for collimating a Newtonian.

A large telescope is not an absolute requirement for deep-sky observing. Most of the objects described in chapters 6 through 9, and hundreds more besides, are visible through a 100-mm reflector from relatively dark suburban settings. This aperture, in my opinion, should be considered a minimum. Better yet is a 150-mm reflector which, as this is being written, costs typically between $300 and $700 depending on its mounting and accessories. This is not by any means a cheap investment — but it is one that will give you a lifetime of joy.

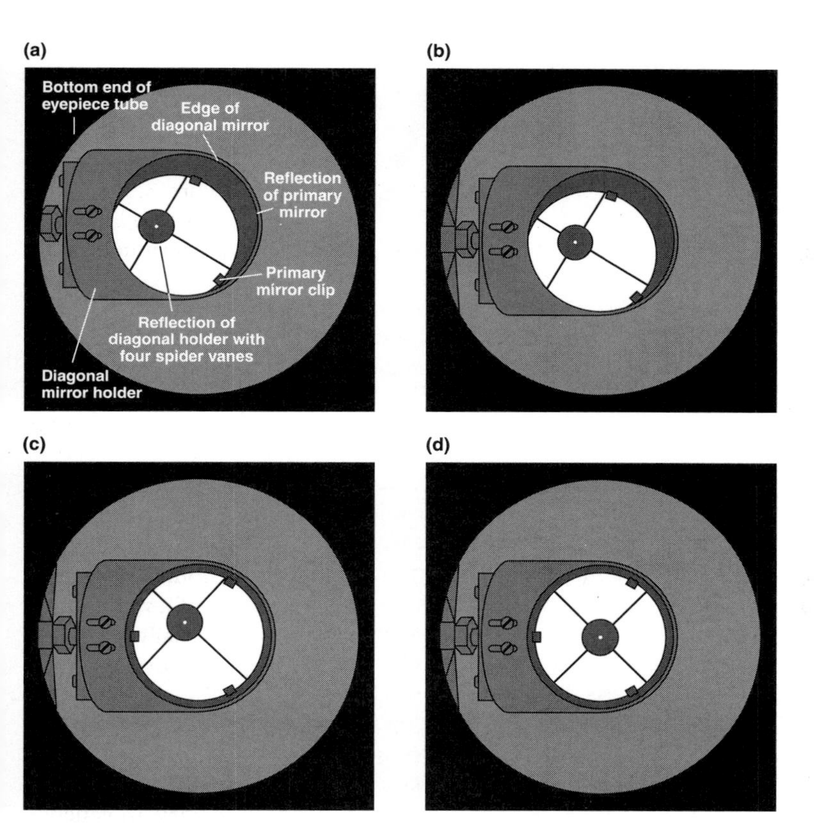

Figure 3.4. *Today's popular short-focal-length Newtonian telescopes must have their collimation checked before each observing session. Here's how to go about doing it. Depicted here is the view through a Newtonian's empty eyepiece holder. In (a) both the primary and secondary mirrors are out of alignment. Adjust the diagonal's central post until the diagonal is centered under the focuser's eyepiece tube, to give a view as in (b). Then turn the diagonal mirror's adjustment screws until the image of the primary mirror is centered, as in (c). Finally, adjust the primary mirror's cell until the reflection of the diagonal is centered in the image of the primary. The view through the collimated instrument will then be as in (d). Diagrams adapted from* Perspectives on Collimation, *a booklet by Vic Menard and Tippy D'Auria.*

Figure 3.5a. *An altitude-azimuth mount telescope features the simplest type of telescope support available. Courtesy Celestron International.*

Catadioptrics

Catadioptric telescopes (sometimes called compound telescopes) come in different designs, but the most popular by far is the *Schmidt-Cassegrain telescope* (Figure 3.3c), or SCT for short. For those who consider ease of portability and convenience of storage the most important criteria, the Schmidt-Cassegrain is a godsend. An SCT is usually only about twice as long as its aperture. No other telescope can fit as large an aperture and as long a focal length into such a short tube assembly. When not in use, the instrument folds away into its footlocker.

Disadvantages? Most experienced observers agree that the optical performance of a Schmidt-Cassegrain is inferior to that of a Newtonian or refractor of equivalent aperture. Because of the comparatively large size of the SCT's secondary mirror, it just cannot match the contrast or "zip" of views through refractors and reflectors. This becomes especially apparent when observing double stars or the planets.

Another problem shared by just about all Schmidt-Cassegrains is "mirror shift." To focus these instruments, the primary is moved back and forth inside the telescope (as opposed to moving the eyepiece back and forth, as with refractors and reflectors). This arrangement is fine in theory, but in practice the focusing mechanism is prone to turning the mirror just a bit. This causes the image to "lurch" slightly whenever the focusing knob is moved.

Telescope mounts

Even a telescope with the finest optics will turn into a useless, quivering mass if it is not supported on a sturdy mounting.

Figure 3.5b. *A Newtonian reflector mounted on a Dobsonian-style altitude-azimuth mount, a simple yet sturdy alternative to heavy, expensive equatorial mountings. Courtesy MorningStar Telescope Works.*

(Figures 3.5a-d). A telescope mount must move the telescope smoothly from one object to the next, allow the telescope to be easily pointed to any part of the sky, and be strong enough to support the telescope's weight while minimizing any vibrations.

Altazimuth (altitude-azimuth) mounts

An altazimuth (altitude-azimuth) mount (sometimes shortened to either alt-azimuth or alt-az) is the simplest type of telescope support available (Figure 3.5a). Its two axes allow the telescope to move both in azimuth (horizontally) and in altitude (vertically), like a camera tripod. In recent years a mutation of the altazimuth has become the preferred mount for large-aperture telescopes intended for visual use only (i.e., not for photography). This is the *Dobsonian mount* (Figure 3.5b), named for John Dobson, an amateur telescope maker from San Francisco. The Dobsonian mount is famous for its simple yet sturdy design made essentially from wood, Formica, Teflon, and some glue and nails.

Left: Figure 3.5c. *A fork equatorial mount telescope. Courtesy Meade Instruments Corporation.* Right: Figure 3.5d. *A German equatorial mount telescope. Courtesy Celestron International.*

Figure 3.6. *Eyepiece quality and selection are both critical for successful deep-sky observing. Today's amateurs have a wide selection from which to choose: the photograph above shows a range of eyepieces offered by a major supplier. Photo courtesy Tele Vue, Inc.*

Equatorial mounts

An equatorial mount (Figures 3.5c and 3.5d) also consists of two perpendicular axes, but instead of being oriented up-down as with an altazimuth mount, the axes are tilted at an angle matching the latitude of the observing site. The net result is a mount that can track the stars in a single east-west motion, rather than the two-step up-down, left-right movement necessary with an altazimuth mount.

There are many benefits to using an equatorial mount, the greatest being the ability to attach a motor drive that allows the telescope to track the sky. Other advantages are that "star-hopping" from one object to another is simplified, and setting circles can be used (see chapter 4). On the minus side of equatorial mounts, however, is that they are almost always larger, heavier, more cumbersome, and more expensive than altazimuth mounts. This is why the simple Dobsonian design is favored for supporting the large Newtonians preferred by experienced deep-sky observers.

Table 3.1 *Suggested exit pupils.*

Target	Exit pupil (mm)
Large star clusters, diffuse nebulae, and galaxies	6–7
Small, bright deep-sky objects	4–6
Small, faint planetary nebulae and galaxies; double stars on nights of poor seeing	2–4
Double stars on exceptional nights	0.5–2

Eyepieces

While no one will argue that a high-quality telescope isn't critical to an observer's success and enjoyment, until recently little emphasis was placed on eyepieces. Today's amateurs have a wide selection of high-quality eyepieces (Figure 3.6) from which to choose. But again, as with telescopes, no single eyepiece design is best for everything.

When deciding which eyepieces are better than others for your intended purposes, you must consider several factors, including magnification, exit pupil, field of view, and eye relief. Base your choice of magnification on what you are trying to see. Widely scattered star clusters and large nebulae are best viewed with low-power, wide-field eyepieces, while smaller deep-sky objects (e.g., planetary nebulae and most galaxies) require higher power. To figure out the magnification that a given eyepiece will provide in a given telescope, simply divide the telescope's focal length by that of the eyepiece. For instance, say you are using a 200-mm (8-inch) f/7 telescope, which has a focal length of 1.4 meters, and a 12-mm eyepiece. First, convert the telescope's focal length to millimeters by multiplying it by 25.4, to get 1,422 mm. Now divide 1,422 mm by the eyepiece's focal length (12 mm). The resulting magnification is 119× (rounding up from 118.5×).

A general recommendation often made in observing handbooks is that if you own a small telescope ("small" here means an aperture of 200 mm or less), you should not use a magnification of more than about 2× per millimeter (50× to 60× per inch) of aperture. By this convention, with a 150-mm (6-inch) telescope you should not exceed a magnification of 360× , and with a 100-mm the maximum is 240×. But this is not necessarily a hard-and-fast rule. Much depends on your local observing conditions and on the instrument itself. Given excellent optics and steady seeing, you may be able to go to nearly 4× per millimeter (90× or even 100× per inch) on some nights, while on others the view may crumble with little more than 1× per millimeter (25× per inch). Large telescopes (those above 200 mm) are rarely able to meet or exceed the "2-power-per-millimeter" rule, regardless of conditions. Instead, they may be able to handle only 1½× per millimeter (nearly 40× per inch), even under steady skies.

The next important factor when selecting an eyepiece is the exit pupil. Recall from the earlier discussion on binoculars that the exit pupil is the diameter of the beam of light leaving the eyepiece and traveling to the observer's eye. To see the exit pupil of an eyepiece,

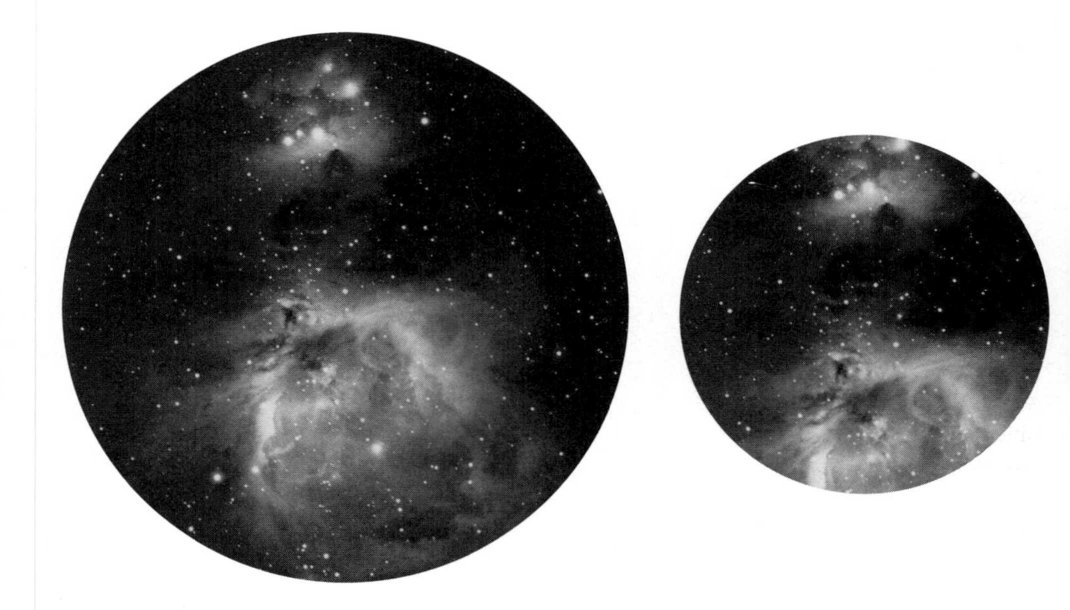

Figure 3.7. *The view through a wide-field eyepiece* (left), *and a narrow-field eyepiece* (right). *Wider views are often more aesthetically pleasing and give a feeling of peering through the porthole of an imaginary spaceship. North is up. Photo of M42 by Jason Ware.*

place it in your telescope and aim the instrument toward a bright surface, such as a light-colored wall or the daytime sky (away from the Sun). Back away a little, and look for the small disk of light that appears in the center of the eyepiece — that's the exit pupil. To work out its exact diameter, simply divide an eyepiece's focal length by your telescope's focal ratio. So, for the 200-mm f/7 telescope and 12-mm eyepiece considered above, the exit pupil is about 1.7 mm. Table 3.1, excerpted from my book *Star Ware,* offers some guidance for matching the optimum exit pupil for particular classes of deep-sky objects. Again, these are only suggestions.

There is also the eyepiece's apparent field of view (Figure 3.7). This is the edge-to-edge angular diameter of the observer's view through the eyepiece. Imagine looking through a long, thin tube, like one from a roll of paper towels. The apparent field of view in this example can be increased by cutting off part of the cardboard tube. Most observers prefer eyepieces with an apparent field of at least 40°, as anything less will produce a tunnel-vision effect. An apparent field greater than 60° gives a panoramic view that is almost like being out in space. Bear in mind, however, that price can rise rapidly as the apparent field of view expands, so temper your desires with a touch of reality. Just how much sky will fit into a field of view is easy to determine by dividing an eyepiece's apparent field of view by the magnification it delivers in a particular telescope.

Finally, a well-designed eyepiece will have good eye relief. Eye relief is the minimum distance from the eyepiece that the observer's eye must be for the entire field of view to be visible. The eye relief with some cheap eyepieces may be only one-quarter of the ocular's focal length. This is much too close for comfortable viewing, especially if you wear glasses. With some modern designs the eye relief can exceed the eyepiece's focal length, making observing more enjoyable. (Table 3.2 surveys the pros and cons of various types of eyepieces.)

Other items

Finders

A finder is a small, low-power refractor mounted on the side of a telescope. Its purpose is to help the observer aim the telescope toward its target. Like binoculars, a finder (including the one shown in Figure 3.8) is specified by a pair of numbers indicating the instrument's magnification and aperture. The most common finders are 6 × 30 (6 power, 30-mm aperture) and 8 × 50 (8 power, 50-mm aperture).

While selecting the right finder for you is a matter of personal preference, most experienced deep-sky observers agree that the

Table 3.2 *Eyepiece pros and cons.*

Eyepiece	Pros	Cons
Kellner	Inexpensive; longer focal lengths offer good eye relief	Suffer from internal reflections ("ghost") more than others, especially when viewing bright objects
Orthoscopic	Sharp, clear images at all magnifications; good contrast; one of the best eyepieces for amateur telescopes	Field of view is typically narrower than other, more modern designs
Erfle	Very wide fields of view, giving outstanding panoramas of the deep sky	The spacious view takes its toll on edge sharpness, which suffers from optical astigmatism and coma
Plössl	Preferred by many deep-sky observers; excellent color correction, good eye relief, and a relatively large apparent field of view	Typically more expensive than orthoscopics
Super-Wide Fields (e.g., Tele Vue Nagler, Panoptic; Meade Super Wide Angle, Ultra Wide Angle)	Extremely wide apparent fields of view with good edge-to-edge image correction; gives the impression of looking out the porthole of a spaceship	Very expensive, possibly more than your telescope

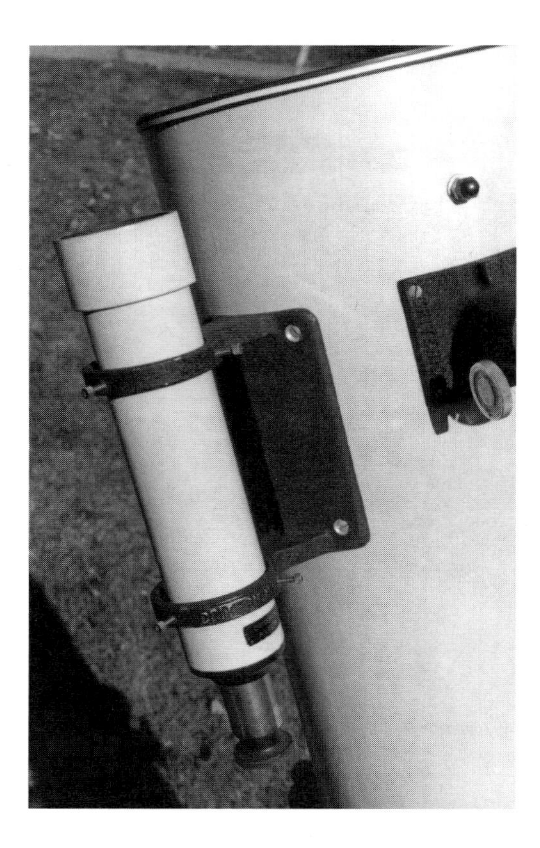

Figure 3.8. *A proper finderscope is crucial for locating the thousands of faint objects that await deep-sky observers. An 8 × 50 finder, such as seen in this photo, is usually considered to be the smallest usable size.*

smallest usable size is the 8 × 50. Anything less simply does not have enough light-gathering ability or magnification. (Incidentally, if your telescope came with a smaller finder, this doesn't mean that you should discard it immediately; rather, use it but understand that 8 × 50 and larger finders can make locating objects easier.)

Many Schmidt-Cassegrain telescopes come with a right-angle finder, incorporating a mirror or a prism at its eyepiece end to divert the incoming light through a 90° angle. While this makes looking through the eyepiece more comfortable, the prism flips everything left-right, making it very difficult to compare the field of view with a star atlas. A straight-through finder gives a view which is upside-down, but it does not swap left and right.

In recent years, one-power aiming devices (i.e., ones that do not magnify the view) have become very popular. The most popular of these new-generation sighting contraptions is the Telrad, designed by the late Steve Kufeld. The Telrad projects a bull's-eye target of three rings onto a clear piece of glass set at a 45° angle. The glass acts as a beamsplitter, reflecting an image of the target rings toward the observer's eye, at the same time letting starlight through from beyond. Other one-power aiming devices work in a similar manner. The purpose of the Telrad is not to replace a magnifying finderscope; rather,

Figure 3.9. *A selection of light-pollution reduction (LPR) filters. From left to right, the Hydrogen-Beta line, Ultra-High Contrast narrowband, Oxygen-III line, and Deep-Sky broadband. Photo courtesy Lumicon.*

it is meant to make aiming a telescope toward the general direction of an object easier.

Light-pollution filters

Perhaps the greatest problem facing today's deep-sky observer is the epidemic of light pollution. Instead of illuminating only their intended targets, many light fixtures scatter photons in all directions — including up. Not only is this a tremendous waste of money and resources, but poorly designed artificial lighting from buildings, streetlights, billboards, and other sources is rapidly swallowing up the stars. Deep-sky objects are especially vulnerable to light pollution since most are faint to begin with. Can anything be done?

Fortunately, many sources of light pollution shine in the yellow region of the visible spectrum, while many nebulae shine mainly in the blue-green and red regions. If the yellow wavelengths could somehow be suppressed without affecting the blue-green and red wavelengths, then the influence of light pollution would be greatly reduced. That's exactly what *light-pollution reduction filters* (or LPR filters) do. LPR filters (an assortment of which is shown in Figure 3.9) are also called nebula filters since they have the greatest impact on the visibility of planetary and emission nebulae. They have only marginal effects on the visibility of star clusters, reflection and dark nebulae, and galaxies, the light from which spans the whole of the visible spectrum.

LPR filters come in three varieties. *Broadband filters,* such as Lumicon's Deep-Sky Filter or Orion's SkyGlow range of filters, pass a wide swath of the visible spectrum and are best suited for astrophotography. They offer little benefit to observers peering through telescopes. *Narrowband filters,* such as Lumicon's UHC or Orion's

Table 3.3 *Some popular star atlases for beginners*

Atlas	Limiting magnitude	Deep-sky objects
Cambridge Star Atlas 2000.0	6.5	866
Edmund Mag 6 Star Atlas	6	over 1,000
Norton's 2000.0	6.5	over 600
1000+	6	over 1,000
Sky Atlas 2000.0	8	2,500
For intermediate to advanced observers:		
Uranometria 2000.0		10,300
Volumes I and II	9.5	(both volumes)
Millenium Star Atlas		Over 10,000
Volumes I, II, and III	11	(all three volumes)

UltraBlock filters, work well for observers trying to spot elusive nebulae from urban and suburban settings. Finally, line filters have an extremely narrow window, allowing only one or two specific wavelengths of light to pass. Lumicon's Oxygen III (O III) line filter, so called because its window is located at the green spectral line of oxygen III, is excellent for glimpsing challenging nebulae.

If you had to buy only one, a narrowband LPR filter would be the one to go for, followed closely by the O III line filter. Both will go a long way in revealing nebulae that are otherwise impossible to see.

Star atlases

All the telescopes in the world won't help if you don't know what to look for and where to find it. That's where a star atlas comes in. A star atlas is a detailed portrayal of the sky showing stars to below naked-eye visibility, as well as hundreds, even thousands of deep-sky objects. There are a number of good star atlases in print today. Table 3.3 lists the best ones.

By now you may be thinking that a deep-sky observer needs a suitcase to lug all this stuff around in. While a full-size suitcase is not necessary, a briefcase will certainly come in handy. Personally, I use two. One holds all of my eyepieces and filters, while the other carries a star atlas, a list of the objects I want to see that evening, three flashlights (a dim red light, a bright red light, and a white light), pencils, a clipboard, observing forms, and a radio. Thus armed, I am ready to tackle the night in style.

When all is said and done, the best stargazing equipment is the kind that gets used often. Sure, one can pour a lot of money into a large telescope, hordes of eyepieces, expensive filters, and books galore. But what good is all this stuff if it only sits in a dusty corner without ever seeing the night sky? On the other hand, if a modest, inexpensive instrument gets used often and gives its owner great pleasure, then it is the best telescope in the world. Which do you own?

CHAPTER 4

Lost in space

Trying to locate objects in the sky that are too faint to be seen with the eye alone can present a great challenge. Most deep-sky observers develop their own methods for zeroing in on difficult targets, but in general, all techniques fall into one of two broad categories: star-hopping and using celestial coordinates. Star-hopping involves going from a naked-eye jumping-off point (typically, a bright star) to a faint target in a series of hops from one star or star pattern to the next, until the quarry is found. It's a great way to get to know the sky and your telescope at the same time. The other method, using celestial coordinates, usually requires a polar-aligned, equatorially mounted instrument. An object's heavenly address is dialed in, and the instrument is swung around to point in the right direction. Let's discuss both.

Star-hopping

In order to play this game of celestial hopscotch, the finderscope must first be lined up with the telescope. It's usually best to align the finder during the day, if at all possible. Center the telescope on a terrestrial target such as a tree or a distant mailbox, then adjust the finder until the views through the finder and telescope match. This way, at night the finder will require only minimal fine-tuning.

With the finder and telescope now pointing in the same direction, look at a star chart and locate the first object you want to look at. Next, survey the area around the object for the nearest naked-eye star that you can recognize in the sky. Aim the telescope in its direction. Return to the star chart and hunt for little patterns among the fainter stars that lie between the naked-eye star and the object of your quest. You might see a small triangle, arc, rectangle, or line; just about any shape is possible. Viewing through the telescope's finder, shift the telescope from the naked-eye star to that pattern of stars, then stop and go back to the atlas. Are you still on track? If so, good; if not, start again. It's important not to rush star-hopping. Take your time and enjoy the thrill of the hunt. Continue going back and forth between the chart and the finder until you are in the object's immediate area. Move the telescope so that the finder is aimed where you believe the object should be, and take a look using your lowest-power eyepiece. The target ought to be in, or at least near, the field of view.

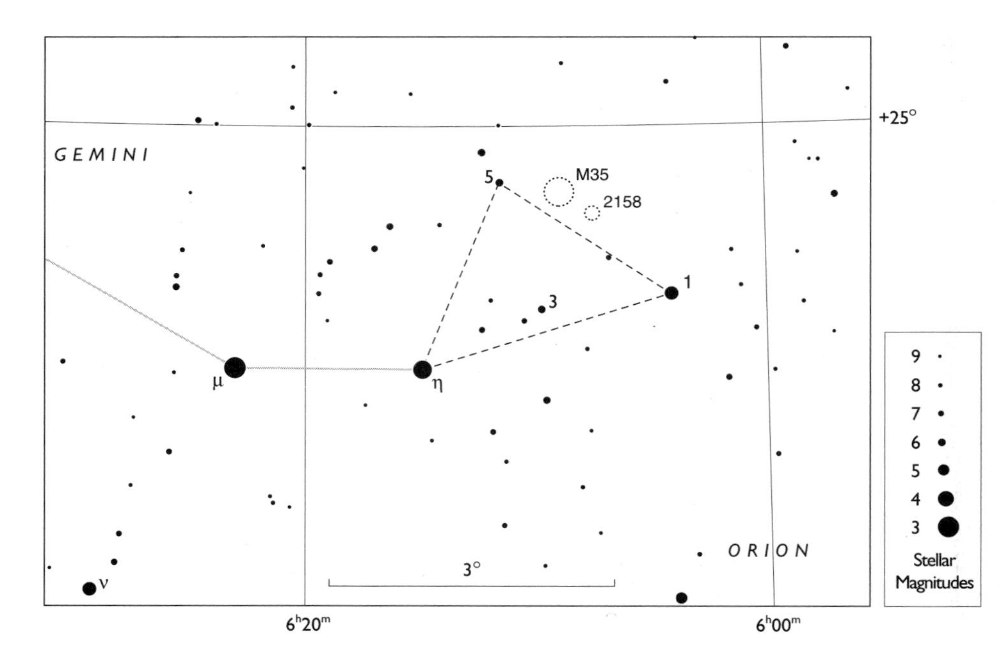

Figure 4.1. *You can get there from here! Using this diagram, you can star-hop to M35, a bright open cluster in the winter constellation Gemini, by way of the two large stars that form the foot of one of the twins of Gemini. Hop from Mu (μ) to Eta (η), then pause to find the triangle formed with 1 and 5 Geminorum. North is up.*

Let's take an example. When I first became interested in deep-sky observing in 1969, I was using a modest (okay, a terrible) 100-mm (4-inch) Newtonian reflector that came with a flimsy altazimuth mounting and a one-power aiming tube that looked more like a drinking-straw than a finder. Back then, I was happy if I could find even the brightest sky objects, such as the Moon and the naked-eye planets; everything else was a challenge.

One object that gave me a great deal of grief was M35, an open cluster in Gemini, the Twins. (You kind of grow into this hobby; today, I can see M35 with the naked eye on dark nights.) The chart in Figure 4.1 shows M35 near the constellation's southwest corner. How would you go about finding it (with a real finder, that is)? First, in the sky trace out the constellation of Gemini. Can you see the two stars that form the foot of the twin on the right? On the chart these stars are labeled Mu (μ) and Eta (η) Geminorum. Aim your telescope at Eta. Notice how two faint stars, one of 4th magnitude, the other of 6th, form a right triangle with Eta. Find that triangle in your finderscope. M35 resides just west of the triangle's faintest star, the one marking the right angle itself. Nudge your telescope until it is aimed at that spot. You may or may not see M35 in the finderscope, but you ought to see the cluster in the telescope's eyepiece. Is it there? Great, you've found it!

Figure 4.2. *Some amateurs prefer using celestial coordinates and setting circles for finding faint objects. Note the finely detailed circles on this equatorial mount.*

M35 is fairly easy to locate compared with some of the other objects listed later in this book. Let's face it — there will be times when you will want to throw up your hands in disgust and quit. But stick with it. Resist the urge to use the mechanical and electronic means for finding deep-sky objects that are outlined below. In my humble opinion, star-hopping is the best way to hone your skills as an observer while learning the night sky.

Using setting circles

Most equatorial mounts are fitted with a pair of graduated rings, one on the right-ascension axis and one on the declination axis, called setting circles. In Figure 4.2 the right ascension circle (bottom left) is graduated in 24 hourly segments, each corresponding to one hour of right ascension, while the declination circle (bottom right) is marked in degrees, from +90° to –90°. Setting circles are designed to let an observer find objects in the sky simply by moving the telescope until the circles correspond to the celestial coordinates of the intended target. Many amateurs are intrigued with the prospect of

Table 4.1 *Suitable stars for setting-circle calibration (epoch 2000.0).*

Star	R.A. h m	Dec. ° '
Spring		
Alphard (Alpha [α] Hydrae)	09 27.3	–08 39
Regulus (Alpha [α] Leonis)	10 08.4	+11 58
Spica (Alpha [α] Virginis)	13 25.2	–11 10
Summer		
Antares (Alpha [α] Scorpii)	16 29.4	–26 26
Altair (Alpha [α] Aquilae)	19 50.8	+08 52
Vega (Alpha [α] Lyrae]	18 36.9	+38 47
Autumn		
Fomalhaut (Alpha [α] Piscis Austrini)	22 57.6	–29 37
Alpheratz (Alpha [α] Andromedae)	00 08.4	+29 05
Hamal (Alpha [α] Arietis)	02 07.5	+23 28
Winter		
Aldebaran (Alpha [α] Tauri)	04 35.9	+16 31
Rigel (Beta [β] Orionis)	05 14.5	–08 12
Sirius (Alpha [α] Canis Majoris)	06 45.1	–16 43

finding deep-sky objects without having to star-hop from one point to another.

Here's the catch. Before setting circles can be used, the telescope mounting's polar axis must be aimed toward the celestial pole. The north celestial pole lies very close to the North Star, Polaris. For our purposes here, aiming the polar axis exactly at Polaris should be adequate, but even that can take a little doing. Use the step-by-step procedure below.

(1) Align the finderscope with the main telescope.
(2) Swing the telescope around until the tube is parallel to the polar axis. Check to see that the declination circle reads +90°. (Most are preset at the factory, though some mounts may require that the circle be adjusted.)
(3) Lock both of the mounting's axes.
(4) By turning the entire mounting left and right (being careful not to move the telescope along its axes), aim the telescope toward the north.
(5) By loosening the mounting's latitude adjustment screw, tip the telescope up or down until you see the North Star in a low-power eyepiece fitted to the telescope.

With the telescope mounting now aimed toward (or at least, near) the north celestial pole, unlock both of the mount's axes and move the instrument toward one of the stars listed in Table 4.1. Once the chosen star is centered in view, turn the right-ascension circle (taking care not to touch the declination circle) until it corresponds to the star's right ascension. Note that while the positions of these reference stars given in Table 4.1 are for January 1, 2000, these values are good for a couple of decades either side of that date.

With your setting circles now calibrated to one of the stars in Table

Figure 4.3. *In recent years, computerized setting circles have eliminated the need for polar alignment before a mounting can be used. Most popular units include built-in libraries containing the coordinates of thousands of objects. Courtesy Orion Telescope and Binocular Center.*

4.1, turn on the clock drive so that the right-ascension circle tracks the object in view; otherwise, you will have to recalibrate it every few minutes. (Incidentally, some telescopes do not have a right-ascension circle that tracks with the clock drive. With such instruments, as well as with all undriven equatorial mounts, you have no choice but to return periodically to a reference star and reset the right-ascension circle.)

Try your luck with the setting circles. Before attempting to locate a faint deep-sky object, try ferreting out one of the other reference stars in Table 4.1. Turn your telescope in right ascension and declination until the dials indicate the proper position, then look through your finder. Is the star centered? Is it at least near the center of the field of view? If it isn't, your telescope is not polar-aligned properly, and you'll have to go back and start the procedure over again.

Unless the polar axis is aimed closely to the celestial pole, you will find that the setting circles' accuracy falls off when moving from one object to another, especially if the two objects are separated by a vast expanse of sky. Then it is best to use a technique called *offsetting*. Before trying to find an obscure target, aim at a bright star that lies nearby. Check and recalibrate your setting circles to the star's coordinates, then move over to the intended target. By calibrating on a known star near the object, then jumping off from there, you minimize the error.

Leave the driving to us

In recent years, digital setting circles, such as the one shown in Figure 4.3, have enjoyed great popularity. These electronic wizards not only make polar alignment easier (in many cases, completely unnecessary), they can also be used on any kind of telescope regardless of mounting. Most are set up simply by aiming at two preselected stars and punching in their proper codes. The computer "brain" inside will then automatically calculate the locations of other objects accordingly. The user need only select which object to observe and key its code into the control unit. The telescope is then rotated until the readout says the intended object is centered in the field of view. Most electronic setting circles include libraries of hundreds or even thousands of celestial objects and allow observers to bounce from one object to another quickly and accurately.

The trickiest thing about using digital setting circles can be figuring out how to attach them to the telescope in the first place. Some units are made to fit easily onto the most popular telescope models, while fitting others requires some mechanical dexterity on the part of the observer. If you are intending to fit digital circles, you would be well advised to contact the manufacturer and ask if there will be any problem adapting the unit to your telescope's mounting.

Personally, I would not recommend that observers, especially those new to astronomy, become dependent on setting circles for finding celestial objects. Make no mistake, though — setting circles do have their place, especially for amateurs making a series of observations of the same objects, such as variable stars, or when light pollution is so severe that finding stars to hop by just isn't practical. For the rest of us, however, my best advice is to learn how to read the sky before learning how to read setting circles, either electronic or mechanical.

"But I don't see anything!"

All observers have had those moments when, even after great care has been taken to aim toward an intended target, it simply isn't there. No matter how often charts and references are checked and rechecked, the object just isn't visible. Is it time to move on to another object? Don't give up so easily. Here are a few secrets that may help render an invisible object visible.

Many observers overlook the need to let their eyes become accustomed to the darkness before searching for faint objects. When we first go out into the dark from a bright room, our eyes begin a two-step process of increasing their sensitivity. The first step happens almost immediately, as the pupil opens, or dilates. The second step is a chemical change in the eye that can take considerably longer to complete: the eye builds up a chemical substance called rhodopsin, or visual purple, which increases the sensitivity of the eye's dim-light receptors, the rods. While most people's eyes achieve partial *dark adaption* in about 20 to 30 minutes, the entire process can take an hour or more to complete. Plan your night's observing program so that brighter objects are viewed first. Wait at least an hour before you begin to search out those faint fuzzies that are just on the brink of visibility.

Figure 4.4. *To find a faint object from a light-polluted site, you may need to go into hiding! A simple "cloaking device" can go a long way in helping you to detect difficult targets.*

Next, comes patience. Take a deep breath (try not to breathe on the eyepiece, however!) and slow down. Just because an object doesn't reveal itself immediately doesn't mean that it won't after a few minutes of concentrated searching. If eye fatigue sets in, move away from the eyepiece and take a short break.

That leads to the next point. Any discomfort will detract from your concentration and introduce "noise" between your eye and your brain, masking faint objects. Believe me, there is nothing more uncomfortable than bending over to look through an eyepiece. Many observers use an observing chair or stool so that they may remain seated while looking through the eyepiece. And, as most mothers would remind us, sit up straight — don't slouch!

If a faint object refuses to be seen when stared at directly, try looking at it with your peripheral vision. This technique, called *averted vision,* has been used successfully by deep-sky observers for years, even centuries. With peripheral vision (that is, slightly looking to one side or the other of the object being observed), light falls on a more sensitive part of the eye's retina. Frequently, objects that are invisible when viewed directly will be seen with averted vision. A good example of this is NGC 6826, the so-called Blinking Planetary in the summer constellation Cygnus, the Swan. This nebula's nickname comes from the fact that, when viewed with averted vision, it appears as a distinct sphere of bluish light, but when viewed with direct vision it disappears entirely! (The Blinking Planetary is described in more detail in chapter 7.)

Some objects still refuse to show themselves even when we use averted vision. In such instances, try centering the object as best you can and tap the side of the telescope tube ever so slightly, imparting a gentle side-to-side motion to the field of view. Marginally visible objects may actually reveal themselves. But again, your touch must be soft, just enough to vibrate the telescope lightly.

Nothing can hide faint objects more effectively than light pollution. As mentioned in chapter 3, filters are available that help decrease light pollution's adverse impact on certain kinds of deep-sky objects. If you are hunting for an especially difficult nebula, then it might just reveal itself when viewed through a light-pollution reduction (LPR) filter. Some LPR filters can work magic with certain types of objects, but they are poor substitutes for clear, dark skies.

Localized light pollution, such as from a street- or porch light, is a little easier to deal with. Though I have never found out why, one of my neighbors insists on leaving the back porch light on all night, every night. To help combat that distraction, I don a cloaking device, simply a piece of dark cloth placed over my head like a photographer's shroud. Many observers use a black turtleneck shirt put on upside-down — instead of putting it on over your head, put your head into the shirt, as shown in Figure 4.4. Then, as you look through the eyepiece, pull the shirt up and over the eyepiece as shown in the photograph. Yes, it looks a little strange (okay, it looks very strange), but I can see about a half-magnitude fainter using this method than I can without it. The only drawback is that, in damp weather, it tends to accelerate eyepiece fogging. To prevent this, try not to hold the cloth too close to you or the eyepiece.

The greatest asset to the observer trying to find and see difficult objects is experience. As your eyes and brain become more experienced at looking for faint, diffuse deep-sky objects, you will find that seeing them will actually become easier. Unfortunately, there is no way to speed up this process, other than just getting out and observing on as many clear nights as possible.

CHAPTER 5

For the record

Ever since I first became impassioned with astronomy some three decades ago, I've kept a detailed logbook of all the celestial objects that I have seen through my instruments. Taking notes at the telescope about each object is a great way for observers to do more than just the quick hunt-and-peck peeking that most are used to. By concentrating on one object, you can detect fine nuances often missed by casual observers. At the same time, taking notes will provide a permanent record of how you develop as an observer.

Notekeeping can be as simple or as complex as you desire. Many ardent observers restrict their notes to simple one-liners, commenting perhaps only on an object's size and brightness, while others prefer detailed studies, which may include comparative comments on the object's general appearance through several different instruments, as well as sketches of the object and notes on how it was found. I employ and recommend a system that is closer to the latter approach. For each new deep-sky object I spot through one of my telescopes, I record the following:

(1) catalog number (Messier, NGC, etc.)
(2) the object's celestial coordinates and home constellation
(3) date and time
(4) telescope aperture and focal length
(5) eyepiece type and focal length
(6) filter used, if any
(7) sky conditions (e.g., transparency, seeing, and faintest star visible to the naked eye)
(8) observing location
(9) general description of the object.

Most of these entries should be self-explanatory, but let's look at a few that might not be. Rating the sky conditions at the time of observation offers an important insight into the quality of the observation. On a night when a thin haze and clouds cover the sky, your notes may record a particular object as especially difficult to see, while on a crystal-clear night the same object may appear obvious.

Sky conditions may be expressed in two different ways: *transparency* and *seeing*. Transparency refers to the clarity of the sky overhead, as it appears to the naked eye. The greater the transparency, the fainter the stars that will be seen. Most observers assess sky clarity

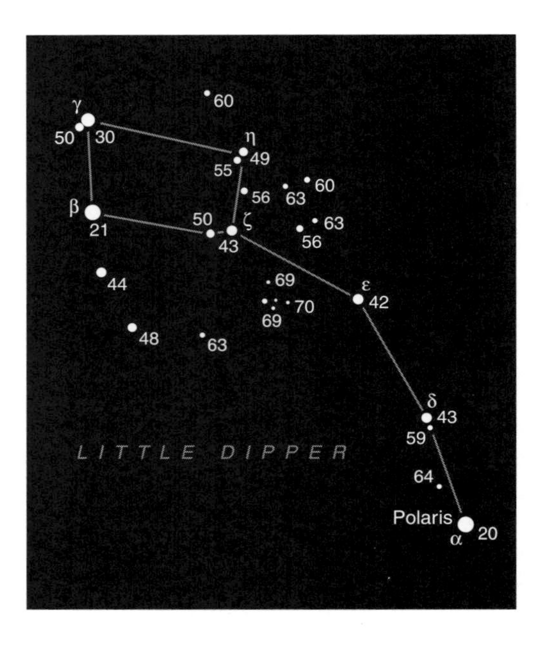

Figure 5.1. *How clear is clear? Many observers in the Northern Hemisphere judge the transparency of the night sky by noting which of these reference stars in the Little Dipper is the faintest visible.*

on a scale from 0 to 10, with 0 indicating completely overcast conditions and 10 denoting perfect ones. Needless to say, there are very few times when your notes will specify either of these two values; most nights they will fall somewhere in between. It is also good practice to record the faintest star visible to the naked eye in a particular part of the sky. Many amateurs find it convenient to use the stars in the Little Dipper as a gauge, since they are in the sky year-round. Figure 5.1 shows the visual magnitudes of these stars, with the decimal points omitted to prevent their being confusing with stars.

Seeing has to do with the steadiness of the atmosphere. How often have you looked at the night sky and seen the stars twinkling rapidly, especially near the horizon? This twinkling effect, called *scintillation* (Figure 5.2), is the result of localized, small-scale density fluctuations in our planet's atmosphere. While pretty to look at, twinkling stars often frustrate observers. Poor seeing (that is, rampant twinkling) interferes especially with observations of close double stars. It is also common practice to quantify seeing conditions on a 0-to-10 scale, with 0 signaling fierce scintillation and 10 signifying very steady skies with no twinkling evident even at high magnification. Again, conditions are usually somewhere between these two extremes.

Make a drawing

Figure 5.4 (shown at the end of this chapter) is a suggested form for recording observations; adopt or change it as you see fit. The big circle at the bottom is for making a drawing of the object under observation; the circle's edge marks the edge of the eyepiece's field of view. It has been said that a picture is worth a thousand words. In observational astronomy, that estimate is low. Trying to describe all the subtleties in a deep-sky object can be a daunting task no matter how broad your vocabulary. A simple drawing will do more to convey the view than even the most thoroughly written description. You don't have to be a student of fine art or an accomplished artist to make accurate sketches of deep-sky objects. All you need are some basic materials and a little perseverance. A shopping list of materials is offered in Table 5.1; the perseverance part is up to you.

It is best to draw deep-sky objects in steps, allowing the sketch to develop along the way. But before the first bit of lead is put to paper, the object must be examined with a wide variety of magnifications to determine which shows it in its best light. Examine it with each of your eyepieces and decide which gives the best overall view. Use this view as the basis for the drawing.

At the same time, determine which way is north in the field of view. Convention has it that a sketch is usually made with north at the bottom, but you decide. You may prefer to orient the drawing so that north is at the top. Whatever you choose, be sure to note where north is somewhere on the drawing. It is also essential to mark east or west, so that later you can distinguish mirror-image drawings from unflipped ones.

Begin the drawing by faintly sketching in the positions of the stars in the field of view, as in Figure 5.3a, locating each one as accurately as possible. Do not feel obliged to draw in every faint star that you see, especially if the field of view is littered with them.

Steps two and three (Figure 5.3b and 5.3c) involve going back over the stars to change the appearance of each to match its apparent brightness, and drawing the outline of the object. Draw the brighter stars larger than the fainter ones. (Some observers also note the magnitudes of individual stars in the field, if they are known,

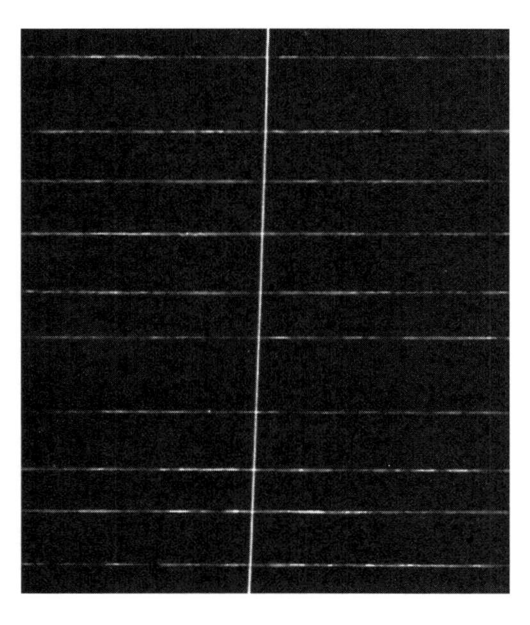

Figure 5.2. *Viewing conditions are critical to the observer's chances of splitting close double stars and discerning fine detail in many other deep-sky objects. In this photo, Finnish photographer Pekka Parviainen has captured the twinkling of Sirius by repeatedly letting the star drift across the field of his camera, raising the aim for each sweep to record a new track. A final vertical sweep of Jupiter — which does not twinkle — produced the continuous white trail that bisects the photo.*

Table 5.1 *The astro-artist's shopping list.*

Clipboard or easel	✔
Pad of smooth, white sketching paper	✔
Soft-leaded artist's pencil (preferably H or HB)	✔
Smudging tool	✔
Wedge-type eraser	✔
Sketching charcoal (optional)	✔

omitting the decimal points to avoid any possible confusion with faint stars.)

Now, ever so lightly draw in the outline of the target. Don't press down on the pencil, but instead let its weight alone keep it on the paper. Double-check to make sure that the object's shape and position on the paper match its true appearance in the eyepiece.

Step four (Figure 5.3d) is a test of bravery: add shading to the object until it matches your visual impression through the telescope. Keep in mind that the drawing is, in effect, a negative image of the object, so the brightest parts of the object must appear darkest on the sketch and vice versa. Rub a small amount of lead or charcoal across the object's outline. Then, using a fingertip, a smudging tool, or a soft eraser, smudge the lead to recreate the subtle darker and brighter regions.

The last step is to refine the drawing's accuracy. Examine the target again with several different eyepieces and add in any detail you previously missed.

Never rush a drawing. There will be times when either your fingers are nearly numb with cold or the mosquitoes are flying about, making it difficult to concentrate. It happens to everyone. Just put the pencil down and take a break. Go inside (but don't switch on a light, because you will lose your dark adaption) and come back out five or ten minutes later. Trying to draw with any accuracy under adverse conditions is not fun. And, after all, fun is why we are here.

Open a file

An approach that is rapidly gaining in popularity is to enter all observational records into a computer. Storing notes in a computer file facilitates the organization, sorting, and analysis of records, allowing such functions to be done much more quickly than is possible with paper records. A scanner can be used to scan drawings into separate files. While computer database programs abound, some astronomical software (such as *The Sky* by Software Bisque) also allows observers to record and store their own notes electronically as well as create customized, detailed star charts. If you like this idea, and also want to compare your findings with others, then investigate the New General Program of Nonstellar Astronomical Objects (NGP for short), created by Dean Williams. (For more about software, see the list at the end of the book.)

Whether you elect to record observations in a notebook or on a

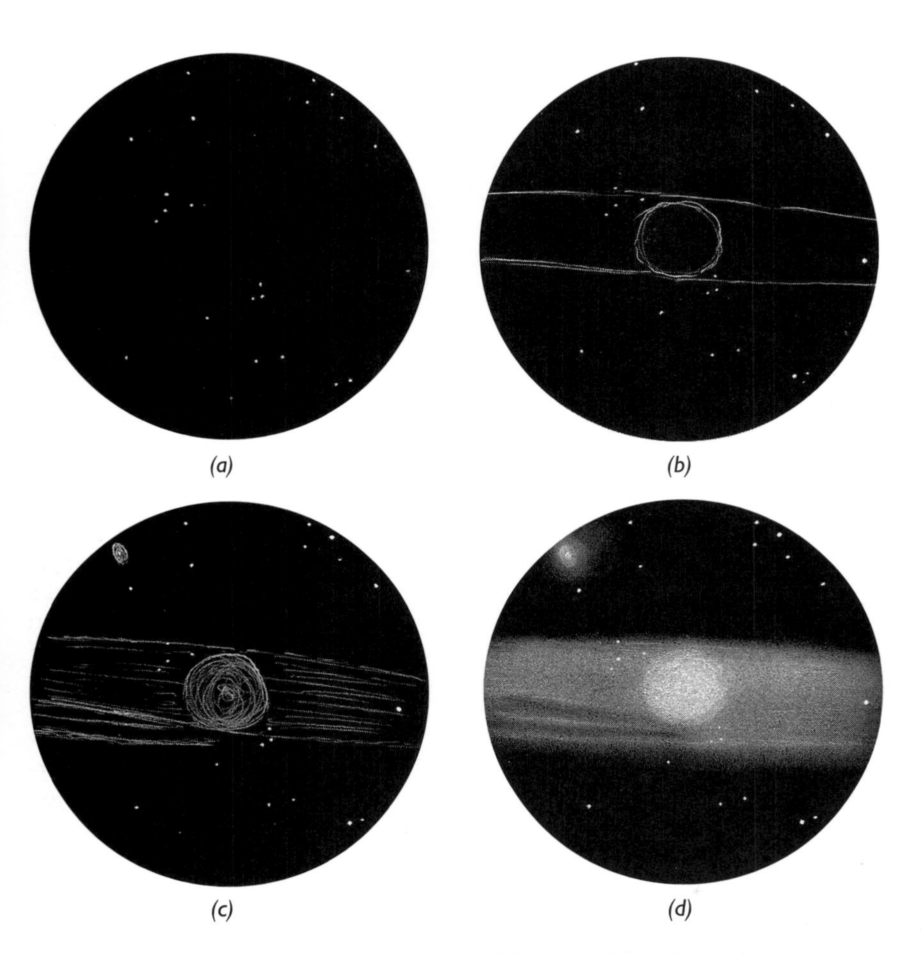

Figure 5.3. *Drawing deep-sky objects is as easy as 1-2-3 — and 4. First, draw in the field stars as dots (a). Faintly draw in the outline of the object and modify the field stars to show their apparent magnitudes (b). Add more details to the basic outline (c). Finally, shade in the object with varying tones of gray, remembering that the drawing is actually a negative view (d). (The drawings above are reverse images of the originals.)*

computerized database, one thing is certain: all other things being equal, a dedicated, experienced observer will see much fainter detail in deep-sky objects than will a casual skygazer. For people with normal vision, spotting faint objects requires no special ability: it is a skill that is learned and fostered only by going outside with a telescope or binoculars. To this day, I can look at objects that I have watched over 30 years and still be surprised by subtleties I hadn't noticed before. How long it takes to develop this ability varies greatly from person to person, but the process can be shortened by making detailed notes and drawings. Doing so forces the observer to slow down and not just look at, but actually *see* the universe.

Date __________ Time __________

Observation Report

Object __________ Constellation __________

R.A. __________ Dec. __________

Observing Site __________

Transparency (1–10) ______ Seeing (1–10) ______ Mag. limit ______

Telescope __________ Eyepiece __________

Filter __________ Misc. Equip. __________

Notes __________

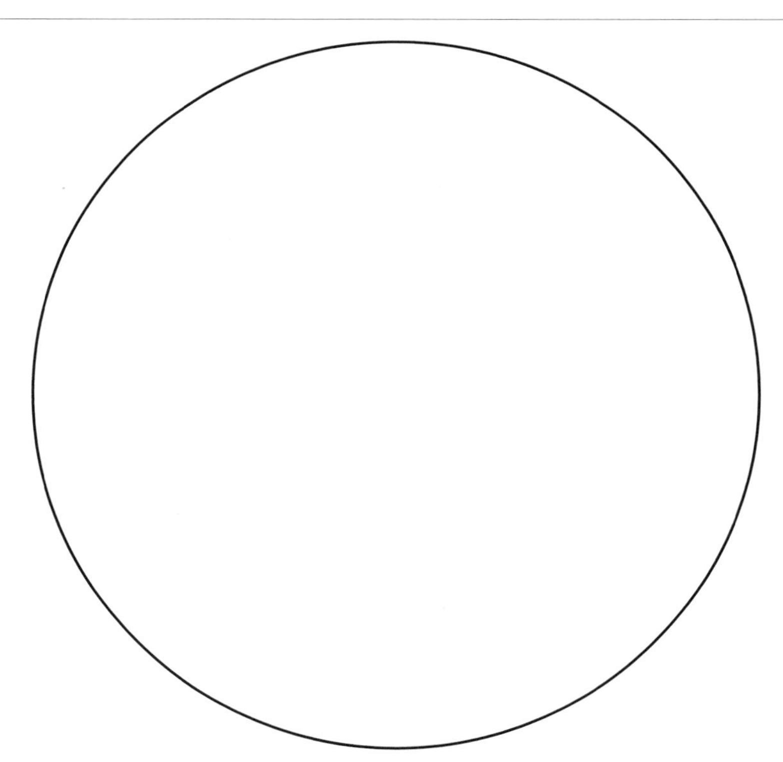

Figure 5.4. *The author's form for recording observations. The "notes" section is for a description of the object as well as essential data such as orientation, magnification, and field of view.*

CHAPTER 6

An extragalactic spring break

The cold nights of winter are a quickly fading memory as the warmth of spring introduces astronomers to a new collection of constellations and associated deep-sky objects. While the summer and winter skies are dominated by nebulae and clusters of stars, our springtime window on the sky (Figure 6.1) is directed away from the Milky Way and out into the farthest reaches of the universe toward hundreds, even thousands, of galaxies.

You don't necessarily need a large telescope to visit the depths of intergalactic space. Most of the galaxies described below can be seen through a 100-mm (4-inch) or 150-mm (6-inch) telescope; some are even visible in steadily held binoculars. But the spring sky has more to offer than just galaxies. Scattered throughout are a variety of interesting double and variable stars, open clusters, and even a rogue globular cluster or two. And for those with a good view toward the deep south, the springtime Milky Way hosts a tremendous selection of deep-sky objects awaiting visitors from the north.

In this chapter and the ones that follow, deep-sky objects are discussed in alphabetical order by constellation, and then by right ascension; their coordinates and other data are listed at the end of the book.

Boötes

The constellation Boötes, representing a herdsman on classical star charts, contributes a lone double star to this survey. A glance at this region in a detailed star atlas reveals a number of galaxies within the constellation's boundaries, though all are too faint for inclusion here. After you complete this introductory study, you might wish to return and try your hand at these fainter targets. Boötes passes the meridian at midnight (what astronomers call *midnight culmination*) in early May and is visible during the early evening hours of June and July.

Xi (ξ) Boötis, discovered in 1780 by William Herschel, is a beautiful pair of colorful stars located 8½° due east of the constellation's most brilliant star, Arcturus. Even the smallest telescopes will show the yellowish 5th-magnitude primary star accompanied by a reddish orange 7th-magnitude secondary sun, set among a crowded field of stars.

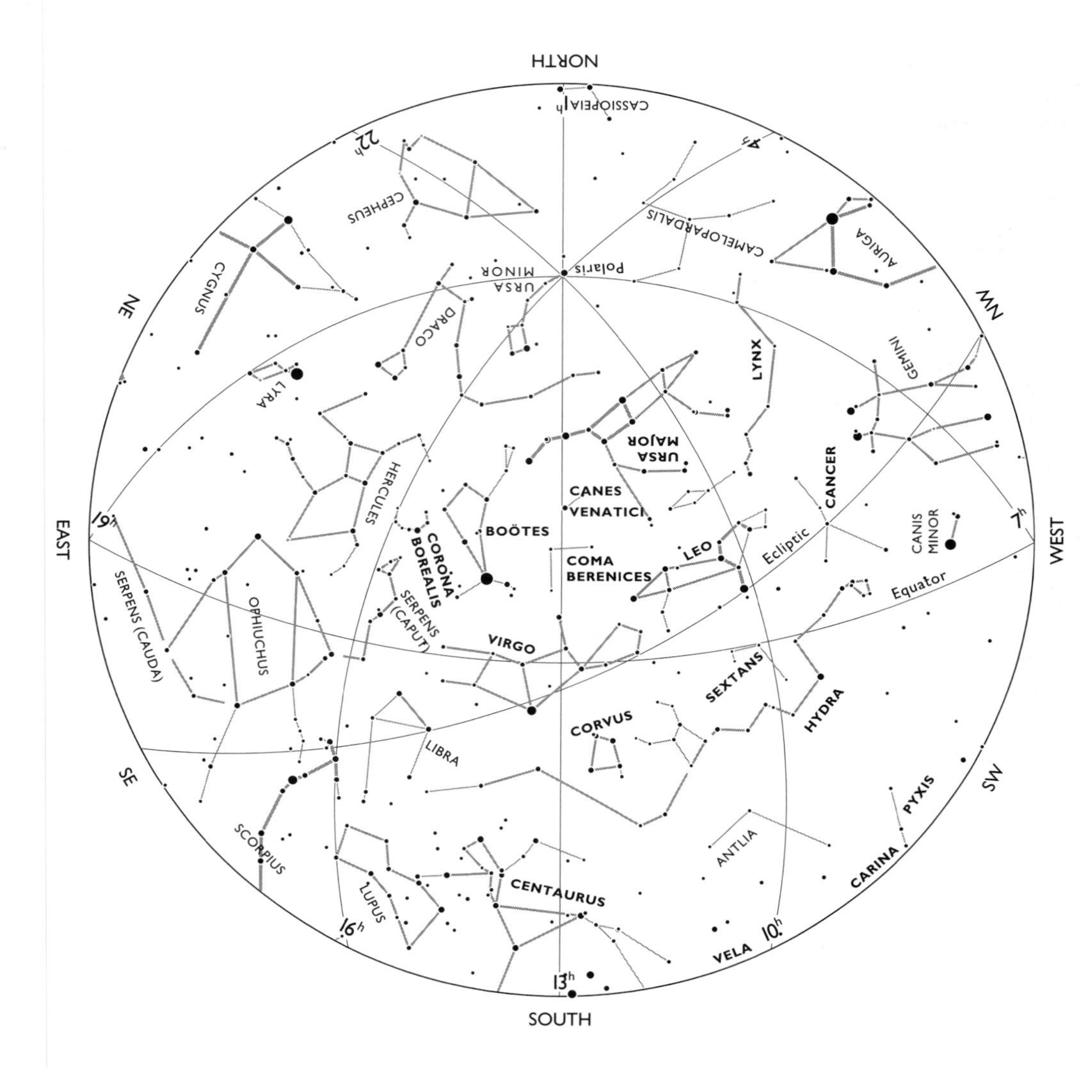

Figure 6.1. *The spring naked-eye sky. Constellations discussed in this chapter are shown in boldface. The chart is plotted for 30° north latitude. For viewers north of that latitude, stars in the northern part of the sky will appear higher above the northern horizon and stars in the south lower (or possibly below the southern horizon).*

Xi Boötis is known to be one of the closest binary-star systems to Earth, with estimates placing it about 22 light-years away. Given this distance and the stars' angular separation, the primary and secondary must be orbiting each other at a distance of about 33 astronomical units. Observations indicate that the secondary circles the primary once in 150 years, causing rapid changes in the system's position angle and separation. As seen from Earth, the companion star passed closest to the primary in 1912, when it was just 1.8" to the southeast. Since then, Xi Boötis B has swung clockwise halfway around Xi Boötis A, reaching a separation of 7.3" in 1984. More than a decade later, they are beginning to close once again, though the separation remains near 6".

Cancer

At first glance the constellation Cancer, the Crab, looks like a void wedged between Regulus in Leo and the brilliant winter stars Castor and Pollux. While Cancer may not be much to look at from a naked-eye standpoint, it holds three of the season's most spectacular deep-sky sights. Cancer passes midnight culmination at the beginning of February and is well placed for early evening observation in April and early May.

Zeta (ζ) Cancri, marking the tip of the Crab's western claw, is one of the season's crowning triple stars. Although Zeta is inconspicuous to the eye, it is easily found by extending a line from Castor through Pollux in Gemini and continuing for twice the distance to the southeast. Backyard telescopes show the 5.6-magnitude primary, Zeta A, matched with two 6th-magnitude attendants. Zeta B, lying about 1" from the primary, completes an orbit every 59.6 years. Zeta C, about 6" from Zeta A, takes a more leisurely 1,150 years to orbit. Collectively, they put on a great show through just about any telescope.

Spectroscopically, Zeta A is classified as a Type-*F* star, while the B and C companions are both cataloged as Type *G,* the same as our Sun, and shine like golden embers. In his classic *Bedford Catalogue,* Admiral William Smyth (1788–1865) described the primary as being yellowish, the closer companion to have an orange tinge, and the third star also as yellowish. What is your impression?

M44 (NGC 2632), known popularly as the Beehive Cluster or Praesepe, is considered by most observers to be the premier open cluster of the northern spring skies (Figure 6.2). Recorded as a "little mist" by the Greek poet Aratus as far back as 260 B.C., the Beehive is located in central Cancer, a little less than halfway between Pollux in Gemini and Regulus in Leo. It can be spotted with the unaided eye and is unmistakable with even the slightest optical aid.

Not all observers will immediately appreciate the grandeur of M44. Its huge 1½° span makes it difficult to fit into the eyepiece field of most telescopes. As a result, the clustering effect becomes lost and the observer may be less than impressed. The saying, "You can't see the forest for the trees," certainly applies to highly spread out open clusters like M44.

The bees in this celestial hive come alive in sparkling style, however, through binoculars and rich-field instruments. Personally, the best overall view of M44 I've ever had is through my 11 × 80 glasses. With these close to six dozen points of light are seen across a field illuminated with the combined glow of another 130 fainter stars. A distinctive trapezoid of four 6th-magnitude suns marks the center of Praesepe, while **Burnham 584** (frequently written as β584), an interesting triple star, lies a little to their south. All three of its components shine with nearly equal brightness and are separated by 45" and 93", respectively.

Figure 6.2. *M44, the Beehive Cluster or Praesepe, in Cancer. Considered by most observers to be the premier open cluster of northern spring skies, its huge 1½° span makes it difficult to fit into the eyepiece field of most telescopes. North is up. Photo by George Viscome.*

M67 (NGC 2682). Sad but true — most amateur astronomers bypass this second open cluster in Cancer in favor of its more famous neighbor, M44. But by ignoring M67, they are missing out on one of the sky's most impressive deep-sky objects. Don't you make the same mistake! M67 is easiest to find by using the trapezoidal head of Hydra, to the south of Cancer, as an arrow. Trace a line from Eta (η) Hydrae northward through Epsilon (ε) Hydrae, and extend it for 5½° to arrive right at M67.

M67 is a crowded horde of as many as 500 stars shining at 10th magnitude and below. This great population creates an impressive sight in all instruments, but especially through 150-mm and larger scopes. In the *Webb Society Deep-Sky Observer's Handbook,* volume 3, Kenneth Glyn Jones of the United Kingdom describes his view of M67 through a 200-mm (8-inch) telescope: "A bright object [with the] brightest star on the north-eastern edge . . . in the central area, there are about 20 stars of 10th to 12th magnitude." With low power, many small- and medium-aperture telescopes display what at first appears to be nebulosity engulfing the cluster. This is merely an illusion of aperture and magnification, for increasing both immediately disperses the imagined "clouds." In their place shine a myriad of stars, many with subtle tints of yellow, orange, and red. These colors tell astronomers that M67 is composed of "mature" stars. In fact, studies point to M67 as being one of the oldest open clusters known.

Canes Venatici

In mythology, the small constellation Canes Venatici, just south of the Big Dipper's handle, represents a pair of hunting dogs. To today's observers, however, the hunting dogs are notable for retrieving an impressive selection of galaxies. Canes Venatici reaches midnight culmination in early April. The constellation passes close to the zenith in the early evening hours of May and June.

M106 (NGC 4258). Let's begin the tour of Canes Venatici at M106, an outstanding example of an Sb spiral. Even under suburban skies, finderscopes will show a dim smudge near a faint field star set just east of the midpoint between Phecda (Gamma [γ] Ursae Majoris, the southeastern star in the Big Dipper's Bowl) and Chara (Beta [β] Canum Venaticorum). Through a 100-mm telescope M106 shows a bright, oval disk, which, to me at least, looks a little like an underinflated football. A 200-mm instrument begins to add some mottled definition to the spiral halo, along with a bright central nucleus.

Many observers comment with some surprise at the small size of M106. Although long-exposure photographs reveal M106 to span about 18' × 8', most visual reports suggest a much smaller extent; the discrepancy is probably due to the low surface brightness of the galaxy's spiral arms. A 305-mm (12-inch) barely reveals these arms, even on exceptional nights, and only the largest amateur telescopes have sufficient light-gathering ability to show the galaxy's full breadth.

M94 (NGC 4736) lies a little northeast of the midpoint between Cor Caroli (Alpha [α] Canum Venaticorum) and Chara (Beta [β] Canum Venaticorum). An Sa spiral, M94 looks nearly circular, with the faint spiral halo enveloping a bright center. Many describe it as looking like an unresolved globular cluster, since any spiral structure in the galaxy's arms is seen only on long-exposure photographs. Although most references and catalogs rate the magnitude of M94 within a couple of tenths of that of M106, it strikes most observers as glowing distinctly brighter. This effect is probably caused by their comparative sizes. M94 exhibits a higher surface brightness than its neighbor since its light is spread across a smaller, 11' × 9' area.

M63 (NGC 5055), a 9th-magnitude Sb galaxy, is nestled in a barren naked-eye field about two-thirds of the way from Alkaid (Eta [η] Ursae Majoris), at the end of the Big Dipper's handle, to Cor Caroli. You'll find it a little north of a distinctive four-star triangle. Often ignored by observers because of its sparse surroundings, M63 is bright enough to be seen with 10× binoculars on clear, dark nights. Small

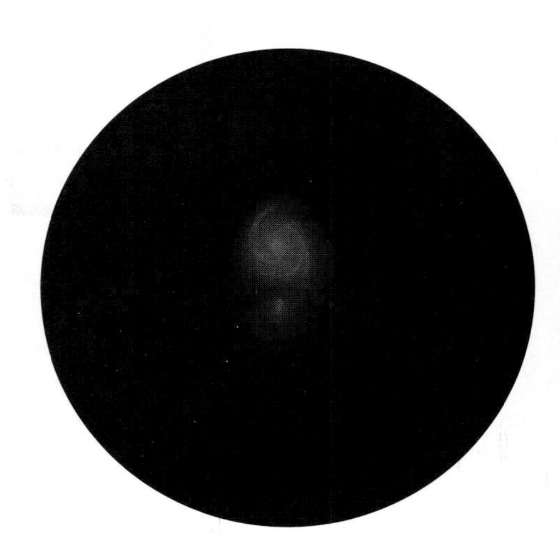

Figure 6.3. Left: *M51, the Whirlpool Galaxy, and its companion NGC 5195 in Canes Venatici. M51 is a classic example of a face-on spiral galaxy. Both galaxies may be spotted through 7× binoculars. Photo by Chuck Vaughn.* Right: Eyepiece Impression 1. *M51, as seen through the author's 333-mm Newtonian and 12-mm Nagler eyepiece (125×). South is up in both images.*

telescopes display it as a silvery sliver of light, brightening, at first slowly, then rapidly, toward an inner core. A 200-mm telescope adds a centralized stellar nucleus, while larger instruments also hint at a mottled surface to the tilted spiral arms. Measuring 12' × 8' in long-exposure photographs, M63 bears magnification well, with the best views coming between 150× and 200×.

M51 (NGC 5194). Of all the countless galaxies scattered across the northern sky, none exhibits as much detail as M51, the renowned Whirlpool Galaxy (Figure 6.3). To find it for yourself, start at Alkaid (Eta [η] Ursae Majoris), the end star in the handle of the Big Dipper. Hop to 24 Canum Venaticorum, a 4th-magnitude point just 2° to the west-southwest, then slide another 2° southwest to a trapezoid of faint stars. M51 lies inside the trapezoid's northeast corner.

The Whirlpool Galaxy is a classic example of a face-on spiral galaxy. Its pinwheel structure, first detected with Lord Rosse's enormous 72-inch (1.83-meter) reflector in 1845, was originally thought to represent a solar system in formation. Only after Edwin Hubble conceived of the true galactic organization of the universe in 1924 were M51 and countless other "spiral nebulae" recognized as remote galaxies.

While it may have taken the 72-inch to discover the spiral arms of M51, today's amateur astronomers can see hints of them in much less. From dark, clear sites, high-quality 150- to 200-mm instruments can readily show the pinwheel-like arms of M51 wrapping around its bright central core; a 305- to 355-mm (12- to 14-inch) telescope yields the same appearance even through suburban skies, as shown in Eyepiece Impression 1. Large amateur instruments

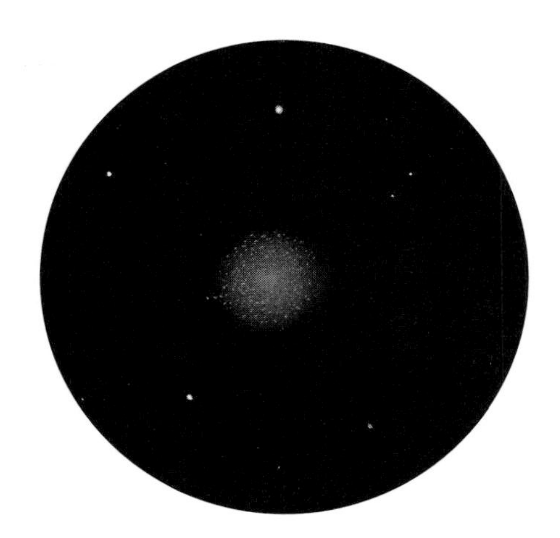

Eyepiece Impression 2. *M3, as seen through the author's 200-mm Newtonian and 12-mm Nagler eyepiece (116×). South is up.*

reveal a distinct mottling to the spiral arms, the result of bright and dark patches of nebulosity.

Extending toward the north of M51 is its famous companion galaxy, **NGC 5195.** Although Charles Messier discovered M51 in 1773, the companion went undetected for another eight years until Pierre Méchain (1744–1805) spied it. Both galaxies are clear in even the smallest amateur telescopes and may be spotted through 7× binoculars by keen-eyed observers. NGC 5195 is an irregular galaxy, appearing as a fuzzy, not-quite-round glow seemingly linked to M51 by a luminous bridge of nebulosity (in reality, this is apparently an optical illusion).

M3 (NGC 5272), the finest globular in northern springtime skies, was discovered by Messier in 1764. The simplest way to find M3 is to aim your telescope just shy of the halfway point from Arcturus (Alpha [α] Boötis) to Cor Caroli, then take a look through your finder. M3 should be within or very near the field of your finder, looking like a fuzzy star just northeast of a 6th-magnitude sun.

While Messier's telescope showed his catalog's third entry as little more than a hazy smudge, Eyepiece Impression 2 is typical of the view through a modern 200-mm instrument, with hundreds of fine stellar points surrounding the cluster's moderately compressed nucleus. Kenneth Glyn Jones described his view of M3 through a 200-mm telescope as "very compact and much brighter at the centre; [at] 121×, beautifully resolved almost to the middle." In a telescope of twice the aperture, stars are resolved across the cluster's face, with several arranged in strings. William Herschel was first to spy these unusual features, describing them as "star chains." Can you duplicate Herschel's find through your own telescope?

Carina

Unfamiliar to most amateur astronomers in more northerly latitudes, the constellation Carina, the Keel, is the southernmost of the four modern constellations formed from the enormous ancient star group Argo Navis (the others being Puppis in the winter sky, and Pyxis and Vela in the spring). Its brightest star, Canopus (Alpha [α] Carinae), is second only to Sirius in brilliance as seen from Earth. Sprinkled among the stars of the Keel is a veritable treasure trove of stunning deep-sky objects. Carina culminates at midnight in early February and is best placed for early-evening observation in late March and April. Keep in mind, however, that only observers south of about 30°N latitude stand much hope of spotting most of these celestial delights.

NGC 2516, technically a winter object, is an impressive open cluster isolated in western Carina just 3° southwest of the 2nd-magnitude star Avior (Epsilon [ε] Carinae). On dark, clear nights NGC 2516 is visible to the naked eye as a soft, gray smudge spanning about one Moon diameter. With even the slightest optical aid, it erupts with starlight. Seven-power binoculars resolve about one-third of the 100 stars that make up NGC 2516, while a 150-mm telescope is capable of showing just about all of them. A trio of orange orbs dominates a scene of otherwise pure-white stars. NGC 2516 spans 15 light-years and is 1,300 light-years from Earth.

NGC 3114, another ravishing open cluster that most northerners miss out on, is located about 6° due east of the 2nd-magnitude star Iota (ι) Carinae. Under dark skies it should also be visible to the naked eye as a 4th-magnitude smudge among the clouds of the southern Milky Way. Three 6th-magnitude beacons rule over the cluster's other hundred or so stellar citizens. Since NGC 3114 spans 35', the best view will be through your lowest-power eyepiece or a pair of binoculars. I could only stare in silent awe when I first saw NGC 3114 through my 11 × 80 binoculars from the Winter Star Party in the Florida Keys some years ago. Other binocularists alongside me whimpered their sadness that both NGC 2516 and 3114 are not visible from back home!

NGC 3293 is another of Carina's many outstanding open clusters. Find it by first locating 3rd-magnitude Mu (μ) Velorum in neighboring Vela. Descend southwestward 6° to 4th-magnitude r Carinae, then another 2° to 4th-magnitude x Carinae. NGC 3293 lies about ¾° to the south of this last star. Once there, you'll find a tightly packed swarm of 50 stars from 6th to 13th magnitude. Binoculars show only the brightest ten or so, surrounded by the gentle glow of unresolved starlight, while a 150-mm telescope will resolve just about all 50. A telescopes twice the size will in addition hint at the faint cloud of nebulosity that engulfs the group. Although vivid in photographs, the nebula requires optimum equipment, skies, and eyes to be spotted visually.

NGC 3372, better known as the Eta (η) Carinae Nebula, is easy to spot with even the slightest optical aid as a dim smudge of light about 11½° west of Crux, the Southern Cross. One of the most amazing objects in the entire sky, this huge glowing cloud puts on an exquisite display regardless of the observer's instrumentation. Dark rifts are easily visible, dividing the nebula into several distinct regions, with the most prominent clearly teardrop shaped.

Entombed within the cloud is Eta Carinae itself, a huge orange star. Bayer recorded it as 4th magnitude in his *Uranometria* star atlas of 1603, as did Edmond Halley (1656–1742) in 1677, when he became the first person to view the region telescopically. By 1730 the star had brightened to 2nd magnitude, only to fall back to 4th over the next half-century. It continued to fluctuate until April 1843, when it suddenly soared to magnitude –0.8. Eta Carinae's brightness continues to fluctuate to this day, but certainly not so dramatically. Presently it is only 6th magnitude, but that does not make it any less interesting. Studies reveal a great abundance of heavier elements in the star, leading many theorists to believe that this is a good candidate for our galaxy's next supernova.

For an overview of the entire nebula, use your lowest-power, widest-angle eyepiece. Even then, however, NGC 3372 will more than fill the field. (Photographs reveal faint strands of cloudiness extending more than 4° across this glorious region.) Switch to a higher magnification for a close inspection of the star's immediate locality. Closely surrounding Eta Carinae is a tiny reddish glow measuring only 3" wide. This diminutive cloud may be seen through moderate-size instruments, but a high magnification is needed for the best view.

NGC 3532. Of the many fine open clusters that surround the Eta Carinae Nebula, most observers favor NGC 3532, about 3° to the nebula's northeast. NGC 3532 is an outstanding gathering of some 400 stars clumped into an area 55' across. More than 60 of them are bright enough to be seen through 7 × 50 binoculars, with hundreds more visible in rich-field telescopes. The cluster as a whole appears wedge shaped, elongated east to west, with many of the stars set in long intertwining lines and curves. One visit to NGC 3532 and you surely will feel as Herschel did when he described it as the most brilliant cluster he had ever seen.

Centaurus

Also riding low in the south on spring nights is the large constellation Centaurus, named for the mythical half-man, half-horse. The Centaur is a great celestial tourist trap, holding a wide variety of wondrous deep-sky targets that are sure to please all who pass its way. Centaurus reaches midnight culmination in early April, and is highest in the early-evening southern sky in late May and June.

Let's begin with the open cluster **NGC 3766,** just north of 3rd-magnitude Lambda (λ) Centauri and sitting right on the galactic equator. A small instrument shows this as a very pretty, compact swarm of stars buried within the nebulous glow of fainter suns, while a 150-mm telescope resolves about 80 stars of 6th to 13th magnitude in an arrowhead formation. The brighter stars — some golden, others bluish — arc across the cluster's northern perimeter. If you have binoculars, pause here a moment and scan the area around NGC 3766. The 4° span between Lambda and 5th-magnitude

Figure 6.4. *NGC 5128 (Centaurus A) in Centaurus is one of spring's most mysterious galaxies. Bisected by a prominent dark lane, it emits unusually strong radio signals. North is up. Photo by R. Dufour.*

Omicron (o) Centauri is one of the most wonderfully star-studded regions in the southern sky.

NGC 5128 is one of the most unusual galaxies to be found anywhere in the sky (Figure 6.4). Photographs record what at first glance appears to be a typical elliptical galaxy, measuring 18' × 14'. Closer inspection, however, reveals it to be bisected by a curious obscuring dark lane. Discussing this galactic curiosity in their book *Constellations,* Lloyd Motz and Carol Nathanson refer to it as the Karate Galaxy because it wears a black belt.

A good southerly viewing site is a must for observing NGC 5128, but even then it can be difficult to find. One way to locate it is to begin at the Corvus trapezoid. Draw a line from Gamma (γ) Corvi southeastward through Beta (β) Corvi and extend it for 16° to 3rd-magnitude Iota (ι) Centauri. From Iota move south-southwest to an L-shaped line of nine 6th- and 7th-magnitude stars. The corner of the L aims directly at NGC 5128, about 2° to the east and just beyond a lone 7th-magnitude field star.

Astronomers have classified NGC 5128 as a type-S0 peculiar galaxy, but remain baffled by the origin and nature of the dark lane that crosses it. Analysis shows that both the galaxy and the lane lie at the same distance, ruling out a coincidental alignment of two unrelated objects. Adding to the mystery is the fact that NGC 5128 is an unusually strong source of radio emissions (as a radio source it is known by the designation Centaurus A). The radio emission comes from two large, optically invisible lobes located to either side of the visible galaxy. Although the debate continues, there is a consensus that NGC 5128 is the result of a merger between a spiral galaxy and an elliptical galaxy that took place within the last billion years.

Observing site and conditions are everything when examining NGC 5128. It is visible from as far north as about latitude 40°N, given a perfect horizon to the south. (One of the few advantages of observing from Long Island as I do is having just such a horizon from the Fire Island Seashore.) On the clearest nights, my 200-mm f/7 reflector shows NGC 5128 as an amorphous sphere. Although my 455-mm (18-inch) Newtonian reveals the dark lane from the same site, any haze along the horizon will render the galaxy invisible with either instrument. From southern latitudes giant binoculars are all that is needed to detect both the 7th-magnitude galaxy and the dark lane. At about 150×, a 150-mm instrument begins to show irregularities in the 6'×1'-wide lane, while at about the same

Figure 6.5. *NGC 5139 (Omega [ω] Centauri) in Centaurus shines at magnitude 3.7, bringing it easily into naked-eye range for observers who are south of about 30° north latitude. Few celestial sights can compare with its raw beauty. North is up. Photo by Luke Dodd.*

magnification a larger telescope will bring out the flared, tattered ends of the lane that reproduce so well in photographs.

NGC 5139. No other globular cluster visible from Earth surpasses NGC 5139, better known as **Omega (ω) Centauri** (Figure 6.5). Omega Centauri is found about 35° due south of the bright star Spica in Virgo and 5° due west of 3rd-magnitude Zeta (ζ) Centauri. Few celestial sights can compare with its raw beauty. Omega Centauri shines at magnitude 3.7, bringing it easily into naked-eye range even on less-than-perfect nights. In fact, Ptolemy included it in his *Almagest* in A.D. 140, while Bayer assigned it the Greek letter omega in his pretelescopic star catalog of 1603.

Even a simple pair of 7× binoculars will begin to reveal the splendor of Omega Centauri, while partial resolution of some of the globular's estimated one million individual stars is clear through 11 × 80 glasses. Telescopes only add to an observer's enjoyment, a 200-mm resolving stars across nearly all of the cluster's huge ½° disk, while larger scopes crack the core of Omega, causing countless suns to pour across the eyepiece field! No matter what the instrument used, your first sighting of Omega Centauri will be remembered for a lifetime.

NGC 5460 is easily located from Omega Centauri by bouncing back to 3rd-magnitude Zeta (ζ) Centauri, then sliding about 2° east-southeast. This attractive open cluster features 40 stars from 8th to 11th magnitude spanning nearly ½°. Binoculars and finderscopes reveal a few of the group's brighter luminaries embedded in a gentle blur of unresolved starlight. A 100-mm telescope will introduce the observer to all of the cluster members, including a striking clique of half a dozen points toward the center.

Alpha (α) Centauri shines like a brilliant beacon low on the southern horizon from the southernmost tier of the United States. Alpha is perhaps best known as the closest star to our solar system, a mere 4.34 light-years away. Deep-sky observers also know Alpha as one of the southern sky's most striking multiple-star systems. Alpha Centauri A, the primary, is a hot Type-*G*0 star that gleams at magnitude 0.0. Alpha Centauri B, positioned about 22" to the primary's southwest, is easy to spy through even the smallest amateur telescope. It shines pure white and, at magnitude 1.2, is about three times fainter than the primary. The third star, called Proxima Centauri, has for nearly a century been thought to be a member of this system. But that association has recently been questioned. Proxima is an 11th-magnitude red dwarf lying 131' — more than 2° — to the primary's southwest.

Coma Berenices

Back north we pause next at the constellation Coma Berenices. In mythology this constellation marked the hair of Queen Berenice, but to modern-day deep-sky observers it is the location of half of the famous Coma-Virgo galaxy cluster. Better known as the Realm of the Galaxies, this broad region of our spring sky offers hour upon hour of galactic challenges for backyard astronomers. The listing below contains just a few select galaxies from within the constellation. Surprisingly, Coma Berenices is also home to a seemingly out-of-place open cluster that is among the sky's finest for binoculars, as well as a pair of telescopic globular clusters. Coma Berenices reaches midnight culmination in early April and is ideally placed for early evening observation in May and June.

Coma Star Cluster. Now here is an open cluster that was made for binoculars! Eighty stars of 5th and 6th magnitude across 5° (that's right, degrees!) belong to the cluster. With 7× wide-field glasses, the view is full of bright stars set in gentle arcs and jagged lines in a pattern resembling a V-shaped flock of northward-flying geese. The large apparent size of the Coma Star Cluster is due primarily to its relative proximity. At a distance of 250 light-years, it is the third-closest open cluster to the solar system; only the Hyades in Taurus and the Ursa Major Moving Cluster are closer.

The Coma Star Cluster is also referred to as Melotte 111, its listing in a catalog of open clusters first recognized and assembled by P. J. Melotte (1880–1961) in the early years of this century. Melotte is credited with identifying numerous other widespread open clusters visible to the naked eye and binoculars. Several double stars highlight the Coma Cluster. Of these, **17 Comae Berenices** is one of the easiest to spot. Even through the smallest opera glass, an observer will have little trouble distinguishing the 7th-magnitude companion from the 5th-magnitude primary thanks to their wide (2.5') separation. Look for the pair just east of the cluster's center.

M99 (NGC 4254) is one of the myriad of telescopic galaxies that belong to the Realm of Galaxies (Figure 6.6). To find it, begin at Denebola (Beta [β] Leonis), the tail star of Leo, the Lion. Scan about 6° due east to 5th-magnitude 6 Comae Berenices, which teams with two 6th-magnitude stars to form a line that extends northeast for about 2°. Aim at the middle star in the line, then slip about 1° southeast to a lone 7th-magnitude sun. M99 lies just southwest of this last star.

When we view M99 we are looking toward a face-on Sc spiral of magnitude 9.8. At first it seems little more than a nondescript glimmer, but on clear, dark nights amateur telescopes will show some tenuous detail within the galactic disk. Most noteworthy is a prominent spiral arm curving toward the southwest (directly opposite that 7th-magnitude field star). Under exceptionally clear conditions, the arm may be glimpsed through a decent 150-mm instrument. A second, less apparent arm extending from the galaxy's bright core toward the northeast is also visible in a decent 200-mm telescope.

While in the area take a moment to visit the spiral galaxy **M98** (NGC 4192), found about ½° to the west of 6 Comae. Shining at magnitude 10.1, M98 appears as a cigar-shaped patch of grayish light highlighted by a bright, circular core.

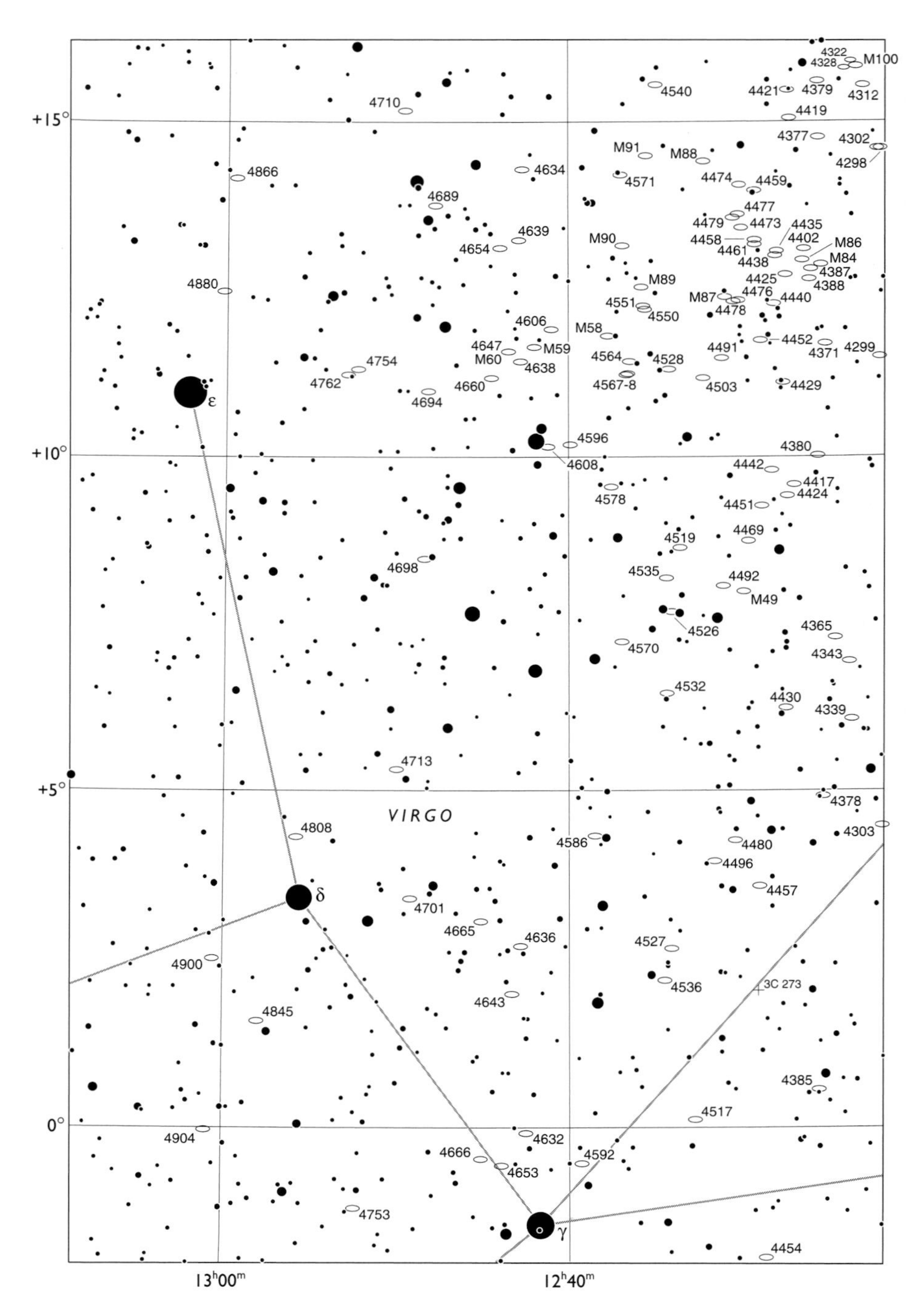

Figure 6.6. *The Realm of Galaxies, spanning the borders of Coma Berenices and Virgo, contains dozens of galaxies that are visible through amateur telescopes. Trying to find your way through what has been characterized as a celestial forest,*

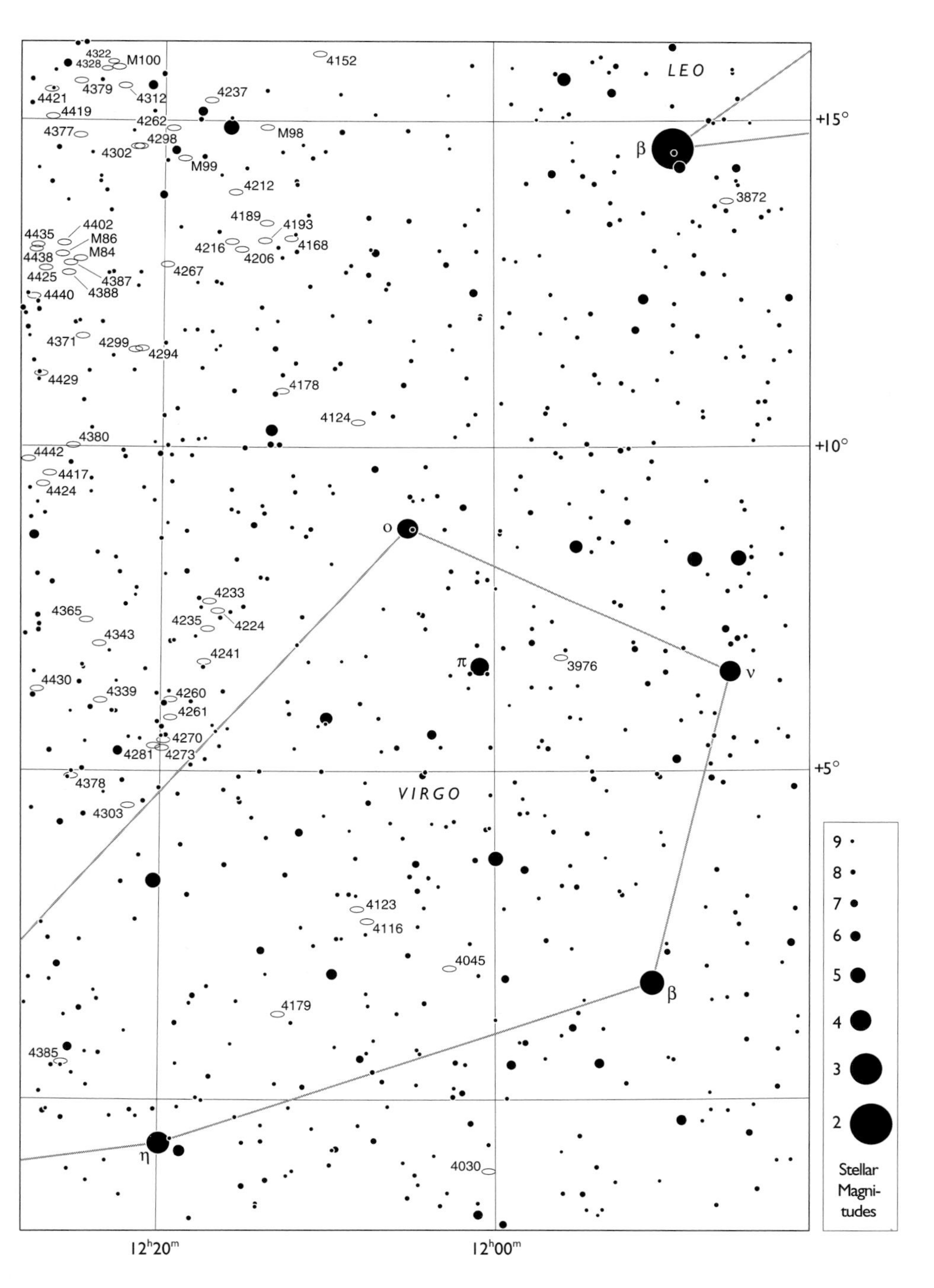

however, can be a daunting task for even the seasoned deep-sky observer. This pair of detailed charts should help you to make your way from galaxy to galaxy. My best advice? Just take it slow — don't rush.

Eyepiece Impression 3. *M64, the Black-Eye Galaxy, as seen through the author's 333-mm Newtonian and 12-mm Nagler eyepiece (125×). Note the black eye itself, an arc of dark nebulosity curving along the galaxy's southern edge. South is up.*

M100 (NGC 4321). Nudge your telescope toward the northeastern member of that line of three stars noted under M99. Continuing ½° farther northeast, you'll strike M100. Another face-on spiral, this galaxy reveals subtle hints of its characteristic pinwheel structure through a 254-mm (10-inch) instrument. Larger telescopes readily show two spiral arms extending from the galactic core. Sharp-eyed observers have recorded that the surface of M100 has an irregular appearance, alluding to large regions of glowing nebulosity and star clouds.

Accompanying M100 is a pair of faint galaxies just to its east. **NGC 4322** and **NGC 4328** shine feebly at about 14th magnitude and appear as little more than dim smudges, each less than 1' across. Only large telescopes stand much of a chance of revealing either.

24 Comae Berenices is a wide double star that is easily separated in most amateur instruments. Look for it at the western end of a curved line of stars that begins at 4th-magnitude Alpha (α) Comae Berenices and includes, from east to west, 38, 36, 33, 32, 27, 25, and lastly 24 Comae. Consisting of a 5.3-magnitude Type-*K* primary and a magnitude-6.6 Type-*A* secondary, 24 Comae is well known for its color contrast. Like Albireo in the summer constellation Cygnus, the brighter star appears distinctly yellow, while the fainter is a pale blue. Just over 20" of dark sky divides the stars from one another, the companion lying almost due west of the primary.

NGC 4565, one of the prettiest examples of an edge-on spiral galaxy anywhere in the sky, is located about 1½° east of 17 Comae Berenices, the aforementioned 5th-magnitude double star on the eastern boundary of the Coma Star Cluster. Even through a small telescope, NGC 4565 looks like a pencil-thin shaft of 9th-magnitude grayish light measuring about 8' × 2' (about half the size of its full photographic extent). Under good conditions a 150-mm will reveal the bulge of the galaxy's central hub along with the faintest suggestion of the characteristic dark lane that cuts across the edge of the spiral-arm halo. Sue French of Glenville, New York, recalls the view of NGC 4565 through her 254-mm f/6 Newtonian: "beautiful. . . quite large, very long and thin, with the dark lane across the center."

M64 (NGC 4826) is one of the most intriguing galaxies in our springtime sky. To pinpoint its location slide about 4° northwest of Alpha (α) Comae Berenices to 6th-magnitude 39 Comae Berenices, then move 3° due east to 5th-magnitude 35 Comae Berenices. M64 lies about 1° to its northeast.

Backyard instruments reveal M64, an Sb spiral, as an oval glow with a noticeably off-center stellar nucleus. M64 shines at magnitude 8.5 and measures 9' × 5' in photographs, though it leaves a smaller impression when viewed visually. But what makes this such an interesting target is the prominent lane of obscuring dust lying north of the nucleus. John Herschel was the first to record this lane, remarking at the time on a "vacuity below the nucleus." The resemblance to a blackened eye is unmistakable through most backyard telescopes, giving rise to M64's popular nickname, the "Black-Eye Galaxy." Eyepiece Impression 3 shows the view through a 333-mm (13.1-inch) telescope.

The globular cluster **M53** (NGC 5024) seems out of place in the spring sky, stranded in the midst of the intergalactic furor of Coma Berenices. Look for this renegade just 1° to the northwest of 5th-magnitude Alpha (α) Comae Berenices, which in turn lies about 15½° more or less due west of brilliant Arcturus in Boötes.

M53 was discovered in 1775 by Johann Bode (1747–1826), who described it as round and pretty "lively," a unique term for characterizing the appearance of a globular cluster. Charles Messier's own words seem to lack the enthusiasm of Bode's, since he referred to it simply as "without stars." Although somewhat colorless, Messier's words paint an accurate picture of M53 when viewed through small backyard instruments. A modern 150-mm telescope is capable of partially resolving the cluster's edges, though none of its members are brighter than 12th magnitude. Larger instruments display a myriad of stars strewn across the globular's circular disk.

While in the neighborhood, also look for **NGC 5053,** a comparatively tough, 10th-magnitude globular located 1° southeast of M53.

Corona Borealis

East of Boötes is semicircular Corona Borealis (Figures 6.7 and 6.8), the Northern Crown. History records that Ptolemy concocted this constellation, though it had been previously known for some time as a wreath. Among the few stellar jewels in the crown are three noteworthy variable stars.

U Coronae Borealis, an interesting eclipsing binary for binocular study, is about 3° west of Theta (θ) Coronae Borealis in the Crown's arc (Figure 6.7). As the fainter companion star orbits the brighter primary once every 3.45 days, we see the stars' combined magnitude fluctuate from a maximum of 7.7 to a minimum of 8.8.

S Coronae Borealis lies in the same low-power field of view as U, less than 1° to that star's southeast. S Coronae Borealis (Figure

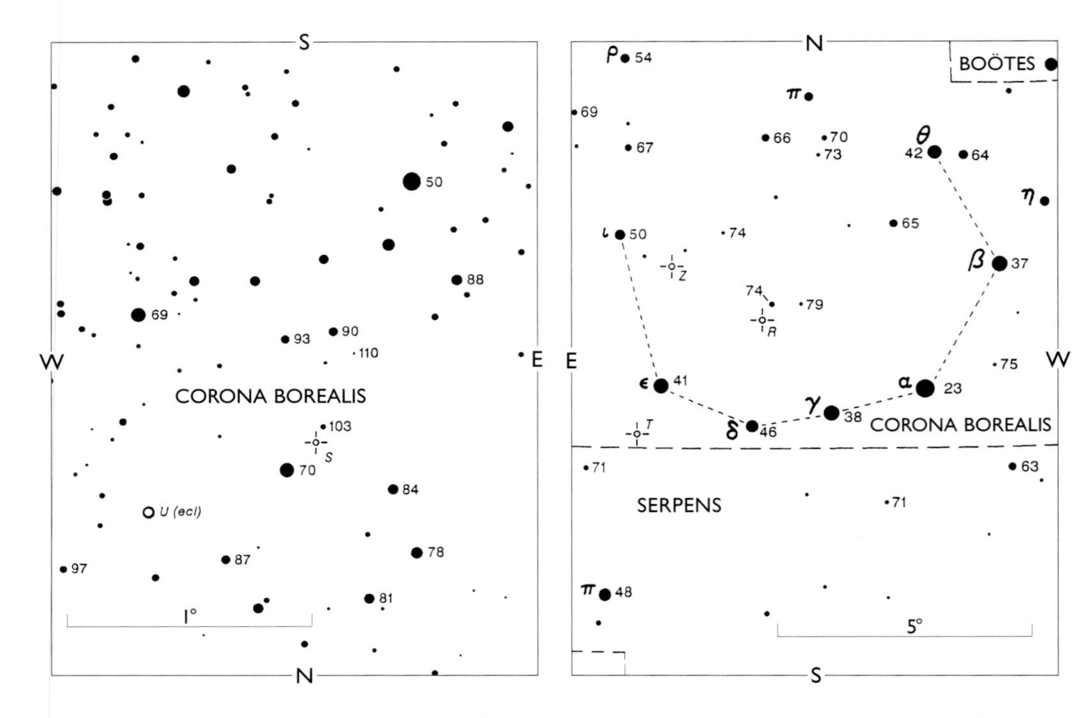

Use these finder charts to estimate the magnitudes of some of the variable stars in Corona Borealis. Left: Figure 6.7 *shows the locations of S Coronae Borealis (below center) and U Coronae Borealis (below and to the left of S). U is further denoted with "ecl" to indicate it is an eclipsing binary. Several suitable comparison stars and their respective magnitudes are also shown; each magnitude's decimal point has been omitted to avoid confusing it with a faint star. (The star labeled "70" is actually magnitude 7.0, for example.) The chart is oriented with south up to approximate the view through an inverting telescope.* Right: Figure 6.8 *plots the locations of R Coronae Borealis (below and left of center) and T Coronae Borealis (at left near the border with Serpens). This binocular view has north up. Both charts courtesy the American Association of Variable Star Observers.*

6.7) is a long-period variable, with a period of just under a year. In that time the star can be seen to vary from magnitude 5.8 to less than 14 and back again. The variable forms a colorful optical double with a white 7th-magnitude star just to its northwest, contrasting nicely with the reddish tint that is typical of long-period stars.

R Coronae Borealis, nicknamed "R Cor Bor" by variable-star enthusiasts, is the constellation's best-known variable. For most of the time it remains at about 6th magnitude and so is visible through binoculars as a point of light southeast of the semicircle's center (Figure 6.8). Once in a while, suddenly and unpredictably, the star will fade to 14th-magnitude in a matter of a few weeks — like a nova in reverse. These unannounced dimmings usually last for several months, though some have continued for years. Then, as unpredictably as it faded, R will brighten from the depths of obscurity back to its "normal" brilliance. This odd behavior is caused by a large obscuring carbon cloud, literally soot, that forms in the star's relatively cool atmosphere. Light is blocked off until the cloud dissipates, whereupon the star returns to normal brightness.

Corvus

For Northern-Hemisphere observers the constellation Corvus, the Crow, never rises very high in the southern spring sky. Yet the distinctive quadrilateral formed by its four brightest stars makes it an easy figure to pick out. To our survey it contributes a seldom-observed planetary nebula, an interesting double star, and a unique asterism. Midnight culmination comes at the end of March, the Crow flying into early-evening view toward the middle of May.

NGC 4361, one of the season's few bright planetary nebulae, is situated about 2½° southwest of Delta (δ) Corvi. Look just a little north of the central point in the Corvus quadrilateral for a pair of 7th-magnitude suns; NGC 4361 lies less than a degree to their north. The lack of bright stars in the immediate area makes the nebula more prominent than it would otherwise be.

NGC 4361 strikes most deep-sky observers as surprisingly large for a planetary, with an apparent diameter estimated to be as large as 45". Describing the view through a 150-mm telescope, Canadian observer P. Brennan wrote in the *Webb Society Deep-Sky Observer's Handbook,* volume 2, that NGC 4361 appeared as a "faint but even nebulosity around a star." A 200-mm instrument will reveal a round or perhaps slightly oval disk of 10th-magnitude light with nary a hint of color. Some keen-eyed observers with 200-mm telescopes can also make out the cloud's progenitor central star shining at about magnitude 13, though most require more aperture to spot its faint glow.

Different people see different things when they look at NGC 4361, even through the same size telescope. For instance, Arkansan Dean Williams describes the view of this planetary through his 200-mm Schmidt-Cassegrain as "round, with no sign of color . . . the central star is visible with averted vision . . . best images at 102× and 140×." On the other hand, observer Glenn Bock, also equipped with a 200-mm Schmidt-Cassegrain, recorded it as having "irregular edges, not circular but with a Crab Nebula–like irregular shape" — a common impression through larger instruments. What do you see?

Delta (δ) Corvi, also known as Algorab, marks the northeast corner of the Crow's quadrilateral pattern. To the naked eye it seems to hold little interest, looking like any other 3rd-magnitude point of light. But as John Herschel and James South (1785–1867) discovered in 1823, Delta is actually a wonderful double star when viewed through just about any telescope. The 9.2-magnitude companion, a dwarf Type-*K* star, may be found just over 24" to the southwest of the Type-*A* primary star. Many observers record the stars' delicate colors as yellow and lilac, though I see them more as yellow and very pale blue white. While the stars' separation and position angle have not changed since their discovery, studies of their common proper

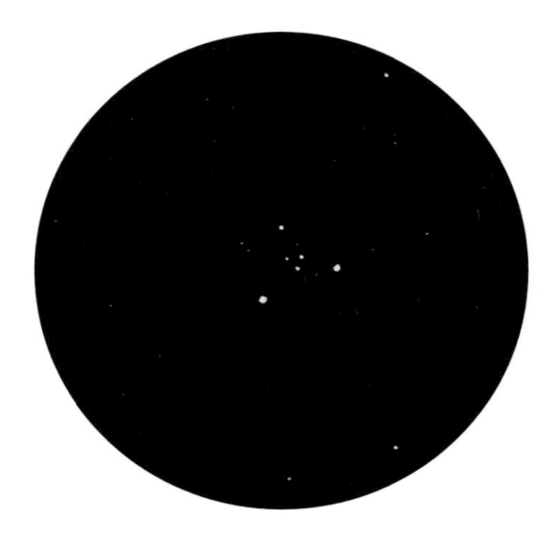

Eyepiece Impression 4. *STAR 20, the Stargate, so-called because of its unique triangle-within-a-triangle pattern of six stars, as seen through the author's 200-mm Newtonian and 24-mm Tele Vue Wide Field eyepiece (58×). South is up.*

motion indicate that their association is real. These suns are thought to be separated from each other by about 900 astronomical units.

STAR 20. Apart from being a pretty double star, Delta Corvi also teams with Eta (η) Corvi to create a wide optical double discernible in binoculars or even with the naked eye. Center your view on Eta, then scan about 3° northward for an isosceles triangle formed by three 6th-magnitude stars and pointing northeast. Notice that the tip of the triangle is aimed at a faint pair of stars that may just be visible in your finder. Those stars are the brightest in STAR 20.

STAR 20 was brought to my attention by several independent observers, including John Wagoner, Father Lucian Kemble, Tom Mote, and Don Farley. All describe a unique triangle-within-a-triangle pattern of six stars that range in brightness from 7th to about 11th magnitude, as shown in Eyepiece Impression 4. Nicknames for the unique pattern range from Kemble's "Delta-Wing Starship" to Mote's "TSP Triangle" (after his first encounter, at the Texas Star Party). I prefer Wagoner's proposal of "Stargate." Call it what you will, STAR 20 is a one-of-a-kind celestial sight.

While in the area, be sure to stop by STAR 21 in Virgo (described later in this chapter), a second asterism only 1° northeast of the Stargate. You may find that they both barely squeeze into the same low-power eyepiece field.

Hydra

Hydra, the Serpent, acclaimed as the longest constellation in the sky, bridges the wide gap between the winter and summer skies. From tip to tip, the Serpent holds several diverse deep-sky treats vying for the observer's attention. The central part reaches midnight culmination in mid-March, though other sections culminate between January and May. Look for Hydra slithering near the southern horizon during the early-evening hours of February through June.

M48 (NGC 2548), the lone open cluster within Hydra, is a splendid splash of stars just inside the constellation's extreme western border. It can be reached by extending a line from Gomeisa (Beta [β] Canis Minoris) through Procyon (Alpha [α] Canis Minoris) and continuing southeast for 14°. As a reference along the way, you'll pass a distinctive line of three stars as well as three other stars set in an equilateral triangle. The cluster shines at 6th magnitude, within the range of most finders and binoculars. Its stars merge into a triangular pattern, with a group near the cluster's center linking to form a bright chain set among fainter orbs. In all, about 80 stars of magnitudes 8 to 13 comprise this attractive early-spring target.

For years, M48 was branded as a "missing" object. Apparently, however, what was really missing was not the object but its correct coordinates. When he discovered M48 in 1771, Messier erroneously reported its location as 4° farther north than it actually is. While no cluster exists at his original plotting, Messier's description closely matches the appearance of NGC 2548, and they are now considered to be one and the same.

NGC 3242, nicknamed the Ghost of Jupiter, is regarded by many deep-sky observers as one of the finest planetary nebulae anywhere in the sky. It rides along the back of Hydra, southeast of the constellation's brightest star, Alphard (Alpha [α] Hydrae). Given relatively dark skies, look about 16° east-southeast from Alphard for a wide, southward-aimed isosceles triangle formed by Lambda (λ), Nu (ν), and Mu (μ) Hydrae. NGC 3242 lies less than 2° due south of Mu, adjacent to an 8th-magnitude field star.

Under dark skies, NGC 3242 is actually bright enough to be seen through large binoculars and finderscopes as a faint, greenish starlike object of about 8th magnitude. A 150- to 200-mm telescope operating between 50× and 75× will reveal the nebula's homogeneous, turquoise disk. By increasing magnification to 100× or greater, these same telescopes might be able to distinguish a bright center surrounded by a faint outer ring. The overall effect reminds many of the CBS "eye" (the logo of the CBS television network). The central star, rated a deceptively bright 12th magnitude, is visible only with difficulty through a 254-mm instrument because of the overwhelming brightness of the nebula itself.

M68 (NGC 4590) also rides on the back of Hydra. Begin your search for this globular cluster at Beta (β) Corvi, the southeastern star in the Corvus quadrilateral. Drop southward 4° to a 5th-magnitude star, then shift ½° to that star's northeast to find M68. When searching for M68, keep in mind that it is never very high above the horizon from midnorthern latitudes (for instance, it never climbs higher than about 15° above the horizon as seen from latitude 40°N). A clear, unobstructed view to the south is therefore a must. Even then, don't be surprised if some searching is required, especially if a thin layer of haze lies near the southern horizon.

Once located, M68 can be partially resolved into individual stars through a 150-mm instrument. A 305-mm reveals innumerable stellar points uniformly packed across an 8' area. Studies classify

Figure 6.9. *M83 in Hydra. This 8th-magnitude galaxy requires careful scrutiny to yield its distinctive features: an irregular texture across its spiral-arm halo, numerous star clouds, and vast regions of nebulosity. South is up. Photo courtesy of Lick Observatory.*

M68 as a Class X globular, implying a loose structure. It is estimated to contain more than 100,000 stars and to lie about 46,000 light-years from Earth.

M83 (NGC 5236) also rides low in the southern sky for observers in the midnorthern latitudes, often causing it to be overlooked in favor of better-placed galaxies. If it were more favorably positioned for observers in the Northern Hemisphere, it would undoubtedly be one of the season's most popular attractions (Figure 6.9). Like M68, M83 is most easily found by beginning at Beta (β) Corvi at the southeastern corner of the Corvus quadrilateral. Head due east for 10° to 3rd-magnitude Gamma (γ) Hydrae, then 5° southeast to a 5th-magnitude double star, itself a nice target. Finally, drop 3° due south to a pair of 6th- and 7th-magnitude stars. M83 lies just a ¼° to their east.

Nicolas Louis de Lacaille (1713–62) discovered this 8th-magnitude galaxy in 1752, describing it as a "small, shapeless nebula." Many of today's amateurs are left with the same impression when viewing M83 for the first time. Only with repeated, careful scrutiny will an irregular texture be seen across the spiral-arm halo. From haze-free southerly vantage points, 305-mm and larger instruments hint at the galaxy's spiral structure as well as numerous star clouds and vast regions of nebulosity. M83 is one of the rare breed of three-armed SBc barred spiral galaxies, and in photographs it offers a hint of a central bar extending from the core.

Leo

No doubt, Leo was one of the first springtime constellations you learned as a novice stargazer. Its unique sickle-shaped star pattern punctuated by the bright star Regulus makes Leo stand out in a region characterized primarily by faint stars. For the deep-sky enthusiast, the area is populated by dozens of galaxies, many of them in pairs and small groups. Leo passes midnight culmination in early March and is well placed for evening observation during May and June.

NGC 2903 lies to the west of Leo's Sickle and is easily visible in telescopes as small as 75 mm (3 inches) in aperture. Begin the hunt for NGC 2903 at the star Lambda (λ) Leonis, just west of the tip of the Sickle. Peering through your finderscope, drop 1½° southward to an east-west pair of 7th- and 8th-magnitude stars. NGC 2903 lies just to the south of the fainter, easterly star. All three form a tiny right triangle that might just squeeze into the field of a low power, wide-angle eyepiece, though moderate magnification is essential for a good look at the galaxy.

NGC 2903 at 9th magnitude is one of the brightest and largest galaxies in the spring sky to be missed by Messier and his contemporaries, though it would have certainly been visible in their instruments. Small telescopes reveal a noticeably oval disk and brighter central hub that stand out well against the weak surroundings. Telescopes in the 200- to 305-mm range add a complexity to this Sc spiral that remains hidden in lesser instruments.

Given clear, dark skies, the spiral-arm disk can be seen to hold more than six dozen bright star clouds and H II regions. The most prominent area, cataloged separately as **NGC 2905,** is located about 1' northeast of NGC 2903's core. To accomplished deep-sky observer Steve Coe, viewing through his 318-mm (12½-inch) reflector at 175×, NGC 2903 appears "bright and mottled across the face with a much brighter core," while NGC 2905 is "a bright spot not far from the galactic hub." The largest backyard instruments begin to show the galaxy's broad spiral arms extending northeast and southwest of the core.

Gamma (γ) Leonis. Let's take a break from our galaxy search to check out a spectacular double star. Also known as Algieba, Gamma

Figure 6.10. *A trio of galaxies in Leo: M65 (top left), M66 (bottom left), and NGC 3628 (far right). M65 and M66 are located within ½° of each other, and both are bright enough to be visible (albeit faintly) through binoculars. East is up. Photo by Martin C. Germano.*

Leonis marks the second star above Regulus in the Sickle. Gamma A shines brightly at 2nd magnitude, while its companion gleams at magnitude 3.5. The pair are separated by 4", with the companion to the primary's south. Gamma A and Gamma B are classified as spectral types *K*0 and *G*7, respectively, explaining the subtle yellow tints noted by some observers.

M96 Galaxy Group. Central Leo is home to many interesting targets, including this gaggle of galaxies. The brightest member is M96 (NGC 3368), a magnitude-9.2 Sb spiral. To find M96 move east-southeast of Regulus to Rho (ρ) Leonis, then turn northeast to 53 Leonis. From 53 look for a 7th-magnitude star between it and 52 Leonis to the north. Center this 7th-magnitude star in the field of your widest-field eyepiece, then slide about 1° due south to M96.

Through amateur telescopes **M96** itself shows a nebulous glow surrounding a brighter, oval heart. Little additional detail becomes visible in larger apertures, though the galaxy's girth grows from about 4' × 2' in 100-mm instruments to over 6' × 4' through 355- to 405-mm (14- to 16-inch) scopes. Because of the galaxy's tilt to our line of sight, the spiral arms prove too difficult to discern.

M95 (NGC 3351) lies just west of M96, almost squeezing into the same low-power eyepiece field as it and M105 (see below). Although M95, at magnitude 9.7, is overshadowed by its brighter neighbor, it can still be seen through binoculars and finderscopes on exceptional nights. Through a 200-mm telescope it appears as a nearly circular patch of grayish light growing brighter toward its center. The core itself is oval and spotlighted by a sharp stellar nucleus. M95, an SBb barred spiral, displays a brighter outer ring measuring about 5' × 4' and surrounding the spiral-arm halo when viewed through large amateur telescopes. The largest backyard instruments also hint at the bar that connects the outer ring with the core.

M105 (NGC 3379) is also within the same low-power eyepiece field as M95 and M96. Visually, this E1 elliptical measures about 4' × 3', smaller than either of its neighboring spirals, and it lies about halfway between them in brightness. M105 offers little visible detail in its nearly circular disk apart from a bright stellar nucleus.

No fewer than 11 other members of the *New General Catalogue* lie within 3° of M96. Of these, the brightest is NGC 3384, just 8' east of M105. **NGC 3384** shines with the equivalent light of a 10th-magnitude star, making it an easy catch in smaller telescopes. My 200-mm f/7 Newtonian reveals it as a slightly smaller and fainter version of M105. A bright stellar core is readily visible in the center of this barred spiral's 3' × 2' disk. Other galaxies within the M96 group include NGC 3299, 3306, 3338, 3367, 3377, 3389, 3391, 3412, 3419, and 3433. All shine at 13th magnitude or brighter, setting pleasant challenges for observers with 200-mm and larger instruments on dark, clear nights.

M65 (NGC 3623) and **M66** (NGC 3627) pair up to create another fine galactic duo in the spring sky (Figure 6.10). Both are located within ½° of each other, about midway between the stars Theta (θ) and Iota (ι) Leonis to the south of Leo's hind triangle. Each galaxy is bright enough to be visible to sharp-eyed observers through 7× binoculars and finderscopes.

Of the two, M66 is the brighter. This broadside Sb spiral displays an oval disk spanning 8' × 4' and is accented by a prominent central core. Instruments from 200 to 254 mm show hints of many bright and dark patches throughout the galaxy's spiral-arm halo. Expanding on this in the *Webb Society Deep-Sky Observer's Handbook,* volume 4, T. Davies notes a "dark lane . . . suspected east of the nucleus" through a 215-mm (8½-inch) telescope.

Occasionally, in the course of this hobby, we see a striking object or event that burns an indelible image into our memories. For me, one of those images is a view I had of M66 a few years ago at the Winter Star Party. Through Florida amateur Tom Clark's monstrous 915-mm (36-inch) Tectron telescope, M66 practically jumped out of the eyepiece at me! But what made this especially memorable was the striking hook of stardust — one of the galaxy's spiral arms — curving eastward away from the southern part of the galaxy's core.

Subsequently, I have spotted this spiral arm through my 333-mm (13.1-inch) Newtonian under Long Island skies, with some hint of it also detected in my 200-mm Newtonian. This is undoubtedly what Davies also saw, though it remained unidentifiable through his instrument. What is the smallest aperture that will reveal this distinctive feature?

M65, just 21' east of M66, is longer, thinner, and about half a magnitude fainter than its neighbor. Although also classified as an Sb spiral, M65's stronger tilt to our line of sight masks the existence of the spiral arms except in the largest backyard instruments. Evident through all 150-mm and larger telescopes, however, is a dark lane of nebulosity that girdles the outer circumference of M65's galactic plane. Try a moderate power for the best view of the galaxy, then switch to higher magnification for a good look at the dust lane.

Forming a triangle with M65 and M66 to their north is **NGC 3628,** an edge-on Sb spiral visible through most of today's amateur telescopes under dark skies (Figure 6.10). Its long, thin disk shines at magnitude 9.5 and spans about 10' × 1' through amateur telescopes. Like many edge-on spirals, NGC 3628 is bisected by a prominent dust lane that passes through the central core and slices close to the galactic halo. In their *Observing Handbook and Catalogue of Deep-Sky Objects,* Christian Luginbuhl and Brian Skiff describe the core itself as "rectangular, 3' × 0'.75, with a circular, more concentrated spot inscribed in it."

All three members of the M66 galaxy group belong to the Leo galaxy cluster, as does the M96 group, though more than 8° separate the two groups.

Lynx

Lynx occupies a barren tract of the early spring's northern sky (leading many stargazers to think of this more as the "missing Lynx!"). Its few naked-eye stars are only dimly perceptible through suburban skies, leading most observers to bypass the constellation in favor of its more prominent neighbors such as Gemini and Ursa Major. Yet this celestial cat holds several challenging objects ready to test an observer's prowess. The Lynx slinks across the midnight meridian in late January, bringing it into the early evening skies during late March and April.

NGC 2419, one of the few globular clusters visible in the early spring sky, frequently goes unnoticed because of its obscure location. To find it start at Castor (Alpha [α] Geminorum). Head 3° north to 5th-magnitude Omicron (ο) Geminorum, then an additional 3° north to another 5th-magnitude star across the border in Lynx. Turn 1° northwest to a lone 6th-magnitude sun, and continue another degree to a close pair of 7th- and 8th-magnitude stars. NGC 2419 lies just 3' to their west.

At first glance NGC 2419 would seem a poor choice to include in a book featuring the "finest" deep-sky objects. Not only does the lack of any bright nearby stars make it difficult to pinpoint, but it glows dimly at 10th magnitude. Even the largest backyard telescopes are unable to show it as anything more than an undefinable glow. What, then, makes it so interesting? When we look at NGC 2419, we are gazing out across more than 300,000 light-years, well beyond all other Milky Way globular clusters and nearly twice as far as our galaxy's two companions, the Magellanic Clouds. This makes NGC 2419 a true independent star system and has led to its being nicknamed the "Intergalactic Wanderer." Something to ponder when you gaze its way on the next clear evening.

NGC 2683, unknown to many deep-sky enthusiasts, is not only the brightest galaxy in Lynx but also one of the brightest in the northern night sky of early spring. Situated about 5° north-northwest of Iota (ι) Cancri, marking the Crab's northern claw, and 2° northwest of 5th-magnitude Sigma^2 (σ^2) Cancri, this fine edge-on Sb spiral is faintly visible in giant binoculars and apparent through nearly all telescopes.

A 200-mm telescope will reveal NGC 2683 as a sharply oval, 10th-magnitude disk accented by a bright core. Larger apertures begin to unveil some dusky mottling along the galaxy's spiral disk. While the full extent of NGC 2683 is recorded as 9' × 3' on photographs, it appears to measure about 6' × 2' in most amateur scopes.

Pyxis

Pyxis, the Compass, which once guided the mythical ship Argo, is a small constellation of faint stars riding the main stream of our galaxy. To our survey, it contributes a unique open cluster. Midnight culmination occurs in early February, with Pyxis placed ideally for early-evening observation, low in the late-March and early-April skies.

NGC 2818. Here's an unfamiliar open cluster that holds a surprise for those who seek it out. Unfortunately, to spot it we have to travel into the unfamiliar territory of the southern spring sky. To me, the easiest way to find it is first to locate 2nd-magnitude Suhail (Lambda [λ] Velorum), which lies about 35° due south of Alphard, the brightest star in Hydra. Look for a short line of three 5th-magnitude stars about 5° north of Suhail. NGC 2818 lies about 1° north of k^2 Velorum, the northern star in that line of three.

At first glance NGC 2818 would appear to be a rather unimpressive collection of about 40 faint stars. "What's all the fuss about?" you may wonder. Have patience, for only after careful examination will you find that there is more here than initially meets the eye. Take a look near the cluster's northwestern edge. Do you see a faint disk of grayish light? That's the dim planetary nebula **NGC 2818A**. The nebula measures about 38" across and appears as a hazy disk through my 333-mm f/4.5 Newtonian here on Long Island. Studies are inconclusive about the relationship between the nebula and the cluster. Are they physically linked with one another or is their pairing just a coincidence, like M46 and NGC 2438 in winter's Puppis? Whatever their true relationship, they form an interesting alliance for observers with moderate to large instruments.

Sextans

Commemorating the sextant, Sextans certainly cannot be accused of dazzling the naked eye. In fact, its brightest star, which lies about 12° due south of Leo's Regulus, shines weakly at 4th magnitude. Yet within this weak constellation is one of the sky's most visually intriguing galaxies. Sextans passes midnight culmination in late February, bringing it into the early-evening skies of May.

NGC 3115. Although Sextans is one of the faintest constellations, locating NGC 3115 is not as difficult as you might suspect. Start not in Sextans, but in neighboring Hydra. Find 4th-magnitude Lambda (λ) Hydrae, lying one "hump" farther east along the back of the Hydra from Alphard. Center Lambda in your finderscope and take a look through the eyepiece. You should see two faint stars that with Lambda form a line aimed toward the north-northwest. Trace this line away from Lambda, stopping at a single 6th-magnitude sun. Curving almost due north of this lone star for about 2½°, keep an eye out for 17 and 18 Sextantis, a pair of 6th-magnitude suns. Pause at these momentarily until you spot a second, fainter pair 1° to their northwest. NGC 3115 lies just 30' to the west from this latter stellar duo and may even be visible in finderscopes under dark skies.

Nicknamed the Spindle Galaxy, NGC 3115 shines with the equivalent light of a 9th-magnitude star. In 75- to 100-mm instruments it measures about 3' × 1', a faint needle of light against a starry backdrop. The galaxy's apparent length grows to about 4' when viewed through a 150- or 200-mm telescope, with a prominent stellar nucleus coming plainly into view. Doubling the aperture again lengthens the faint extensions of the spindle and boosts the apparent size to about 5' × 2', though the galaxy's disk remains perfectly flawless, with no suggestion of any dust lanes or spiral shape.

Ursa Major

The Great Bear, Ursa Major, is certainly the best-known circumpolar constellation of the northern sky, its seven brightest stars forming the Big Dipper. The Bear also hosts a wide array of deep-sky treats for backyard telescopes. Double stars, an errant planetary nebula, a curious asterism, and galaxies galore all await observers who dip into the region. Although Ursa Major is visible throughout the year from much of the Northern Hemisphere, it is traditionally associated with the spring sky. The center of the constellation passes midnight culmination in the middle of March, with the constellation highest in the early evening during mid-May and into June.

NGC 2841. Not as popular among observers as some of the other Ursa Major galaxies, NGC 2841 is nonetheless bright enough to be seen in 75- and 100-mm telescopes. To locate this type-Sb galaxy, look toward 3rd-magnitude Theta (θ) Ursae Majoris, the bear's front knee. From there scan toward the west-southwest, pausing at the 6th-magnitude double star 37 Lyncis. The galaxy lies a scant 22' to this double's southeast.

Once spotted, NGC 2841 shows itself as a 9th-magnitude, cigar-shaped glow accented by a bright, oval core. The galactic halo measures about 5' × 1.5' visually, extending northwest-southeast. Just

Eyepiece Impression 5. *M81 (top) and M82 (bottom), as seen through a 200-mm Newtonian and 24-mm Tele Vue Wide Field eyepiece (58×). Even though it is about a full magnitude fainter than its neighbor, M82 offers more structural detail. South is up.*

beyond the galaxy to the northwest lies a 10th-magnitude star, while a 13th-magnitude point appears to be implanted in the galaxy's halo. Although both stars are merely foreground objects lying within the Milky Way, they have been mistaken for supernovae by unwary observers. In a 305-mm telescope or larger, there may also be a hint of a dark lane along the galaxy's major axis, but otherwise NGC 2841 appears homogeneous in form.

M81 Galaxy Group. Mention early spring to me, and one of the first things that crosses my mind will be **M81** (NGC 3031), an outstanding Sb spiral (Figure 6.11). While, at 7th magnitude, it is bright enough to be visible in 7 × 50 binoculars, M81 frequently gives amateurs trouble since it is so far from any handy reference stars. Here is just one way to find it. A star atlas shows that M81 lies on a long line extended to the northwest from Phecda (Gamma [γ] Ursae Majoris) through Dubhe (Alpha [α] Ursae Majoris), with both the galaxy and Phecda more or less equidistant from Dubhe. Gazing through your finderscope, star-hop (more like star-bound) from

Figure 6.11. *M81 (top) and M82 (bottom) in Ursa Major. M81 is one of early spring's signature galaxies. M82's most prominent feature is a jagged rift of darkness across its center. South is up. Photo by Alfred Lilge.*

Phecda to Dubhe, then on toward M81. If you don't see M81 at first, keep a watch for 24 Ursae Majoris, which marks the 90° angle of a small right triangle whose longest side extends toward the southeast and almost directly at M81, just 2° away.

Once spied, M81 will put on a memorable performance. As depicted in Eyepiece Impression 5, a 200-mm telescope shows a bright, oval patch with a prominent core. On exceptional nights, the same aperture will also bring out some vague irregularities in the spiral halo of M81, while telescopes of 305 mm and larger add a few bright and dark blemishes across the galaxy. The largest backyard instruments bring out the best in M81 by adding hints of the galaxy's two great spiral arms wrapping around the core, with the southern arm more prominent. All in all, however, M81 appears much less detailed than other bright spirals.

M82 (NGC 3034), a peculiar galaxy seen edge-on from our vantage point, is located 37' north of M81 and in the same low-power eyepiece field (Figure 6.11). Even though it is about a full magnitude fainter than its neighbor, M82 offers more structural detail. The most prominent feature of this faint sliver of grayish light is a jagged rift of darkness that rips across its center. There are several other dark filaments, the more conspicuous of them toward the galaxy's eastern edge and angled away from its major axis. Sharp-eyed observers with medium- to large-aperture telescopes might also spot a fan-shaped glow protruding from the galaxy's center toward the south.

Two other galaxies within reach of most amateur instruments also belong to the M81 Galaxy Group. Southwest of M81 lies **NGC 2976,** an oval-shaped Sc spiral whose low surface brightness requires a 150-mm or larger instrument. **NGC 3077,** lying just 4' from an 8th-magnitude star to the east-southeast of M81, is much easier to find. Even a 75-mm telescope reveals its nebulous 10th-magnitude disk as a soft glow about 1' across. A 200- to 254-mm telescope expands the galaxy's apparent size to about 4', and brings into view a pronounced core.

NGC 3079, a splendid example of an edge-on spiral galaxy, is tucked away in Ursa Major to the west of the Dipper asterism. To find it draw an imaginary line from Phi (φ) and Upsilon (υ) Ursae Majoris; these two stars should just squeeze into your finder's field of view. On either side of the line's midpoint lie two 5th-magnitude stars, creating a north-south diamond pattern. From the diamond's eastern star slide about 1° south until you come to a small triangle of 8th- and 9th-magnitude stars. NGC 3079 lies within this triangle.

A 150-mm telescope reveals NGC 3079 as an 11th-magnitude streak of grayish light with a brighter core. Through his 200-mm Schmidt-Cassegrain telescope Arkansas amateur Dean Williams saw NGC 3079 as "very elongated . . . with hints of significant structural detail in the form of tenuous brightenings and darker areas. I had some difficulty picking it out at 63×, where it showed as just a tiny diffuse patch, but 102× shows the great elongation and internal

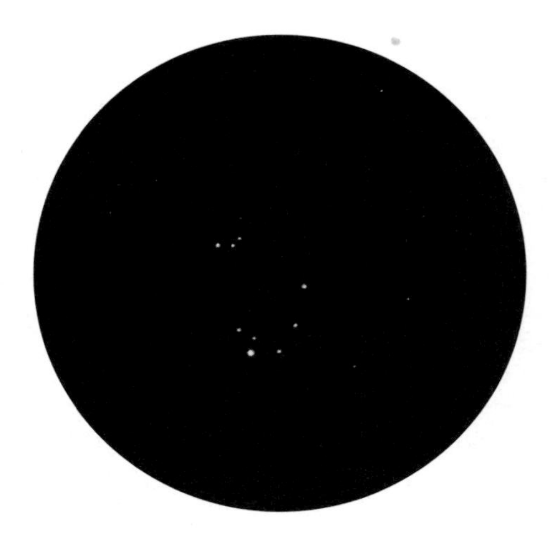

Eyepiece Impression 6. *STAR 19, the Broken Engagement Ring, as seen through the author's 200-mm Newtonian and 24-mm Tele Vue Wide Field eyepiece (58×). South is up.*

detail." High magnifications reveal a faint star embedded near the galaxy's northern tip, as well as faint hints of mottling across the disk, while larger instruments also show **NGC 3073,** a fainter galaxy just west of the triangle's brightest star.

STAR 19. Here's a neat little asterism that I bumped into one night while stalking galaxies in western Ursa Major. To find STAR 19, scan about 1½° west of Merak (Beta [β] Ursae Majoris) for a pair of 6th- and 7th-magnitude stars. Center your telescope on the fainter, southern star in the pair and take a look. As illustrated in Eyepiece Impression 6, a low-power eyepiece will show this to be the brightest of nine suns that form a broken oval about ¼° in diameter. The overall effect reminds me of a ring, with the brightest star representing a diamond. From the shape it's in, however, the ring must have been through quite a battle, so it only seems appropriate to call it the "Broken Engagement Ring."

M108 (NGC 3556) and **M97** (NGC 3587). The Sc spiral galaxy M108, found about a degree and a half to the southeast of Merak (Beta [β] Ursae Majoris), is a fine sight through medium- and large-aperture telescopes (Figure 6.12). The galaxy's plane is presented to us nearly edge on, causing the 10th-magnitude disk to resemble a nebulous cigar nestled among an attractive field of stars. Although there is no sign of any central bulge or nucleus, M108 abounds in faint detail that only a slow, concentrated scan will reveal. My 333-mm f/4.5 Newtonian at medium power makes M108 look "lumpy," with many bright knots, dark patches, and two faint Milky Way stars strewn across the galaxy's 8' major axis. The brighter of these two field stars (12th magnitude) overlaps the western tip of the spiral-arm halo, while the other (13th magnitude) lies near the galaxy's center. Both are frequently reported as supernovae by unsuspecting amateurs.

While in the neighborhood, be sure to stop by planetary nebula M97, the renowned Owl Nebula. The Owl is perched just 48' southeast of M108, possibly within the same low-power field of view

Figure 6.12. *A celestial odd couple in Ursa Major: M97, the Owl Nebula (upper left) and the galaxy M108 (lower right). M108 resembles a nebulous cigar. M97's two dark ovoids require a 200-mm instrument to reveal themselves. East is up. Photo by Martin C. Germano.*

(Figure 6.12). A 200-mm telescope begins to add a little personality to the Owl by showing hints of two dark ovoids either side of center. These are the famous "eyes" that gave M97 its nickname. Larger instruments add further definition to the eyes, but you'll need a mighty big telescope to reveal the nebula's 16th-magnitude central star.

Mizar (Zeta [ζ] Ursae Majoris), marking the bend in the handle of the Big Dipper, teams with Alcor (80 Ursae Majoris) to form the most famous naked-eye double star in the entire sky. In reality, however, the team of Alcor and Mizar is illusory since they are actually nowhere near each other in space.

Mizar, on the other hand, is a true binary star — one of the most frequently observed binaries in the northern sky. History tells that Mizar was the first double to be discovered telescopically when the Italian astronomer Giambattista Riccioli (1598–1671) noticed its duality in 1650. Since the discovery, the companion star Mizar B has moved little relative to Mizar A. Although there is no doubt of the stars' true relationship, the companion's orbit must take thousands of years to complete. Mizar is believed to lie about 59 light-years away. If so, then the two stars must be separated by about 260 astronomical units, or nearly seven times the distance from the Sun to Pluto.

Currently Mizar B appears as a 4th-magnitude point of light about 14" to the 2.3-magnitude primary's southeast, creating a striking pair of beacons through all telescopes. Can you detect any coloration to the stars of Mizar, as some past observers have? They have been described as "brilliant white and pale emerald" by Richard Hinckley Allen, in his *Star Names: Their Lore and Meaning*, and white and greenish by the Reverend Thomas W. Webb, in his *Celestial Objects for Common Telescopes*. Most observers perceive them both as whitish, which is in keeping with their spectral classification as Type-*A* stars.

M101 (NGC 5457). Even though its low surface brightness makes it infamously difficult to locate, especially to those new to deep-sky observing, M101 holds a wealth of detail for patient visitors. Indeed, patience is a prerequisite to finding this huge face-on spiral galaxy in the first place. Here's a little help. Focus your attention on Mizar (Zeta [ζ] Ursae Majoris). Shift from Mizar to Alcor (80 Ursae Majoris), then 2° farther east until you come to 81 Ursae Majoris. This last star marks the western end of a string of four 5th- and 6th-magnitude stars. Follow this line to 86 Ursae Majoris at the eastern end, then look to the northeast for three 7th-magnitude stars set in a right triangle. M101 lies just east of the midpoint along the triangle's hypotenuse.

When initially encountered, M101 will probably strike you as little more than an amorphous, featureless glow that seems to vanish slowly into the background. However, on closer inspection this initial blandness blossoms into great intricacy. But be forewarned that with large, faint galaxies like M101 there is no substitute for telescope aperture, optical quality, and sky transparency. You'll have to wait for that special, once-in-a-season clear night for M101 to reluctantly reveal its true nature. Although M101 spans close to ¼°, don't be afraid to use a medium- or even high-power eyepiece to distinguish individual features.

Under exceptional skies, a 200-mm telescope reveals a subtle spiral pattern within the exceedingly faint halo. The brightest arm curves away from the core to the north and unfurls toward the east. Following the arm, look for several bright patches marking either stellar associations or H II regions. Toward the end of this arm, seemingly detached from the galaxy, is a bright clump of light cataloged separately as NGC 5462. Larger instruments add another delicate spiral arm curving away from the core's south. Again, several bright associations may be seen along its length. Many of the other condensations within M101 have also received individual NGC entries: NGC 5447, 5449, 5450, and 5451 lie west of the core;

NGC 5453, 5455, and 5458 are found to its south; while NGC 5461, the aforementioned 5462, and 5471 are to its east.

Vela

Sailing along the southern horizon during the early spring is the little-noticed constellation Vela, the sails of the mythical ship Argo Navis, at one time the largest constellation in the sky. Vela is one of four modern constellations formed from it, along with Puppis, the Stern; Carina, the Keel; and Pyxis, the Compass. Vela, Carina, and Pyxis are spring constellations, while Puppis is detailed in chapter 9, on the winter sky. Since the Sails never rise very high in our skies, Vela must be visited when it lies on or near the meridian, when it is at its farthest above the horizon. Vela passes through the meridian at midnight in mid-February and is well placed for viewing during April and early May evenings.

Gamma (γ) Velorum, the brightest star in Vela, is a wonderful five-star system for backyard telescopes. The system's 1.9-magnitude primary star is accompanied by 4.2 magnitude Type-*B* companion about 41" to the primary's southwest. A third, magnitude-8.2 star lies 62" southeast of the primary, while Gamma D shines at 9th magnitude and is separated from Gamma A by 94", to the southeast. Completing the system is Gamma E, a dim 12.5-magnitude speck 2" to the southeast of Gamma D.

Gamma Velorum A is itself especially interesting as it is an intensely hot Wolf-Rayet star. Although they were once classified spectroscopically as Type *O*, Wolf-Rayet stars now have their own spectral class, Type *W*, and are well known for their high intrinsic luminosity. At its estimated distance of 520 light-years from Earth, Gamma A must have an absolute magnitude of –4.

NGC 2547 is found some 2° south of Gamma Velorum. Over 80 stars make up this open cluster, with more than a dozen shining brighter than 9th magnitude. Many of the suns in this dazzling cast appear to form lines and curves; all told, they strike me as a crooked cruciform lying on its side.

IC 2391. This beautiful open cluster, lying just north of Delta (δ) Velorum, can be seen with the unaided eye on a good evening, and is easy in binoculars. My 11 × 80 glasses show a large, coarse gathering of many bright stars, including several doubles. Its brightest member, magnitude-3.6 Omicron (o) Velorum, looks like a brilliant sapphire set into a glistening backdrop. Keep in mind that most telescopes will limit the cluster's attractiveness since it measures almost 1° across.

NGC 2669 may be found by scanning about 1° due east of IC 2391. This little-known open cluster is comprised of about 40 stars, the brightest half-dozen forming a trapezoidal pattern. On its own, NGC 2669 is not especially impressive, but add its magnificent environs, and you have a wonderful star-studded show for binoculars.

NGC 3132 is one of the brightest planetary nebulae in the entire sky (Figure 6.13). It is about 11° northeast of Lambda (λ) Velorum,

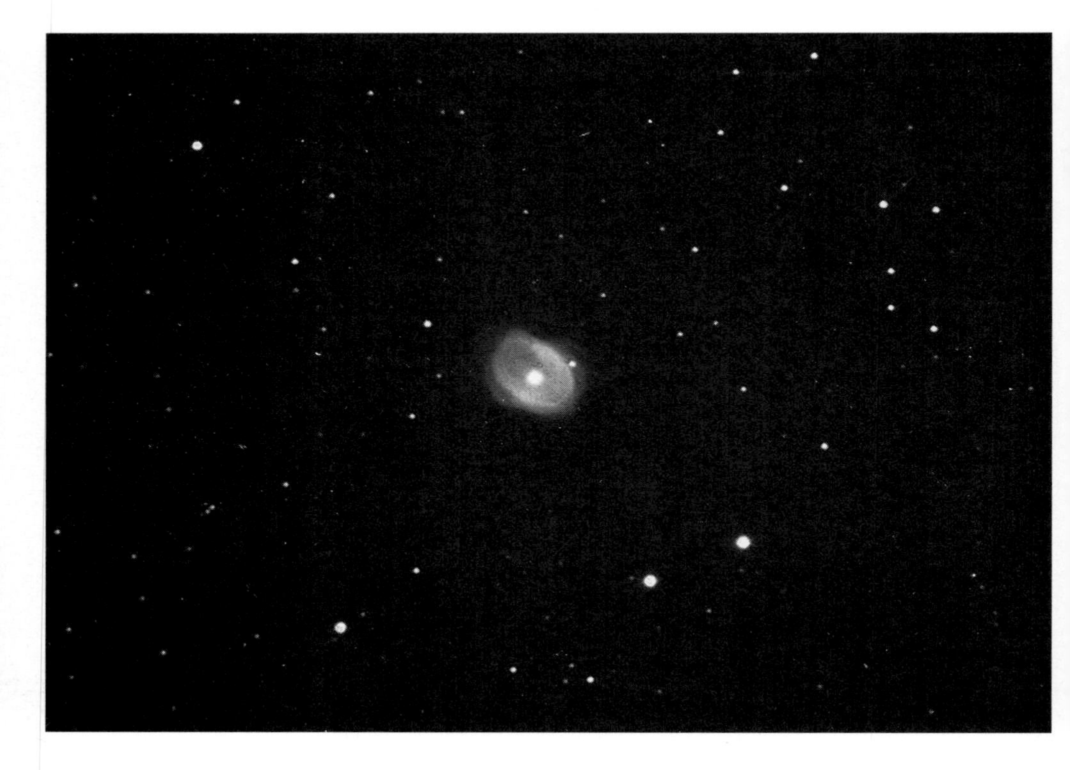

Figure 6.13. *NGC 3132, the Eight-Burst Nebula in Vela, is one of the spring sky's brightest planetary nebulae. Its subtle features require large-aperture instruments, but its slight bluish tinge is apparent in most telescopes. South is up. Photo courtesy Harvard College Observatory.*

close to the border with Antlia. NGC 3132 shines conspicuously at 8th magnitude and is comparable in size to the Ring Nebula in Lyra. The unusual multiring structure of this cloud, best seen in long-exposure photographs and larger backyard instruments, has led astronomers to nickname it the "Eight-Burst Nebula." Although these complex features may not be readily apparent through your telescope, a slight bluish tint should be. Its central star glows at 10th magnitude and is visible in 100-mm and larger telescopes.

NGC 3201, the only bright globular cluster in Vela, rides very low in the southern sky from midnorthern latitudes. Look for its 7th-magnitude smudge set between a pair of 7th-magnitude suns, about a third of the way between 3rd-magnitude Mu (μ) Velorum and 4th-magnitude Psi (ψ) Velorum. To observers south of about 20°N latitude, it should be visible with just about any optical aid. NGC 3201 is listed as a Class X globular, implying a loose stellar concentration. Many of the 13th-magnitude stars that surround the cluster's core spring into view in a 150- or 200-mm telescope, as do a clumping of stars along the cluster's northern edge. Strangely, even though this

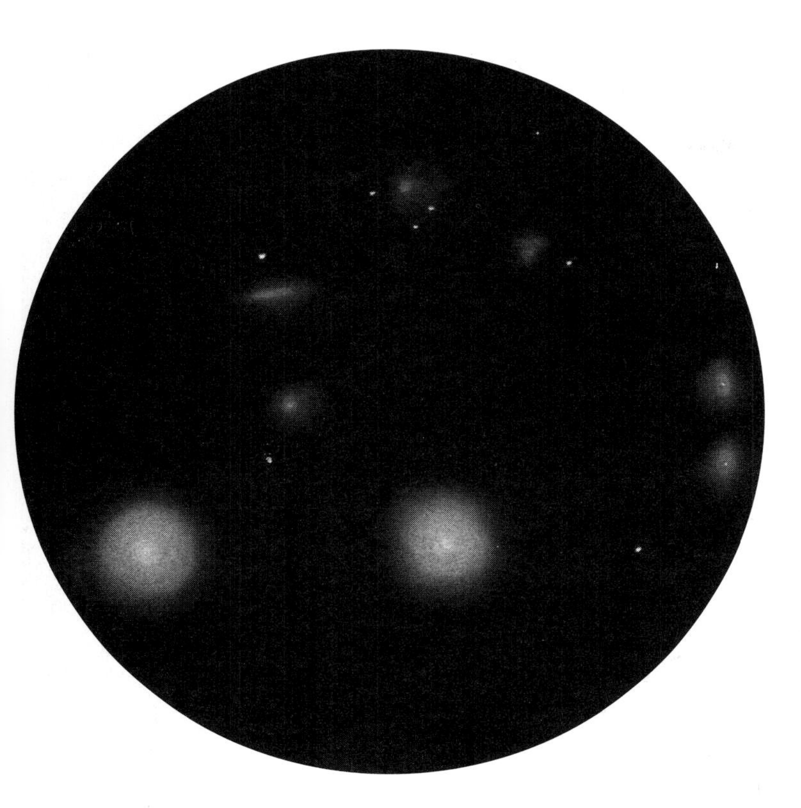

Eyepiece Impression 7. *Spring skies can be a galaxy hunter's dream come true. This small portion of the Realm of Galaxies in Virgo contains M84 (left) and M86 (right), surrounded by a host of fainter galaxies. Turn the book upside down to see how the galaxies are "looking back" at you! Seen through the author's 333-mm Newtonian and 24-mm Tele Vue Wide Field eyepiece (63×). South is up.*

effect has been well documented by many visual observers, photographs show no evidence of it.

Virgo

Home to the southern half of the Realm of Galaxies (Figure 6.6), the constellation Virgo, the Virgin, holds a lifetime of fascination for deep-sky observers. You can quite literally spend the entire season within the constellation's boundaries and still not run out of new objects to peruse. This is a galaxy-hunter's dream come true. Virgo reaches midnight culmination in the middle of April, bringing it into the early evening during May and June.

M84 (NGC 4374) and **M86** (NGC 4406). Right in the thick of the Realm is this pair of almost identical elliptical galaxies. The challenge is to pick them out from the crowd. To isolate M84 and M86, move eastward 6° from Denebola (Beta [β] Leonis), the easternmost star in Leo's triangular hindquarters, to 5th-magnitude 6 Comae Berenices. Now, trace a line southeastward for 3°, passing a 7th-

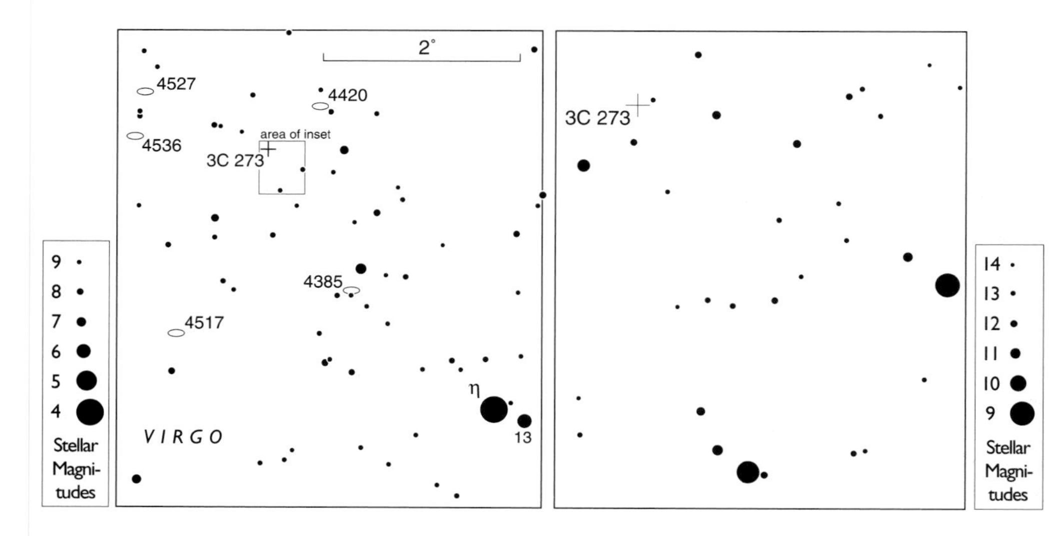

Figure 6.14. *This pair of finder charts will help you locate quasar 3C 273, the most distant object visible through most amateur telescopes. Use the wide field chart (left) when aiming with your finderscope. Once you are pointed in the right direction, switch to your telescope and a medium-power eyepiece. Then, using the detailed chart (right), which shows stars as faint as magnitude 14, zero in on the quasar itself. Look for a faint stellar point of pale blue light. That will be 3C 273. Just be careful not to confuse it with faint field stars.*

magnitude star along the way, to M84 and M86. Eyepiece Impression 7 shows how the galaxies look like two spheres of light punctuated by bright, nonstellar nuclei. Closer inspection will show M84 (the westernmost of the pair) to be a little brighter than its neighbor, while M86 appears slightly elongated. M84 is classified as an E1 elliptical and M86 as an E3.

Surrounding these bright systems are many fainter galaxies within the grasp of a good 200-mm telescope. To the north of M86 is **NGC 4402,** a spiral seen nearly edge on — look for a long, thin needle of 12th-magnitude light. South of M84 and M86 lies **NGC 4387.** Measured on photographs as 2' × 1', this faint elliptical will challenge even large telescopes on less-than-perfect nights. Finally, **NGC 4388** forms a triangle with M84 and M86. Measuring about 3' × 0.5' in a 200- to 254-mm telescope, this edge-on galaxy displays a conspicuous stellar nucleus when examined with high power. To Sue French of Glenville, New York, M84, M86, NGC 4388, and NGC 4402 form an intergalactic face. Consider M84 and M86 to be the eyes, NGC 4388 the nose, and NGC 4402 the thin mouth. Here's looking at you!

The brightest galaxy near M84 and M86 is **M87** lying 75' to their southeast and just south of a 9th-magnitude Milky Way star. In 1918, Heber Curtis (1872–1942) of Lick Observatory discovered a strange luminous jet of material bursting from M87's core. Later, the jet was found to be a source of radio waves and x-rays, and to emanate from the very core of the galaxy. Radio observations have since found several hundred galaxies and quasars with associated jets. The jets are thought to act as conduits for the transport of matter and energy away from the cores of these objects. In 1994, observations with the Hubble Space Telescope provided strong evidence that the core of M87 is powered by a black hole.

Amateur telescopes reveal that the galaxy's brilliant not-quite-stellar nucleus is engulfed in a fainter round mist that fades slowly toward its edges. Although M87 stands up well to scrutiny at high magnification, backyard telescopes just don't have enough oomph to show the jet.

While in the area, be sure to look for two fainter galaxies, 11th-magnitude **NGC 4478** and 12th-magnitude **NGC 4476,** both to M87's southwest.

3C 273. Let's pause from our galactic tour to find the most distant object visible through most amateur telescopes: the quasar 3C 273. Begin your trip to the depths of the universe at the naked-eye star Eta (η) Virginis; the quasar lies a little more than 3° northeast of the star (Figure 6.14). Viewing through your finderscope, slide about 1½° east-northeast of Eta to a northward-pointing triangle of 8th-magnitude stars. Continue another 1½° to the north, past a lone 8th-magnitude star, to another 8th-magnitude star. Switch to a low-power, wide-field eyepiece, and slip to the east about ¾°. Keep on the lookout for a ½°-wide triangle of 9th- and 10th-magnitude stars. The quasar lies just north of the triangle's 10th-magnitude, northeast-corner star, adjacent to a 13th-magnitude star.

3C 273, the 273rd entry in the *Third Cambridge Catalogue of Radio Sources,* is the brightest quasar in the sky. At magnitude 12.8 it is within the grasp of a 150-mm telescope, but only on nights when the air is dry and the sky is crystal clear. Its blue color should help set it apart from stellar neighbors.

The spectrum of 3C 273 reveals that it is receding from us at a tremendous rate — about one-sixth the speed of light! Hubble's law states that the faster a galaxy is receding from us, the farther away it is. The high rate of recession of 3C 273 implies that it must be about two billion light-years away and therefore extremely luminous — about 1,000 times brighter than an average galaxy. Imagine: the light we see coming from 3C 273 tonight left there when single-cell life forms ruled the Earth!

STAR 21. If you are feeling a little homesick for our galaxy, try your luck with the asterism STAR 21, just northeast of STAR 20, the Stargate (described under Corvus). As with the Stargate, cast off from Delta (δ) and Eta (η) Corvi, which together mark the northeast corner of the Corvus quadrilateral. From Eta, move about 3° northward to a northeast-pointing isosceles triangle formed by three 6th-magnitude stars. The tip of the triangle is aimed directly at the Stargate, which lies about 1° to the northeast. Continue another 1° northeast to a 7th-magnitude sun, the brightest in STAR 21.

Figure 6.15. *M104, the Sombrero Galaxy, in Virgo. This well-known galaxy looks like a Mexican hat only when viewed north-up! Photo by Martin C. Germano.*

Tom Mote of San Antonio, Texas, and John Barra of Peoria, Illinois, were the first to introduce this pattern to me. Mote calls it "Little Sagitta" after its resemblance to that summer constellation, while Barra teams it with the Stargate to form the "Double T Clusters." As shown in Eyepiece Impression 8, I see STAR 21 as more like a northward-swimming shark, the brightest star marking a portion of the head or mouth and fainter stars forming an emaciated body. I therefore call it "Jaws." With a little imagination you'll even see a dorsal fin!

M104 (NGC 4594). Although it is believed to be a bona fide member of the Coma-Virgo Realm, M104 lies about 25° due south of the cluster's center (Figure 6.15). Many observers use Delta (δ) Corvi as their jumping-off point to M104. As just described above, hop northward from Delta to the Stargate asterism (STAR 20), then leap toward Jaws (STAR 21) across the border in Virgo, and finally shift 20' east to M104.

M104, also shown in Eyepiece Impression 8, is well known among deep-sky observers for its prominent central core, broad spiral-arm rim, and conspicuous dark lane. The overall visual effect is that of a

Eyepiece Impression 8. *M104 (left) and STAR 21, Jaws (right), as seen through the author's 200-mm Newtonian and 24-mm Tele Vue Wide Field eyepiece (58×). The asterism's two brightest stars, seen toward the bottom of the view, mark the shark's open mouth, while a line of fainter stars extending southward make up the thin body. A lone star to the west (right) marks the tip of the dorsal fin. North is up.*

Mexican hat, leading early observers to nickname this the "Sombrero Galaxy." When viewed through a 100- or 150-mm telescope, the Sombrero's dark lane may be seen cutting across the south side of the nucleus. The added light-gathering ability of a 200-mm instrument extends the lane fully across the galactic "brim" of M104.

As you can see, springtime is indeed galaxy time, a season when deep-sky observers reach beyond the Milky Way toward the hordes of more distant galactic systems. With so many diverse galaxies from which to choose, it is easy to see why the spring is many an astronomer's favorite time of year.

CHAPTER 7

Summer's galactic getaway

While the spring sky was chock full of distant galaxies, warm summer evenings are bathed in the gentle glow of our own galaxy, the Milky Way. Scattered across the summer sky are hundreds of open and globular clusters, huge rifts of bright and dark nebulosity, petite planetaries, double stars, asterisms, and even a select few galaxies, all beckoning us to join them.

Before starting out become familiar with the season's many bright stars (shown in Figure 7.1) and use them as guideposts to help orient yourself to the sky. Once acclimated, begin your deep-sky venture with the brighter objects mentioned here. Later, continue with some of the more difficult targets. Before you know it, you'll be sailing the vast ocean of our Milky Way like an old salt.

Ara

Ara, the Altar, is a far-southern constellation occupying a small region of the summer sky south of more prominent Scorpius. While it contains many interesting open clusters and diffuse nebulae, Ara is best known as home to one of the closest globular clusters to Earth. Keep in mind, however, that unless you are located south of about 30°N latitude, the horizon will probably come between you and the Altar. Ara transits the meridian at midnight in the middle of June, bringing it into our August evening skies.

Of Ara's open clusters, **NGC 6193** stands out from the crowd. It is situated in the northwest corner of the constellation and is most readily found by sliding about 7° to the south-southwest of Zeta (ζ) Scorpii in the Scorpion's hook-shaped tail. Once there, look through your finder or binoculars for a gentle haze surrounding a 5th-magnitude point of light.

With more aperture and magnification, NGC 6193 appears as a large smattering of about 30 stars scattered across 15'. Many of the cluster members are arranged in long, looping stellar chains. The brightest is the multiple star **h4876**. Only two or three of this system's six components are likely to be resolved in most amateur telescopes.

Faint nebulosity surrounds NGC 6193. The brightest portion of this cloud, to the cluster's southwest, is designated **NGC 6188**. Photographs of NGC 6188 always remind me of a "headless" Horsehead Nebula — a long, thin streak of dim nebulosity. Spotting it can prove even more challenging than seeing its famous likeness in Orion.

IC 4651. Of the open clusters in Ara, this is my favorite. Look for

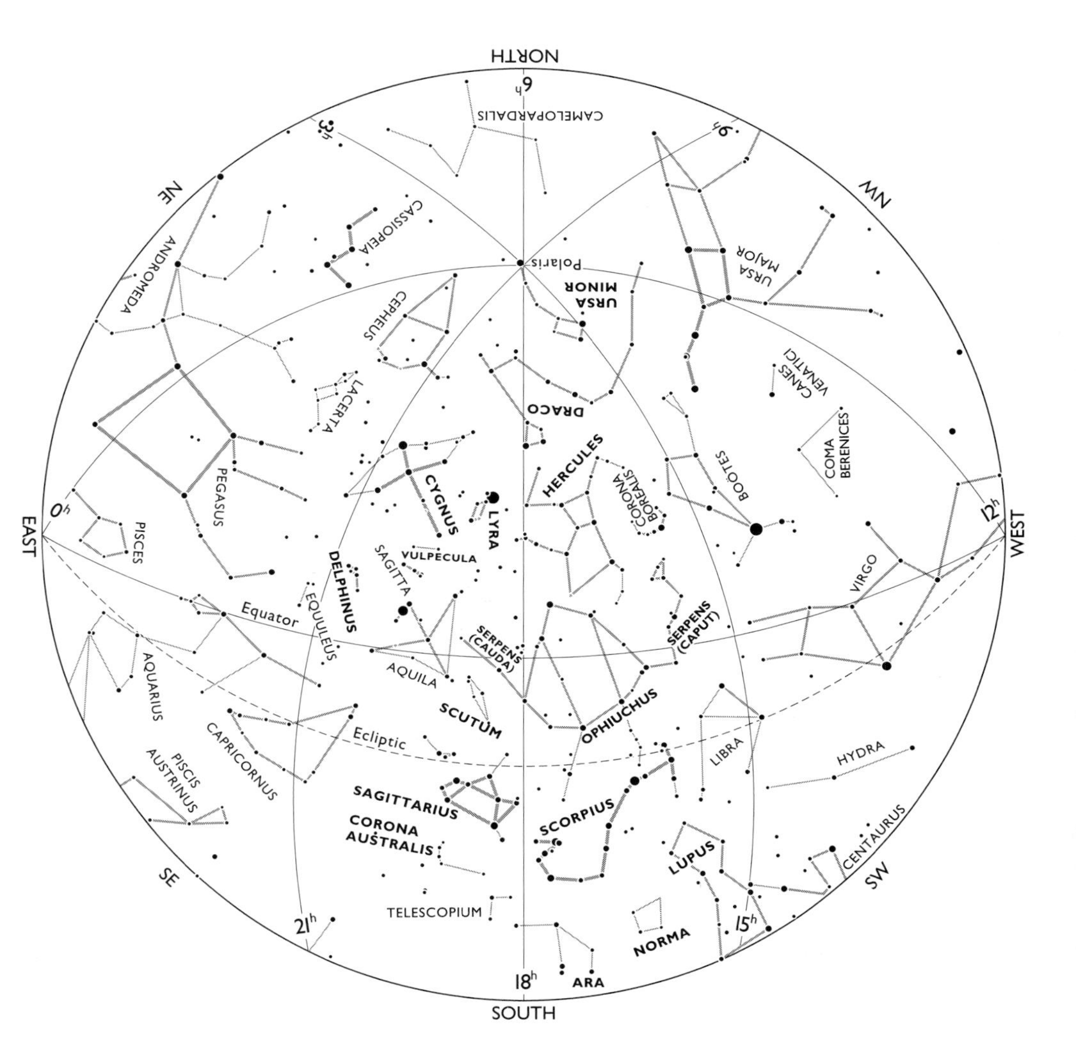

Figure 7.1. *The summer naked-eye sky. Constellations discussed in this chapter are shown in boldface. The chart is plotted for 30° north latitude. For viewers north of that latitude, stars in the northern part of the sky will appear higher above the northern horizon and stars in the south lower (or possibly below the southern horizon).*

it about 2° due west of 3rd-magnitude Alpha (α) Arae. Although binoculars show IC 4651 as little more than a muddled haze, telescopes display a dense fog of faint stars. Most of the 80 blue-white suns in the cluster are between 9th and 10th magnitude and are confined to a region 12' across, so the nicest show is delivered by medium magnifications.

At an estimated distance of 8,200 light-years, **NGC 6397** is one of the closest globular clusters. It lies about 3° northeast of 3rd-magnitude Beta (β) Arae and within the same binocular field as 5th-magnitude Pi (π) Arae. Through binoculars NGC 6397 is a small ball of cosmic fluff against an attractive field of stars. A 100-mm (4-inch) telescope will begin to resolve the cluster into a myriad of stars, while larger scopes will easily crack its stellar vault. Viewing with a 100-mm f/10 refractor from the Winter Star Party in the Florida Keys some years ago, I noted what can best be described as a gap between the stars along the globular's western edge. I have never read about such a feature in other deep-sky handbooks, though some observers have described this globular as "triangular." What do you see?

Corona Australis

The Southern Crown, Corona Australis, is formed from an arc of 3rd- and 4th-magnitude stars directly east of the tail of Scorpius. From midnorthern latitudes the constellation scrapes the horizon and suffers greatly from the haze of summer nights. From a good observing site farther south, however, it is worth a look at the Crown's globular cluster. With Corona Australis reaching midnight culmination in early July, the best time for seeing the Crown in the early evening is late August and early September.

NGC 6541, a 6.6-magnitude globular, is most readily found by casting off from Theta (θ) Scorpii and heading southeast toward 96 Scorpii. Continue another 2° northeast to a second 5th-magnitude star, where you will also find NGC 6541 just 20' to its southeast. NGC 6541 is one of those rare globulars that is an attractive sight through just about all optical instruments. Binoculars and small telescopes show it as a tiny piece of celestial cotton, perhaps giving the slightest hint of a mottled surface. A 150-mm (6-inch) is probably the smallest instrument capable of resolving any individual stars, while larger scopes easily resolve the cluster into a myriad of points.

Cygnus

While the saying goes that "man's best friend is his dog," one of the deep-sky observer's best friends is the Swan, Cygnus, and its cornucopia of wonderful deep-sky objects. Here we can only scratch the surface of what Cygnus has to offer amateur astronomers. Begin with the objects described here, then expand the list with your own entries. Cygnus flies past midnight culmination in late July, and is best seen in the early evening during late August and September.

Beta (β) Cygni. Of all the double stars sprinkled across the heavens, none is more beautiful than Beta (β) Cygni, or Albireo (Figure 1.1). Albireo marks either the base of the Northern Cross or the tip of the Swan's beak, depending on how you look at it. To the naked eye it looks like just another 3rd-magnitude point of light, but a telescope, or even steadily held high-power binoculars, reveals Albireo as one of summer's most colorful stellar jewels. Albireo A, a Type-*K* star, is a gleaming 3rd-magnitude gem that shines with a golden radiance, while its companion, Albireo B, is a Type-*B* star that appears as a dazzling sapphire.

When the renowned double-star observer F. G. W. Struve first examined Albireo in 1832, he measured the apparent distance between Albireo A and B to be 34". Although the stars have changed little in

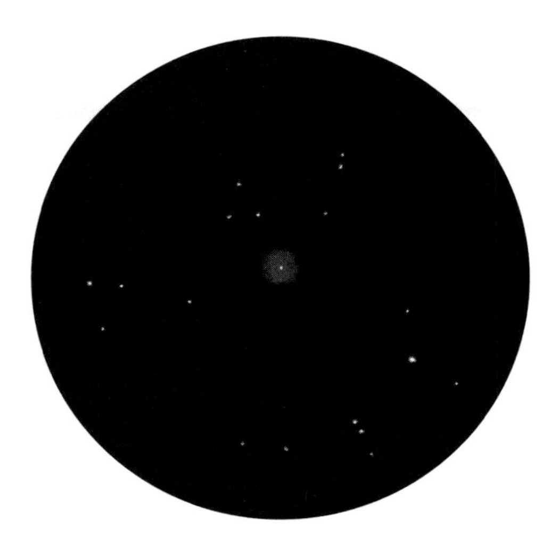

Eyepiece Impression 9. *NGC 6826, as seen through a 355-mm Schmidt-Cassegrain telescope at 326×. Drawing by Dave Kratz. South is up.*

separation since their discovery, most binary-star authorities believe that they form a true binary system. Albireo B is estimated to be at least 4,400 astronomical units from Albireo A. At that tremendous distance its period of revolution must be nigh on 100,000 years.

Some astronomers take exception to this conclusion. In 1984, J. Kemp, D. Kraus, and I. Beardsley concluded that, based on observations made with the 610-mm (24-inch) telescope at the University of Oregon, Albireo may just be an optical double — a chance alignment of two widely separated stars. In fact, Albireo B may be as much as 50 percent farther from Earth than Albireo A. Clearly, more work and analysis remain to be done.

NGC 6819. For the next selection, scan along the Milky Way toward Sadr (Gamma [γ] Cygni), then head west for 8° until a triangle of 6th- and 7th-magnitude stars comes into view. The open cluster NGC 6819 is just southwest of the triangle's brightest component. Small amateur telescopes show NGC 6819 as a hazy glow bespangled with a few dim points of light. A 150- to 200-mm (6- to 8-inch) instrument will partially resolve the cluster into a dense throng of faint stars unevenly distributed across 10'. Larger instruments continue to improve resolution, and more than 100 cluster members are visible through 305- to 355-mm (12- to 14-inch) telescopes.

Many observers have commented on the asymmetric appearance of NGC 6819. For instance, in the *Webb Society Deep-Sky Observer's Handbook,* volume 3, Guy Hurst describes the cluster as roughly U-shaped as seen with a 254-mm (10-inch) telescope. By contrast, in their book, *Observing Handbook and Catalogue of Deep-Sky Objects,* Luginbuhl and Skiff liken it to the letter K or X.

NGC 6826, one of the season's prettiest and most curious planetary nebulae, is near the western wing of Cygnus. From Iota (ι) Cygni, at the wing's tip, scan about 1½° southeast to Theta (θ) Cygni (itself a pretty optical triple through binoculars and finderscopes). Now travel east-northeast about 45' to 5th-magnitude 16

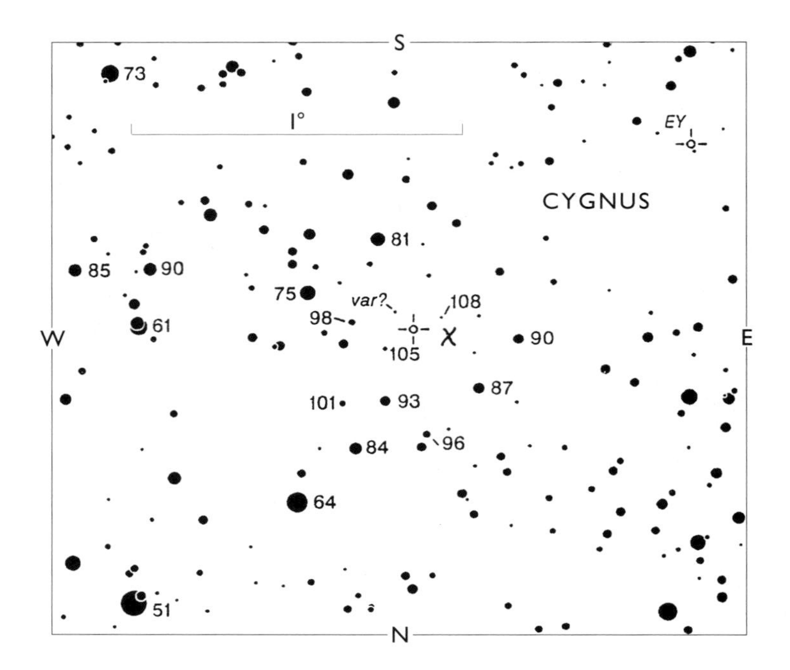

Figure 7.2. *Use this finder chart to try your luck at estimating the magnitude of the variable star Chi (χ) Cygni. See photo on page 14. Several suitable comparison stars and their respective magnitudes are also shown, though each magnitude's decimal point has been omitted to avoid confusing it with a faint star. (The star labeled "75" is actually magnitude 7.5, while the star labeled "108" is magnitude 10.8, and so on.) The star labeled "var?" may be variable; further study is required. The chart is oriented with south up. Courtesy the American Association of Variable Star Observers.*

Cygni, then another 30' eastward to find the planetary.

NGC 6826 has been given the unusual nickname of the Blinking Planetary because of the disappearing act it performs through smaller amateur telescopes. When viewed with averted vision NGC 6826 shows a distinctive bluish or greenish disk measuring some 30" in diameter, as shown in Eyepiece Impression 9. Its 10th-magnitude central star is also easy to see through a telescope as small as 50 mm (2 inches). But when you look directly at the planetary, it disappears! Where does it go? Of course, it doesn't actually vanish. What happens is that the light from the central star overpowers the nebula when both are viewed with the less sensitive central part of the retina. When their combined light falls along the more sensitive area to one side of the center, both nebula and star can be distinguished independently.

Chi (χ) Cygni is a long-period variable star noted for its broad magnitude range (Figures 1.4 and 7.2). Look for its orangish glint along the Swan's long neck, about a quarter of the way from Sadr (Gamma [γ] Cygni) to Albireo (Beta [β] Cygni). At greatest light

Left: Eyepiece Impression 10. *NGC 6888, known as the Crescent Nebula, is the brightest of many nebulae found near the center of the Northern Cross. This drawing shows the view through the author's 200-mm Newtonian equipped with a 24-mm Tele Vue Wide Field eyepiece (58×) and an O III nebula filter.*
Right: Figure 7.3. *Compare the drawing to this photograph of the Crescent Nebula taken by Preston Scott Justis. In both cases, south is to the upper left.*

Chi reaches magnitude 3.3 and becomes visible to the naked eye. It then fades at a fairly constant rate, until it bottoms out at 14th magnitude, with the entire cycle taking 406.93 days. Thanks to its prominent location near the zenith on midsummer nights, Chi is a favorite variable of observers with binoculars and telescopes alike.

NGC 6888. Of the many patches of nebulosity adjacent to the center of the Northern Cross, NGC 6888 is the brightest and easiest to find (Figure 7.3). From Sadr (Gamma [γ] Cygni), scan south-southwest 2° to 5th-magnitude 34 Cygni, then 1° west-northwest to a small diamond of 7th- to 9th-magnitude suns. A careful check through a telescope shows these stars to be engulfed in the faint wisps of NGC 6888.

This cloud is known to many as the Crescent Nebula for its arc-like appearance, which is apparent in Eyepiece Impression 10. The cloud's brightest section lies between the diamond's northern and western stars and may be seen in a 75- to 100-mm (3- to 4-inch) telescope, while the fainter extension usually requires at least a 150-mm for it to be detected. A subtle though distinct mottled texture is revealed in a 254-mm, while the nebula's full curve is usually visible only through 305-mm and larger instruments and in photographs. Experience shows that the view is greatly enhanced by an O III nebula filter.

STAR 26. If you're a bird watcher, you may well have seen a yellow-bellied sapsucker or a red-winged blackbird. But have you ever

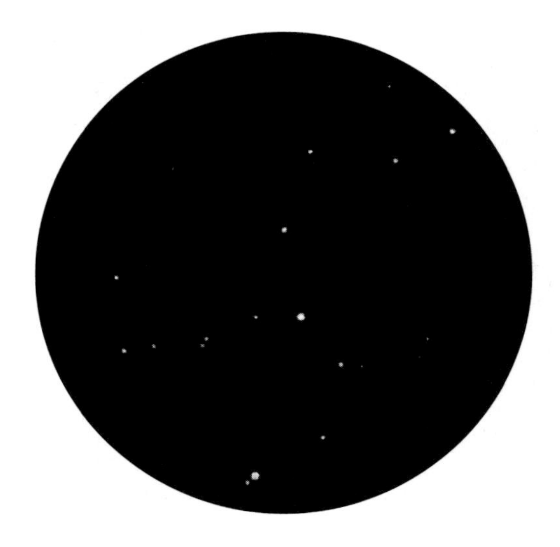

Eyepiece Impression 11. *STAR 26, the Red-Necked Emu, as seen through the author's 200-mm Newtonian and 24-mm Tele Vue Wide Field eyepiece (58×). The Emu's head is to the right. Its long neck and body curve toward the left, with its tail extending to the north to end at the star 29 Cygni. South is up.*

seen a red-necked emu? Probably not, since the Red-Necked Emu is not found here on Earth; rather, it is found accompanying the Swan across the summer night sky. John Barra of Peoria, Illinois, was first to spot the Emu using a 200-mm reflector. You can spot it too by beginning at Sadr (Gamma [γ] Cygni). Move 2½° south-southwest along the "neck" of Cygnus to 34 Cygni, then another 1½° to 29 Cygni. This latter star, a wonderful multiple sun, marks the Emu's tail.

The profile of the Red-Necked Emu is drawn sideways in the sky, with its feet toward the east and its head cocked to the west, as shown in Eyepiece Impression 11. All of the stars in the Emu's tail, feet, triangular body, and head shine with a blue-white luster. Only a lone reddish star marking part of its neck appears noticeably different — hence the name. Try your luck capturing this rare bird on the next clear night.

The **Veil Nebula** complex is unquestionably one of the best-known telescopic sights of the summer sky. Believed to be the fragmented remnant of an ancient supernova, the Veil always proves challenging through nearly all amateur telescopes.

Brightest of the three main portions of the Veil network is **NGC 6992,** the eastern loop (Figure 7.4). Look for it just southwest of a small triangle of 7th- and 8th-magnitude stars about halfway between Epsilon (ε) and Zeta (ζ) Cygni, the stars that form the Swan's eastern wing. Measuring around 1° in length, NGC 6992 is easiest to find with a low-power, wide-field eyepiece. Once spied, however, switch to an eyepiece of moderate focal length to glimpse the cloud's delicate gossamer structure. My most memorable encounter with NGC 6992 took place several summers ago from Ron Woodland's house in the Berkshire hills of western Massachusetts. Peering through his 445-mm (17.5-inch) Newtonian, the nebula's intricate filamentary structure was as breathtaking as in any photograph — but I was actually seeing it live!

Figure 7.4. *NGC 6992, the brightest of the three main portions of the Veil Nebula in Cygnus. Under dark, crystalline skies many of the twists and knots within the Veil's clouds can be seen through telescopes as small as 200-mm aperture. South is to the upper left in this photograph by Martin C. Germano.*

NGC 6960, the westernmost section of the Veil complex, is a long, thin nebulous band of grayish light extending to the north and south of the bright star 52 Cygni. While studies indicate that the star is merely an unrelated foreground object, it does curiously mark a change in the nebula's appearance. As shown in Eyepiece Impression 12, the area to the star's north is far easier to see than the southern region. In fact, the southern half of NGC 6960 may go completely undetected unless 52 Cygni is first removed from the field of view. As with NGC 6992, tangled rifts and eddies can be detected through medium and large backyard telescopes with moderate-power eyepieces. Nebula filters (especially O III filters) help to enhance this

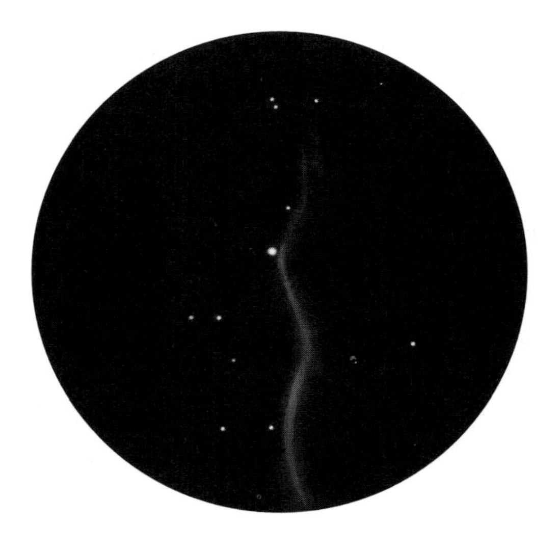

Eyepiece Impression 12. *NGC 6960, as seen through the author's 200-mm Newtonian, 24-mm Tele Vue Wide Field eyepiece(58×), and O III filter. South is up.*

fine detail. But a large telescope is not absolutely necessary for seeing NGC 6960, as I. Genner of Newcastle-upon-Tyne in the United Kingdom has demonstrated. In the *Webb Society Deep-Sky Observer's Handbook*, volume 2, he recounts how the nebula was "difficult but certain with averted vision" through a 75-mm telescope.

Several fainter sections of the Veil complex challenge even the most dedicated deep-sky observers. The faintest of the three main parts of the Veil is a long, thin, wedge-shaped nebulous glow between NGC 6960 and NGC 6992. Not included in the *New General Catalogue*, it has come to be known as **Pickering's Wedge.** You might also try to search out **NGC 6974** and **NGC 6979.** The former appears as a small, curved cloud to the northwest of a pair of 7th- and 8th-magnitude stars just west of NGC 6992. The latter, lying farther to the northwest, is a smaller, formless glow measuring about ½°. Spotting any of these requires exceptionally transparent skies.

NGC 7000, the famous North America Nebula, is always the subject of hot debate at summer star parties. Many observers find this one of the most difficult objects in the sky to spot, as it defies detection in nearly all telescopes; others feel it is one of the easiest nebulae, since it can be seen with the naked eye if one knows exactly where to look. Both schools of thought are correct. The North America Nebula may indeed be seen with no optical aid at all as a breakaway Milky Way cloud about 3° due east of brilliant Deneb (Alpha [α] Cygni). Yet its large dimensions and low surface brightness make it rarely visible in a narrow field of view.

Binoculars are the instrument of choice for viewing NGC 7000. The "East Coast" is easily detected through 7× glasses, while "Florida" and the "Gulf of Mexico" are both distinct through 11×80 binoculars. The "West Coast", extending east toward 62 Cygni, is much more difficult to distinguish against the starry surroundings.

Defining the East Coast of this celestial continent is a dark nebula

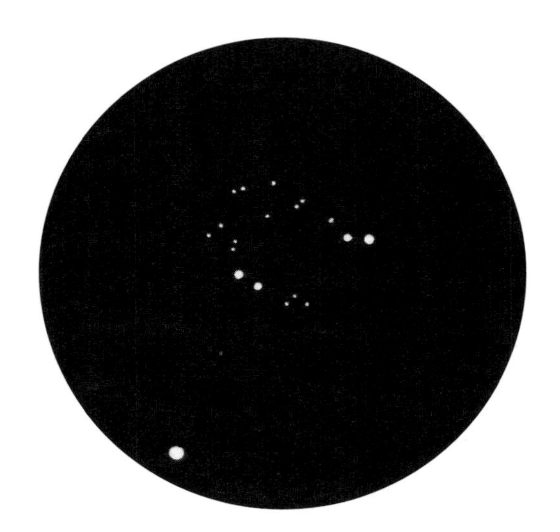

Eyepiece Impression 13. *STAR 28, the Horseshoe, was first recorded more than a century ago by the famed deep-sky observer Thomas Webb. This modern rendition was sketched by Dan Hudak through a 200-mm telescope. South is up.*

known as **Lynds 935** (LDN 935), though it might be more appropriate to call it the "Atlantic Ocean Nebula." LDN 935 is a large dark patch measuring 150' × 40'. Its large apparent size makes it all but impossible to spot through most telescopes, but the wide fields of view of binoculars and rich-field telescopes cause it to stand out quite well.

Now, look near where Hudson Bay should be. Can you see a dark rift extending northward? If so, then you have seen the dark nebula **Barnard 352.** *Sky Catalogue 2000.0* (volume 2) specifies it as 20' × 10' in extent.

Once you have successfully pinned down the North America Nebula, try your luck with the open cluster **NGC 6997,** which is positioned against NGC 7000 approximately where the Great Lakes are in the North American continent. In all, 40 faint stars have been identified as members of this 7'-wide group. Estimates place both the cluster and the nebula at approximately 1,600 light-years from Earth.

The western border of LDN 935 marks the edge of **IC 5070,** the Pelican Nebula, lying between NGC 7000 and Deneb. Many amateurs hastily assume that the Pelican is an "impossible" object, but this may not always be so. Though IC 5070 itself is all but invisible, binoculars can distinguish its eastern border (that is, the Pelican's eastern-facing profile) against the comparatively star-free backdrop of LDN 935.

STAR 28. As a different kind of observing project, Daniel Hudak of Canfield, Ohio, has conducted a systematic survey of the many fine star fields mentioned by Rev. Thomas W. Webb in his classic 19th-century work *Celestial Objects for Common Telescopes*. This one, resembling a horseshoe of stars, is especially attractive. To discover it for yourself, slide about 3° east-northeast of Deneb (Alpha [α] Cygni) to 5th-magnitude 60 Cygni, then 2° farther northeast to 5th-magnitude 63 Cygni. STAR 28 is found just ½° south of 63.

As shown in Eyepiece Impression 13, Hudak's drawing of the horseshoe displays an arc of two dozen stars ranging in brightness from about 7th to 11th magnitude and measuring about 20' across. Webb described this star field as "a curious horseshoe and magnificent galaxy field." Hudak adds that this is an "excellent object; despite being set upon a star-packed Milky Way background, the horseshoe stands out."

M39 (NGC 7092) rides the plane of our galaxy about three-quarters of the way between Deneb and 4th-magnitude Pi² (π^2) Cygni. Under dark skies you just might be able to see it with the naked eye as a 5th-magnitude hazy patch floating in the Milky Way. Most sources agree that LeGentil discovered M39 in 1750, though *Burnham's Celestial Handbook* (volume 2) references a statement by J. E. Gore that M39 "was noted by Aristotle as a cometary-appearing object" as far back as 325 B.C.

With its 30 stars covering an area of sky as large as the full Moon, M39 is best appreciated through giant binoculars and small rich-field telescopes. My best view has always been with a 24-mm wide-angle eyepiece in a 108-mm (4¼-inch) f/4 rich-field Newtonian reflector that my daughter and I built a few years ago. At 20×, this telescope transforms M39 into about two dozen blue-white stellar sapphires poured into a triangular pattern. Be forewarned that the cluster may lose some of its appeal when viewed with "normal" telescopes (i.e., ones with a focal ratio greater than about f/6) or at higher power. For instance, in the *Webb Society Deep-Sky Observer's Handbook*, volume 3, Kenneth Glyn Jones describes his impression of M39 through a 200-mm telescope as a "very large group but rather sparse," even at 40×.

IC 5146, the Cocoon Nebula, lies near a short queue of three 7th- and 8th-magnitude stars about 4° southeast of the bright open cluster M39. There we find a faint, nondescript glow surrounding a sparse open cluster of stars. Photographs show that the nebula is traversed by dark lanes in much the same way as the Trifid Nebula in Sagittarius, but only instruments of 305-mm and larger stand much chance of revealing these features visually.

IC 5146 is located at the eastern end of the long, thin dark nebula **Barnard 168.** Given dark skies, the obscuring cloud is surprisingly easy to see through giant binoculars as a dark lane stretching west of the Cocoon. It stands out in striking contrast against IC 5146 and the bright surrounding Milky Way star field.

Delphinus

Lying just to the east of the Milky Way, the tiny constellation of Delphinus, the Dolphin, contributes three interesting deep-sky sights to our survey. Several fainter targets also lay in wait within Delphinus, inviting you to return again and again. Delphinus passes midnight culmination in early August and can be found in the early-evening September sky as summer fades into autumn.

STAR 9, situated to the south of the Dolphin's diamond-shaped body, was first recorded in Webb's *Celestial Objects for Common Telescopes* in 1859 and subsequently listed in my *Touring the Universe Through Binoculars*. The brightest star in the pattern is 6th-magnitude Theta (θ) Delphini, an orange sun that lies at the end of a curve of five stars bending toward the northwest. Adding to the scene is a tight equilateral triangle of 7th-magnitude suns just to the east of Theta. In all, two dozen stars of 9th magnitude and brighter are found here. Like many open clusters, the wide expanse of STAR 9 is best appreciated through binoculars and low-power rich-field telescopes.

Gamma (γ) Delphini marks the northeast corner of the Delphinus diamond. Small telescopes reveal Gamma to be a double star, with a 4.3-magnitude primary attended by a 5.2-magnitude partner. Since 1830, when the pair was first measured by F. G. W. Struve, their separation has closed from nearly 12" to about 10", with the companion now to the primary's west. Many observers comment on both of the stars' golden color, while a few have referred to the companion as greenish.

STAR 27. I first came upon this little asterism, nestling just inside the constellation's eastern border, at the 1993 Stellafane amateur telescope-maker's convention in Springfield, Vermont. You can find it too by heading 5° due east of the Dolphin's diamond-shaped body to a southward-pointing right triangle of three 6th- and 7th-magnitude stars. Fit a wide-angle eyepiece, and extend the triangle's hypotenuse to the northeast for about 45' until you see a small triangle of 9th-magnitude stars. You've now found the treasure of the Dolphin — those three stars along with a cloud of a dozen fainter stellar diamonds overflowing from the triangular treasure chest make up STAR 27. The view through a 100- or 150-mm backyard telescope is especially impressive, since larger instruments tend to dispel the treasure's misty appearance.

Draco

Draco, the Dragon, winds halfway around the northern circumpolar sky. Although all of its stars appear faint to the unaided eye, the Dragon holds several interesting deep-sky objects in its lair, including one of the season's finest planetary nebulae and an interesting asterism. Dozens of galaxies also populate Draco, though none are bright enough to be included in this brief survey. While portions of Draco are visible throughout the year, the constellation's center passes midnight culmination in mid-May and is highest in July and August evening skies.

NGC 6543 is one of summer's most remarkable planetary nebulae. To find it, aim your telescope halfway between Delta (δ) and Zeta (ζ) Draconis, a pair of 3rd-magnitude stars about 12° north of Draco's distinctive four-sided head. There, you ought to see a trapezoid of four faint stars. NGC 6543 lies about 1° west of the pattern's southwestern star, toward a lone 8th-magnitude sun.

Although at lower magnifications it appears uniform in texture (as illustrated in Eyepiece Impression 14 and Figure 7.5), NGC 6543 begins to reveal a swirling complexity through larger telescopes at

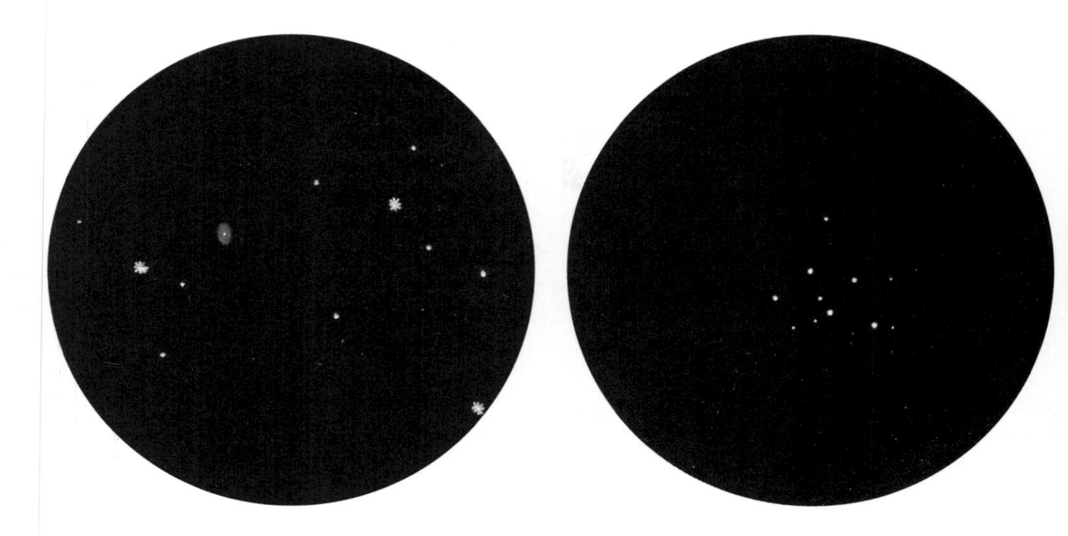

Left: Eyepiece Impression 14. *This drawing of planetary nebula NGC 6543 in Draco was made by Dave Kratz through a 200-mm Schmidt-Cassegrain telescope at 133×. South is up.* Right: Eyepiece Impression 15. *STAR 25, the Little Queen, bears an unmistakable resemblance to the constellation Cassiopeia in this drawing made through the author's 200-mm Newtonian and 24-mm Tele Vue Wide Field eyepiece (58×). South is up.*

high magnifications. Many references call it the Cat's-Eye Nebula, a name that stems from its greenish cast, oval shape, and bright central star. The effect is enhanced by a dark, circular void in the nebula directly adjacent to the central star.

STAR 25, a charming asterism independently discovered by Raymond Maher Jr. of Port Elizabeth, New Jersey, and Attila Kovács of Vác, Hungary, bears more than a passing resemblance to the constellation Cassiopeia. This "Little Queen" is best found by extending a line from Kochab (Beta [β] Ursae Minoris) to Eta (η) Ursae Minoris and continuing on to a triangle formed from Phi (φ), Chi (χ), and Upsilon (υ) Draconis. From Chi, shift eastward about 75' to STAR 25, a small clump of 6th-magnitude and fainter stars.

As shown in Eyepiece Impression 15, the five brightest stars in STAR 25 create a distinctive 10' × 20' "W" pattern that is visible even in 7 × binoculars on dark nights. Telescopes reveal that 18 stars ranging from 6th to 11th magnitude contribute to this pretty asterism.

Hercules

Some deep-sky observers may know the legendary strongman only as the home of the Great Cluster, the globular M13. While there is no denying that M13 is one of the northern sky's true showpieces, Hercules holds much more of interest besides, including another outstanding globular cluster, a number of fine double stars, a nice planetary nebula, and even a pair of little-known asterisms. Hercules crosses the midnight meridian in mid-June and can be found standing

Figure 7.5. *NGC 6543, nicknamed the Cat's-Eye Nebula, is Draco's finest deep-sky treasure. This photograph by Martin C. Germano is oriented with south up.*

near the zenith (albeit upside-down from our Northern-Hemisphere perspective) during late July and August evenings.

STAR 23, first described in Webb's *Celestial Objects for Common Telescopes,* was pointed out to me by Ohio observer Daniel Hudak. It is easily located about 75' southwest of Zeta (ζ) Herculis at the southwest corner of the Hercules Keystone. Looking at Zeta through your finder, you ought to see a 7th-magnitude sun accompanied by a couple of fainter ones. They all belong to STAR 23.

In all, 13 stars down to 11th magnitude create this asterism. While Webb described them as a "recurving line of small stars," it is simpler to characterize the pattern as a backward letter "S" or possibly a rattlesnake. As shown in Eyepiece Impression 16, the rattler's head is marked by the curve's brightest sun, while the tip of the rattle is a 9th-magnitude sun. Even though the remaining 11 stars in the snake's body are only of 11th magnitude, the overall pattern stands out well.

M13 (NGC 6205). Since its discovery by Edmond Halley in 1714, the globular cluster M13 has been widely known as the grandest of its kind north of the celestial equator. In dark skies, a keen eye will be able to spot it without any optical aid as a faint glow about a third of the way from Eta (η) to Zeta (ζ) Herculis in the Keystone. Observers can expect to see a few of M13's estimated 1 million stars with a 100-mm telescope at moderate power. At twice the aperture, M13 is resolved into a huge globe of tiny points, as shown in Eyepiece Impression 17, like a pinch of sugar dropped onto a black-velvet backdrop. Exceptional sky conditions may cause these stars to appear to stand out, almost three-dimensionally — quite an effect.

If you look carefully, you may notice that many of the outer cluster members form chains, or lines, radiating outward from the core. This "spiderlike" impression, first noticed by John Herschel, has been confirmed by many observers since. It's also possible to see three diverging dark lanes, looking almost like a three-blade

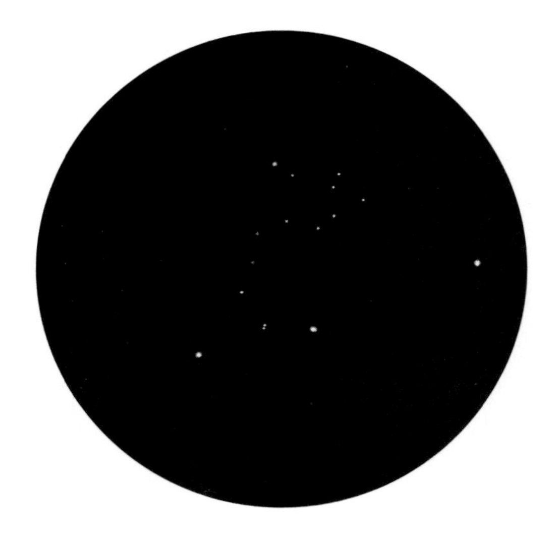

Eyepiece Impression 16. *STAR 23, the Backward S, as seen through the author's 200-mm Newtonian and 24-mm Tele Vue Wide Field eyepiece (58×). South is up.*

propeller, to the south of the cluster's core. Sharp-eyed observers with a 150-mm telescope or larger may also notice a faint smudge of light about 20' northeast of M13. That's **NGC 6207,** a 12th-magnitude spiral galaxy. Can you spot it?

NGC 6210. Discovered by F. G. W. Struve during his 19th-century survey of double stars, the planetary nebula NGC 6210 is bright enough to be seen in telescopes as small as 75 mm in aperture. To find NGC 6210, center your telescope on 51 Herculis, a 5th-magnitude star north of the halfway point between the stars Kornephoros (Beta [β] Herculis) and Delta (δ) Herculis. Scan 2° southwest toward Beta until you see a small triangle of 7th- and 8th-magnitude stars. NGC 6210 is right next to the triangle's westernmost star.

Even the smallest backyard instruments reveal NGC 6210 as a homogeneous disk about 14" in diameter. Nearly all observers will notice the nebula's distinctive color immediately, which is described variously as blue, green, aqua, or turquoise. NGC 6210 holds up well under higher magnifications, so don't hesitate to push your telescope to its limit. Depending on the telescope and seeing conditions, the best views will be enjoyed at 150× or more. Although the central star shines at about 13th magnitude, it is difficult to detect because of the nebula's high surface brightness. You might spot it slightly off center in the nebula through 305-mm or larger instruments.

Alpha (α) Herculis, or Rasalgethi, is not just the constellation's brightest star, it is also an outstanding double star for backyard telescopes. Even small instruments reveal the brighter primary, an orangish 3.5-magnitude star, teamed with a pure white 5.4-magnitude companion some 5" to the southeast. In addition, Alpha Herculis A is an irregular variable that fluctuates from 3rd to 4th magnitude, while Alpha Herculis B is a spectroscopic binary.

M92 (NGC 6341) is one of the unsung glories of the heavens, as observers frequently ignore this globular cluster in favor of its

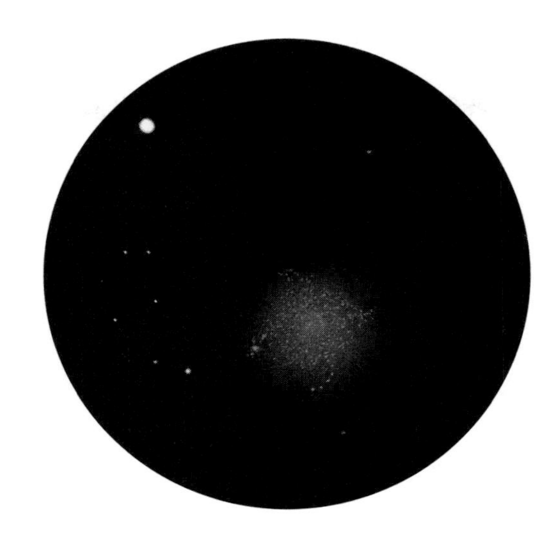

Eyepiece Impression 17. *M13, the Great Cluster in Hercules, as seen through the author's 200-mm Newtonian and 26-mm Plössl eyepiece (55×). South is up.*

famous neighbor, M13. But once they see it, and discover that it is so easy to find, most observers quickly add it to their permanent must-see list. To discover this unappreciated treasure, head 4° north-northwest from Pi (π) Herculis at the northeast corner of the Hercules Keystone to a distinctive triangle of 5th- and 6th-magnitude stars. From there it's just a matter of taking a 3° northeastward hop, past an arc of three 7th-magnitude stars, to M92.

Binoculars show it as a fuzzy patch of 6.5-magnitude light, while a 150-mm will partially resolve stars around the cluster's edges. The largest amateur telescopes can achieve full resolution and may also be able to confirm some of the peculiar dark patches seen through Lord Rosse's giant 72-inch (1.83-meter) reflector more than a century ago.

95 Herculis, one of the prettiest binary stars in Hercules, is the westernmost member of a quadrilateral of 4th- and 5th-magnitude stars about 11° east-southeast of Delta (δ) Herculis. The binary's 5th-magnitude components are separated by a little more than 6" and can be resolved through a 75-mm telescope. Many observers comment about their subtle colors, with the brighter star shining blue and the fainter one usually appearing yellowish. Others see things a little differently. For instance, John Herschel called them "bluish-white and reddish" in 1824, F. G. W. Struve saw them as "greenish-yellow and reddish-yellow" around 1830, while the well-known English amateur William Dawes (1799–1868) felt them to be "apple green and cherry red." Little wonder, then, that in the light of these widely divergent descriptions some observers have suggested that these stars may actually change in color. While a real variation in the stars' color is unlikely, it is interesting to note how color perception can vary from one person to the next, especially when it comes to the subtle colors of stars.

STAR 24 is another noteworthy star field mentioned in Webb's *Celestial Objects for Common Telescopes* and suggested by Daniel

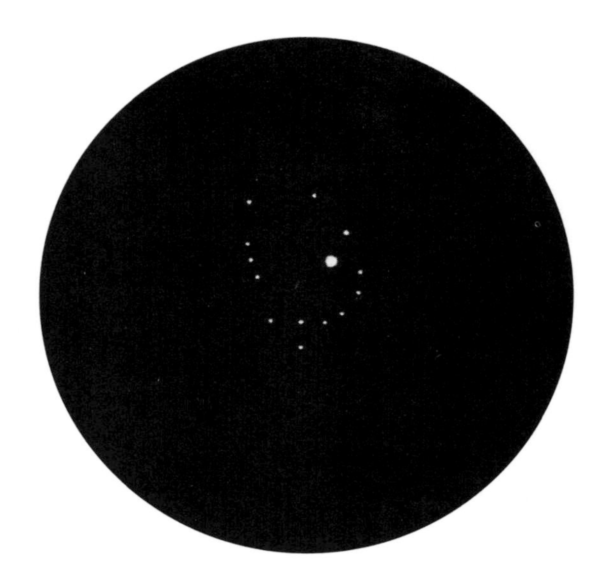

Eyepiece Impression 18. *STAR 24, as seen by Dan Hudak through a 200-mm telescope. South is up.*

Hudak. To find STAR 24, first find Mu (μ) Herculis to the southeast of the Keystone. Move about 4° east-southeast and center on 89 and 100 Herculis, an east-west pair of 5th-magnitude suns separated by about 3°. In between, you ought to spot an arc of three 7th-magnitude suns. Center your telescope on the arc's southernmost star.

Once there, a telescope as small as 100-mm aperture will show a view similar to that portrayed in Eyepiece Impression 18: a conspicuous ring (or "wreath," according to Webb) of 11th-magnitude stars encircling an off-center 7th-magnitude sun. Like those in STAR 23, none of the stars in STAR 24 are terribly bright, yet all combine to make an easily recognizable figure.

Lupus

Between springtime's Centaurus and summertime's Scorpius is the comparatively obscure constellation of Lupus, the Wolf. Obscure or not, thanks to the Milky Way skirting its eastern boundary, Lupus hosts many fine deep-sky objects. Unfortunately, because of its far-southern position, the Wolf is liable to remain elusive unless you observe it from south of about 30°N latitude. Lupus reaches midnight culmination in mid-May and rides highest in July's early evening skies.

NGC 5822 is unquestionably the finest sight within Lupus. Lying about one-fifth of the way between Zeta (ζ) Lupi and Alpha (α) Centauri, it is a delightful open cluster for binoculars and rich-field telescopes. Viewing through 11 × 80 binoculars from the Florida Everglades (25°N latitude), I recorded the group as a rich gathering of 9th-magnitude and fainter suns set against the bright glow of other, unresolved cluster members.

Across the border in the constellation Circinus, 1° south of NGC 5822, is **NGC 5823,** a second, smaller open cluster that also makes quite a good target for smaller apertures. Most binoculars will show little more than a faint glow, while telescopes reveal a few scattered 10th-magnitude points of light within.

Lyra

Although it covers an area of less than 16° × 22°, Lyra, the mythical lyre of Orpheus, contains some of the season's most notable objects. Best known of the lot is M57, the famous Ring Nebula, but the Lyre also holds other celestial treats for those who want to delve a little deeper. For observers near 40°N latitude, Lyra, marked by the brilliant star Vega, passes near the zenith on its trek across the sky. Midnight culmination is reached in early July, bringing the Lyre high into the evening skies of August and September.

Epsilon (ε) Lyrae, better known as the "Double Double" quadruple-star system, may be faintly glimpsed with the unaided eye just northeast of dazzling Vega. Since the component stars are separated by 3.5', near the naked-eye resolution limit for most people, sharp-eyed observers can resolve Epsilon into a pair of 5th-magnitude stars with the unaided eye. The stars are easily resolved using almost any optical aid; the northern component is designated Epsilon^1 (ε^1), and the other Epsilon^2 (ε^2).

If you own a telescope of at least 75-mm aperture, try viewing the Epsilon twins at about 100× . If the seeing is good, each should resolve into a tight pair. Epsilon^1 consists of a 6th-magnitude secondary orbiting a 5th-magnitude primary once every 1,165 years or so; they are separated by more than 150 astronomical units. Here on Earth, this distance translates into a 3" separation, with the companion to the brighter star's northwest.

The southern pair, Epsilon^2, is a little tighter than Epsilon^1 and therefore slightly more difficult to resolve. They are isolated from one another by 2", with the 5.5-magnitude secondary nearly due east of the brighter, 5.2-magnitude primary. Like Epsilon^1, these two stars are separated by more than 150 astronomical units, but with an orbital period of about 585 years, only half that of Epsilon^1.

M57 (NGC 6720), the Ring Nebula, is the sky's most famous planetary nebula (Figure 7.6). One reason why it is so well known may be that it is so easy is find, about a third of the way from Sheliak (Beta [β] Lyrae) to Sulafat (Gamma [γ] Lyrae) along the bottom of the Lyre. Visible in giant binoculars as a faint point of light, M57's characteristic smoke-ring shape is visible at about 100× through telescopes as small as 50 mm. Although the Ring seems smooth and even in small instruments, a 200-mm telescope or larger will uncover some bright and dark variations in its surface texture, as suggested in Eyepiece Impression 19.

One of the great challenges facing deep-sky observers is trying to glimpse the 15th-magnitude central star that created the Ring so many millennia ago. Faint to begin with, the star is made more difficult to see by the overwhelming brightness of the Ring itself. As a result, a 305-mm is probably the smallest telescope that can show the central star, and then only under ideal conditions. In fact, even through my 455-mm (18-inch) f/4.5 Newtonian from my suburban

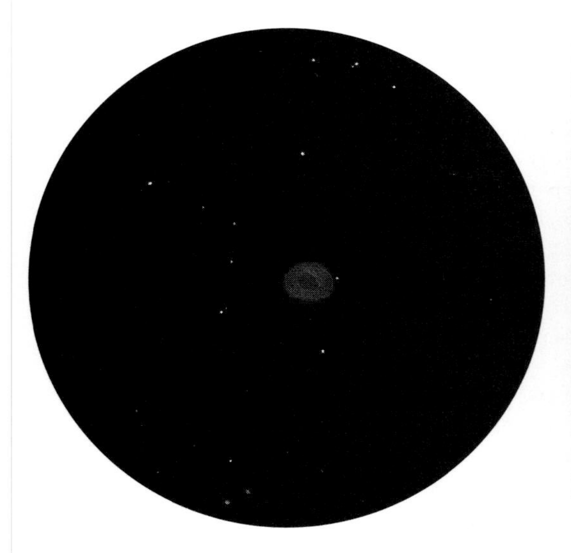

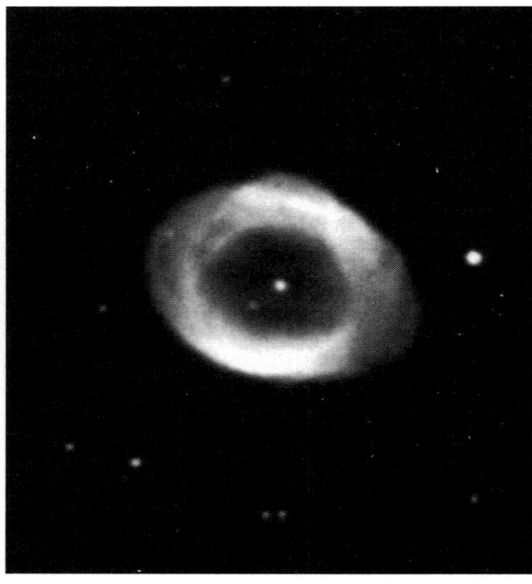

Left: Eyepiece Impression 19. *M57, the Ring Nebula in Lyra, is the sky's most famous planetary nebula. While the central star is absent, compare the nebula's "texture" in this drawing made through the author's 333-mm Newtonian and 7-mm Nagler eyepiece (214×) with the photo at right.* Right: Figure 7.6. *M57, as photographed by Kim Zussmann. Don't be fooled by the brightness of the central star in photographs, which can give observers a false sense of its prominence. South is up in both illustrations.*

backyard, I can only suspect the presence of the central star by using averted vision.

M56 (NGC 6779). Overshadowed by its famous neighbor, the Ring Nebula, this globular cluster is another of the unsung glories of the deep sky. Happily, M56 is easy to find, about halfway between Sulafat (Gamma [γ] Lyrae) and Albireo (Beta [β] Cygni), just southeast of a 6th-magnitude field star.

Resolving it can prove more difficult than finding it. Depending on the quality of its optics, a 150-mm telescope can show some of the cluster's stars, giving M56 what some observers describe as a "grainy" appearance. A 200-mm increases resolution, though an experienced eye and moderately high magnification are still required. Larger telescopes resolve cluster stars right across the core, though again the cluster lacks the typical strong central core that is so apparent in most other globulars.

Norma

West of Ara and east of Lupus and Circinus is another constellation of the deep south — Norma, the Carpenter's Square. Norma covers an area rich in open clusters but poor in bright stars. In fact, you will find most of the objects listed here more easily by beginning at stars in neighboring Ara and slowly working your way into Norma. Norma passes midnight culmination in late May and is highest in the evening skies of July and early August. However, only observers fortunate to be south of about 30°N latitude stand much chance of spotting its southern gems.

NGC 6067 is a fine open cluster nestled in a rich star field. Unfortunately, locating it can prove frustrating. Your best bet is to start at 3rd-magnitude Zeta (ζ) Arae and scan westward more or less blindly for 6° to 5th-magnitude Kappa (κ) Normae. The cluster lies a scant ½° to Kappa's north. The light from its 100 stars blends to 6th magnitude, making NGC 6067 easily visible with binoculars, small telescopes, and perhaps even the unaided eye on superb nights. Individually, the stars shine at 8th magnitude or fainter, with several set in close pairs.

NGC 6087, another of Norma's attractive open clusters, may be found by first locating 3rd-magnitude Eta (η) Arae. Looking through a finderscope, slide 6° west-northwest to an east-west line of three 5th-magnitude stars. (For reference, the middle star in this line is Iota^1 (ι^1) Normae, while the westernmost is Iota^2.) NGC 6087 lies just east of the line's easternmost star. Forty suns belong to this swarm, the brightest being S Normae. A golden Cepheid variable, S Normae fluctuates between magnitudes 6.1 and 6.8 over a 9.75-day cycle. The other stars of NGC 6087 shine at 8th magnitude or fainter and stretch across 12'.

NGC 6152 is a coarse open cluster situated near the constellation's eastern border. Look for it about three-fifths of the way from Zeta (ζ) Arae to the wide optical double formed by Gamma^1 (γ^1) and Gamma^2 (γ^2) Normae. Spanning one Moon diameter, NGC 6152 holds 70 stars. None shine brighter than 11th magnitude, making them a bit tentative through binoculars. A 100- to 150-mm telescope, on the other hand, will reveal the cluster as a dazzling array of faint stardust.

Ophiuchus

One of the season's largest constellations, Ophiuchus, the Serpent-bearer, contains more globular clusters and planetary nebulae than just about any other constellation in the sky. Yet globulars and planetaries are not all that Ophiuchus holds in store for deep-sky observers. The huge constellation's pentagonal body reaches midnight culmination in the middle of June, bringing it into the early evening skies of late July and August.

M10 (NGC 6254) and **M12** (NGC 6218). Central Ophiuchus contains two spectacular globular clusters: M10 and M12. Let's begin with the former. Draw a line connecting Eta (η), Zeta (ζ), Epsilon (ε), and Delta (δ) Ophiuchi along the "bottom" of the constellation. Looking through binoculars or a finderscope, extend a perpendicular line from Zeta (at the line's center) toward the northeast. Along that line you should spot 5th-magnitude 23 Ophiuchi, set in a small triangle, and farther along, 5th-magnitude 30 Ophiuchi. M10 lies about 1° to this last star's west, while M12 lies another 3½° farther northwest, on the other side of a 6th- and 7th-magnitude pair of stars.

Amateur telescopes have a fairly easy time resolving separate suns in M12, owing to the cluster's low star density. A 254-mm is more than capable of showing stars across the entire object. In the 19th century Adm. Smyth noted several "knots" of stars toward the cluster's center, while Lord Rosse thought he saw through his huge telescope a suggestion of spiral structure.

M10 shines about as brightly as its neighbor, yet its tighter structure makes it more difficult to resolve. A 200-mm has little trouble revealing stars around the edges of M10, but the nucleus remains an unresolvable blaze. At twice the aperture, resolution of the cluster's individual stars is much better. The distance to M10 is given as 18,000 light-years, comparable to that of M12. If these estimates are correct, then the clusters are within 4,000 light-years of each other.

M62 (NGC 6266) rides the Scorpius-Ophiuchus border. Starhoppers should be able to spot the 7th-magnitude glow of this globular cluster just west of a triangle of faint stars a little less than halfway from Epsilon (ε) Scorpii to Theta (θ) Ophiuchi. Charles Messier was the first to cross paths with M62, in 1771, but its true cluster identity was not established until William Herschel observed it years later. Although many references claim the cluster stars to be of 11th magnitude, they are too difficult to spy through a 150- or 200-mm telescope, even under superb sky conditions. John Herschel was the first of many observers to note M62's asymmetric appearance: there is a distinctive northward "bulge," which is apparent both visually and in photographs.

About a dozen other globulars, including M9 and M19, reside in this region of Ophiuchus and neighboring Scorpius. Each deserves at least a moment's attention.

Pipe Nebula. Spanning a total of 11 square degrees, the Pipe Nebula is one of the largest dark-cloud formations visible from the Northern Hemisphere. Look above the tail of Scorpius, not far to the south of Theta (θ) Ophiuchi. To the naked eye the nebula appears as a 2° × 3° rectangle (the Pipe's bowl) with a long, thin appendage (its stem) extending 5° to the west. Although the edges of the Pipe appear sharp and straight visually, they look rather ragged and irregular in photographs. It was this photographic appearance that led Edward Emerson Barnard to record the Pipe as several separate catalog entries. The bowl was assigned **B78,** while the stem is listed as **B59, B65, B66,** and **B67,** from west to east. Listed as having an opacity of 6, on crystal-clear nights the Pipe is quite distinct to the eye alone as well as through wide-angle binoculars.

Barnard 72. Slithering out of the northwest corner of the Pipe Nebula's bowl, about 1½° north of Theta (θ) Ophiuchi, is the dark nebula B72, better known as the Snake Nebula. With an opacity rating of 6, on dark nights B72 stands out well against its surrounding star fields as a distinctive S-shaped lane. Giant binoculars are adequate to reveal its slender form, while telescopes add depth to the overall effect.

NGC 6369, nicknamed the "Little Ghost" planetary nebula, is easy to scare up as long as you have an unobstructed view to the south. First locate 3rd-magnitude Theta (θ) Ophiuchi, about 12° east of Antares (Alpha [α] Scorpii). Hop just over 1° northeast of Theta to 44 Ophiuchi, then another 1° east-northeast to 5th-magnitude 51 Ophiuchi. The Little Ghost floats ½° west-northwest of this last star.

Looking like a miniature Ring Nebula, NGC 6369 is an exciting planetary nebula to view at moderately high magnification. When the air is especially clear, its pale blue hue may be detected through a 100-mm telescope, while larger instruments show the bluish color even under less-than-perfect conditions. A faint, ghostly halo measuring fully 1' surrounds the planetary's brighter inner ring. While the largest backyard instruments may be used to capture this latent feature photographically, it is probably too subtle to be seen visually. The nebula's central star is also a tough catch even through the largest amateur instruments, as it glows weakly at 15th magnitude.

IC 4665 is a coarse collection of stars set 1½° northeast of 3rd-magnitude Beta (β) Ophiuchi. Although not as rich as some of the other open clusters in this survey, it stands out surprisingly well thanks to its spartan surroundings. A finderscope or binoculars will display about 10 stars, while a 150-mm telescope increases this number threefold. Since IC 4665 measures 41', the best telescopic views will be through low-power, wide-field eyepieces; but before you leave the cluster behind, try switching to higher magnification to spot a trio of double stars set near its center.

NGC 6572 is the brightest of the nearly 50 planetaries within the boundaries of Ophiuchus. To spot it, first locate 72 Ophiuchi, a 4th-magnitude star that forms an equilateral triangle with Alpha (α) and Beta (β) Ophiuchi to its west. From 72, hop southward to 71 Ophiuchi, then the same distance again southwest to a widespread pair of faint suns. NGC 6572 lies about 1° directly to their south.

Even the smallest telescopes display this planetary's tiny bluish green oval disk, though it may appear starlike at first. As magnification climbs, so will the nebula's size, but don't expect to glimpse its central star easily through a small instrument; the progenitor star is rated at 13.6 magnitude. I have never spotted it through my 333-mm f/4.5 Newtonian. As is the case with so many other planetaries, this is probably down to the high surface brightness of the engulfing nebulosity — an interesting twist to the phrase "light pollution." It's not that the star is exceptionally faint but rather that the contrast between it and the surroundings is so low. (The Ring Nebula's central star is probably the best-known example of this effect.)

70 Ophiuchi is one of the fastest-changing binary-star systems visible through amateur telescopes. You'll find it faintly with the naked eye, along with nearby stars 66, 67, 68, and 73 Ophiuchi, in a V-shaped asterism collectively set about 5° southeast from Cheleb (Beta [β] Ophiuchi). Incidentally, many note the resemblance of these five stars to the head of winter's Taurus, the Bull, an analogy that lead Abbé Poczobut in 1777 to call the pattern "Taurus Poniatovii," in honor of Polish King Stanislaus Poniatowski. The three brightest stars — 67, 68, and 70 — form the Bull's head, while fainter 66 and 73 mark the tips of its horns.

What makes 70 Ophiuchi noteworthy is how rapidly its components, magnitudes 4.2 and 6.0, change positions relative to one another. In 1989, the stars were separated by only 1.5", with the faint star at P.A. 233° — southwest of the bright star. In 1996, the companion has swung around to P.A. 160° — now to the bright star's south-southeast — and separated by 2.9".

An interesting long-term observing project would be to make a drawing of this celestial dynamic duo tonight, then put it aside. Come back in a few years and see how things have changed. By 2000, for instance, the stars' separation will increase to 3.9", with the fainter star to the bright star's southeast, at P.A. 147°. A decade later, the pair will be almost 6" apart, the companion at P.A. 131°.

A keen eye may see an orange tinge to both stars, indicative of their Type-*K* spectral classification. Both are known to be 16 light-years away, with their distances apart ranging from 11 to 33 astronomical units.

Pavo

Pavo, the Peacock, flies below most of our southern horizons, with only a portion of the constellation visible from the southern tier of the United States. Of course, observers near or south of the equator will have a bird's-eye view of the constellation's deep-sky wonders. Pavo passes midnight culmination in mid-July, and is best placed for early-evening observation in early September.

NGC 6752 is an outstanding globular cluster that goes little observed from the Northern Hemisphere. Beginning again at Lambda (λ) Pavonis, head 2° north to 5th-magnitude Omega (ω) Pavonis, then turn east for 1½° to reach the globular. Rated at magnitude 5.4, NGC 6752's slightly yellowish disk should be plainly evident with just about any optical aid.

Were it not for its far-southern declination, this would undoubtedly be one of the most popular globulars of the season. Partial resolution is possible through a 100-mm instrument, while 150 mm or larger gives a grand view. As with just about all globulars, NGC 6752 bears magnification well, so don't be afraid to raise your telescope's power as far as its optics and the seeing will allow.

Sagittarius

To naked-eye observers Sagittarius resembles a teapot more than the legendary archer-centaur that the ancients saw. To deep-sky observers, however, the constellation is more of a playground than anything else. When we gaze its way, we are staring straight toward the center of our galaxy, so — not surprisingly — we find a tremendous variety of deep-sky objects scattered throughout the area. You'll discover that the Archer has something hidden within its borders for just about everyone. The question is not what to look at, but what to look at first, for regardless of what object you pick it is sure to be a winner. Sagittarius passes midnight culmination in early July, bringing it into our August and early-September evening skies.

NGC 6440 and **NGC 6445.** Here is a challenging deep-sky-object team. Although they are located along the Milky Way, both can be a little tough to zero in on since there are no nearby naked-eye stars. I always begin at Theta (θ) Ophiuchi, some 12° east of Antares. From Theta, hop 1° northeast to 44 Ophiuchi, then another 4½° to 58 Ophiuchi, and finally about 1½° farther northeast to a pair of 7th- and 8th-magnitude stars. Our deep-sky duo resides about 1° farther north.

First up is NGC 6440, a small, 10th-magnitude globular cluster. A 200-mm reveals a mass of unresolved stars, while a 254-mm or larger at a high power resolves some individual suns. Just northeast of NGC 6440 is the planetary nebula NGC 6445. In most backyard telescopes it looks like a whitish disk about 34" across. That's quite large for a planetary nebula and thus assists us in telling it apart from surrounding stars.

There is some disagreement over the exact magnitude of NGC 6445. *Sky Catalogue 2000.0* describes the planetary as having a photographic magnitude of 13.2; the *Webb Society Deep-Sky Observer's Handbook,* volume 2, lists it at magnitude 9.5, while through my 200-mm reflector I have estimated its magnitude as 10.5. How bright does it look to you? One thing observers all agree on, however, is that regardless of the size of your scope, the nebula's central star is nowhere to be found. It shines at a dismal 19th magnitude.

M23 (NGC 6494) is an open cluster found about 11° northwest of the Teapot's spout. To find it, cast off again from Theta (θ) Ophiuchi. Move a little over 1° northeast from Theta to 44 Ophiuchi, then hop another 5° northeast to 58 Ophiuchi, and finally 4° more northeast to M23. On exceptional nights, M23 might just be visible to the naked eye as a slightly brighter spot along the lane of the Milky Way, while through binoculars it appears as a nebulous glow spanning about ½° and marked by a few faint points of light. The best telescopic views of M23 are, not surprisingly, with low-power eyepieces. These show a beautiful field of glittering stars, 120 in all. Some observers have reported that the group appears fan-shaped, though that image has never occurred to me. Perhaps you have a better imagination.

NGC 6520 is a dense open cluster not far from the Teapot's spout. From Alnasl (Gamma [γ] Sagittarii), the star at the spout's tip, continue northward for 2½°, passing 4th-magnitude W Sagittarii, to a 7th-magnitude field star; NGC 6520 is just to the southeast. Although the field literally overflows with starlight, almost masking the cluster's existence at first glance, NGC 6520 will reveal itself even through 7× binoculars. A 100-mm telescope at low magnification shows it as a small, elongated patch of interstellar fluff oriented northwest-southeast. Higher powers and larger apertures are needed to resolve the cluster into a rich bevy of 60 faint stellar beauties

Figure 7.7. *This wide-field photograph captures one of the most spectacular star fields in the entire sky. Dominating the view is M8, the Lagoon Nebula (bottom), and M20, the Trifid Nebula (top right), both in Sagittarius. Compare the view of M8 with the drawing on the next page. This figure is oriented with north toward the top left. Photo by Robert Bickel.*

swimming in the stream of the Milky Way.

M8 (NGC 6523), the Lagoon Nebula, is considered by most observers to be the premier bright nebula of the summer sky (Figure 7.7). On dark, moonless nights it is visible to the unaided eye as a hazy island in the Milky Way, about 6° north of the Teapot's spout. Look for it surrounding several stars, including 5th-magnitude 7 Sagittarii.

Small amateur telescopes show a glowing cloud of great intricacy sliced in two by a dark lane, or "lagoon," of obscuring dust. Through large backyard instruments the view is truly awe inspiring, with pockets of dark nebulosity sewn between patches of bright clouds. The Lagoon Nebula spans the width of three full Moons, making a wide field of view a must to take it all in. The view through my 333-mm Newtonian, portrayed in Eyepiece Impression 20, shows that M8 more than fills the field and is breathtaking in its complexity. On exceptional evenings, it is even possible to catch something of the reddish color that is so prominent in photographs.

Charles Messier described M8 as "a cluster which looks like a

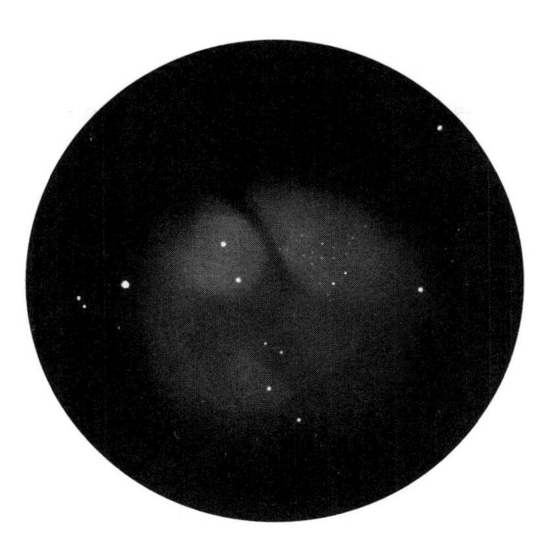

Eyepiece Impression 20. *M8, the Lagoon Nebula, as seen through the author's 333-mm Newtonian and 24-mm Tele Vue Wide Field eyepiece (63×). South is up.*

nebula in an ordinary telescope . . . but in a good instrument one observes a large number of small stars." His description is especially noteworthy since the open cluster **NGC 6530** lies within M8. This is a loose cluster of about two dozen 7th- to 9th-magnitude stars. Collectively, they remind Dave Kratz of Poquoson, Virginia, of a wine goblet.

The Lagoon Nebula contains an abundance of dark nebulosity. William Herschel described this vast network of clouds as "an extensive milky nebulosity divided into two parts," while his son John noted "a collection of nebulous folds . . . including a number of dark, oval vacancies." One of the Lagoon's most obvious dark patches is cataloged as **Barnard 88** and lies just north of a slim triangle of 8th-magnitude stars just above 5th-magnitude 9 Sagittarii. Most observers note its resemblance to a "negative" comet, with a grayish "tail" arcing southward away from a round, black "coma."

Barnard 89 is a long, thin, dark cloud located on the eastern edge of the Lagoon's associated star cluster, NGC 6530. To find this subtle region, first focus your attention on the cluster, then look to the east for a lone 7th-magnitude star. A careful scan between the two should turn up B89 as a triangular void with two "wings" extending north and south.

Finally, **Barnard 296** is a fan-shaped "dent" along the south-central edge of M8. Through my 333-mm reflector, its presence becomes obvious only when a narrowband nebula filter is used to heighten contrast. Still, the exact extent of B296 is difficult to pin down in this and most other amateur telescopes.

M20 (NGC 6514), the Trifid Nebula, is located about 2° northwest of M8. Like M8, Messier also interpreted this object as a star cluster. William Herschel saw the nebulosity cut into thirds by dark lanes, leading to the Trifid nickname we use today. The clouds of the Trifid are rather faint and difficult to discern through small instruments on hazy summer nights. In the *Webb Society Deep-Sky Observer's Handbook*, volume 2, David Allen pointed out that M20 is "a difficult object [through a 300-mm telescope] at the altitude it reaches in Britain." Observers in Canada and the northern United States will no doubt agree.

While most bright nebulae are best seen through low-power eyepieces, M20 is one of a rare breed that handles magnification well. In fact, many observers of the Trifid favor a telescope-eyepiece combination that yields between 100× and 150×. Some of my best views of the Trifid through my 455-mm Newtonian have been with a 12-mm Nagler eyepiece. In the Nagler's monstrous field of view, the soft, gossamer wisps of the Trifid stand out magnificently against a dazzling stellar field.

M21 (NGC 6531). Since I first viewed M21 through a 60-mm (2.4-inch) refractor over 25 years ago, it has become one of my favorite clusters on a midsummer's night. One reason may be the company it keeps, for M21 lies less than 1° northeast of the Trifid Nebula, M20. They form a sparkling pair of celestial celebrities when viewed with giant binoculars and rich-field telescopes. M21 looks good in just about any amateur telescope. Through small instruments it is easily resolved into about two dozen points of light with a string of stars near its center. Larger scopes increase the star count to about 60, with several close pairs visible within the group's 13' span.

Barnard 87 is nestled just inside the western edge of Sagittarius, about 8° northeast of the Scorpion's tail. Using your lowest-power eyepiece, aim at a point about halfway between the open cluster M7 and Gamma (γ) Sagittarii, keeping an eye out for a 12' void floating among the stars. One inventive mind saw in the cloud a resemblance to a parrot's head in profile, and christened it the Parrot's Head Nebula. Photographically, the likeness is quite striking with the bird's beak curving to the east from the circular head. A lone 9th-magnitude star marks its imaginary eye. Visually, however, my comparatively uninventive mind detects B87 as a triangular wedge with rounded corners. What do you see?

M24. Until recently, M24 was thought to be one and the same as NGC 6603. Yet, when Messier's description of his catalog's 24th entry is compared with that of NGC 6603, it is clear that they cannot be the same object. Messier's description of M24 reads: "a large nebulosity in which there are many stars of different magnitudes . . . Diameter 1° 30'." By contrast, Dreyer described NGC 6603 as a "remarkable cluster, very rich and very much compressed, round, stars of [12th] magnitude and fainter." Can these two very different descriptions possibly be of the same object?

Dreyer's portrait closely matches the visual impression of NGC 6603 through many backyard telescopes. Notice how the four brightest stars in the M24 star cloud are set in a pattern resembling a sideways kite. Aim at the easternmost star in the kite (mark the kite's "top"), then shift 13' to the north to spot NGC 6603. Telescopes in the 150- to 200-mm range will show it as only a faint glow

Figure 7.8. *M17, known popularly as the Omega or Horseshoe Nebula for its arc-like shape, is one of the finest objects in the constellation Sagittarius. As described in the text, note how the nebula's brightest portion, along the top, bears a resemblance to a distended "2." South is up. Photo by Preston Scott Justis.*

measuring 5' across and accented with a few feeble points of 14th-magnitude light. Although most definitely an open cluster, its misty appearance has led more than a few observers to conclude erroneously that it is a globular cluster. It is only by close examination at high magnification through a large-aperture instrument that the true nature of NGC 6603 comes through.

In his book *Messier's Nebulae and Star Clusters* Kenneth Glyn Jones first suggested that M24 is actually the entire region that surrounds NGC 6603. Also known as the Small Sagittarius Star Cloud, it would seem to fit Messier's description perfectly. Through binoculars, the whole area disintegrates into countless points of light of many magnitudes. Dominating the scene are **Barnard 92** and **Barnard 93,** a pair of dark nebulae seen in silhouette against the northwest corner of the star cloud. B92 is relatively easy to see, even through light-polluted suburban skies, as an oval void measuring about 12' × 6'. Most amateur telescopes reveal only a lone 10th-magnitude star in front of the dark cloud. Although it is nearly equal in size to B92, B93 proves more difficult to spot because of its lower contrast with the background. Unlike B92, which has sharp borders that clearly delineate it from its surroundings, B93 slowly diffuses into its environs.

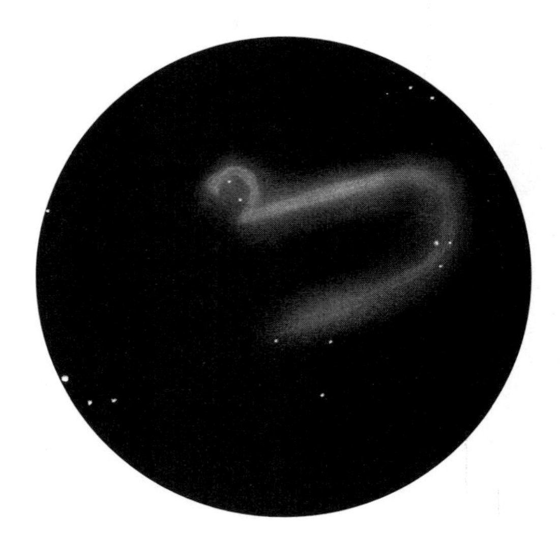

Eyepiece Impression 21. *M17, the Omega Nebula, as seen through the author's 333-mm Newtonian, 24-mm Tele Vue Wide Field eyepiece (63×), and narrowband nebula filter. South is up.*

M17 (NGC 6618), one of the sky's outstanding bright nebulae (Figure 7.8), lies almost upon the northern border of Sagittarius, a little more than 9° north-northwest of Kaus Borealis (Lambda [λ] Sagittarii). Even from suburban backyards, its soft, gossamer glow should be visible in 7× binoculars.

The Swiss astronomer Philippe de Chéseaux (1718–51) discovered M17 in the spring of 1764, beating Charles Messier by just a few months. Messier was later to describe his catalog's 17th entry as "a train of light without stars . . . in the shape of a spindle," indicating that he saw only the brightest portion of the cloud. Observing through larger instruments, William Herschel was the first to appreciate the full extent of M17. He likened its form to the Greek capital letter Omega (Ω), leading to M17 being nicknamed the Omega Nebula or Horseshoe Nebula.

The Omega Nebula is a spectacular sight through nearly all binoculars and telescopes. Giant binoculars and small telescopes reveal an intricate cloud in the shape of an extended number "2." In fact, it is the cloud's long, curved "neck" that has led some to dub this the Swan Nebula. With increasing aperture the Swan slowly disappears into a huge, glowing, semicircular arc of light. A 200-mm instrument just begins to expose M17's fainter regions curving up and away from the bottom bar of the "2," while, as shown in Eyepiece Impression 21, the horseshoe figure and its numerous rifts and swirls are clearly seen in 305- to 355-mm instruments. Adding a narrowband nebula filter, as was used for Eyepiece Impression 21, brings out an interstellar spectacle whose magic cannot possibly be conveyed in words or pictures.

M25 (IC 4725) is also most easily found by stepping off from Kaus Borealis (Lambda [λ] Sagittarii) and heading about 6° north. With a finderscope, binoculars, or even the eye alone, M25 ought to be discernible as a 5th-magnitude smudge adrift in the Milky Way.

Figure 7.9. *M22 in Sagittarius is one of the sky's greatest unsung sights. Many observers feel that it is the equal of, or even superior to, better-known M13 in Hercules. South is up in this photo by Gary Jones and Mike Warnell.*

As with so many open clusters, M25 is best appreciated through a low-power, wide-field eyepiece. Telescopes reveal a rich congregation of stars both bright and faint scattered without apparent form across the field. Estimates of the stellar population of M25 range from 30 to 80, with individual stars shining as brightly as 6th magnitude. One of the most interesting members of M25 is **U Sagittarii,** a Cepheid variable that fluctuates between magnitudes 6.3 and 7.1 with a period of 6.75 days.

NGC 6645. Move 2½° farther north of M25 to encounter this next open cluster, a striking group of faint stars sprinkled across 15'. The brightest of the 40 or so stars that call this cluster home shine no greater than 11th magnitude, causing telescopes smaller than 100 mm to show only a nebulous mass highlighted by perhaps one or two points of light. Instruments of 200-mm and larger reveal a dense swarm floating among a rich star field. Two arcs of stars appear to extend from either side of the cluster's round core, painting a picture in my mind of a bird in flight.

M22 (NGC 6656) ranks as one of the sky's showpiece globulars, closely resembling the Hercules Cluster M13 in size and structure (Figure 7.9). It's easily located, about 2½° northeast of Lambda (λ) Sagittarii, the top of the Sagittarius Teapot.

Small backyard telescopes show 5th-magnitude M22 as a round, "grainy" disk with a mottled appearance. A 100-mm telescope is probably the smallest instrument that will begin to resolve some stars around M22's fringe, while a 200-mm is more than capable of displaying hordes of 11th-magnitude and fainter stars across the cluster's core. Take a careful look at the cluster. Do you notice

Figure 7.10. *NGC 6822, known as Barnard's Galaxy, is a member of the Milky Way's Local Group of Galaxies, but is still one of the toughest objects listed in this book. South is up. Photo by Martin C. Germano.*

anything peculiar about its shape? Unlike most globulars, M22 is not perfectly round. Its northeast-southwest bulge is noticeable in 20 × 80 binoculars and might be visible through less powerful instruments. The total population of M22 may well exceed half a million stars.

M55 (NGC 6809). Just finding this globular cluster, tucked away in an empty portion of eastern Sagittarius, can prove a formidable challenge for many star-hoppers. Try your luck by beginning at Ascella (Zeta [ζ] Sagittarii) in the Teapot's handle. Star-leap about 5° due east to a pair of 6th-magnitude stars oriented northwest-southeast. From these you might be able to spot another 6th-magnitude star about 4½° to their southeast. Look for M55's 7th-magnitude glow about three-quarters of the way from the pair to that lone star.

M55 is the least concentrated globular in Messier's catalog as well as in our survey. Reports of partial resolution in 75-mm instruments have been recorded, though this would seem a tough task from mid-northern latitudes. A low-power 200- to 254-mm telescope will resolve many individual points of light sprinkled across the nucleus, the view giving the impression of a sparsely populated cluster. Most of M55's stars are 17th magnitude and below, so they are seen only through either very large instruments or in photographs.

NGC 6818 and **NGC 6822.** Nicknamed the Little Gem, NGC 6818 is an easy planetary nebula to see using just about any backyard telescope. Finding it, on the other hand, can be another matter. From Nunki (Sigma [σ] Sagittarii) in the Teapot's handle, slide northward to an arc formed by Pi (π), Omicron (ο), and Xi (ξ) Sagittarii. Follow the arc's curve to the northeast, pausing at Rho[1]

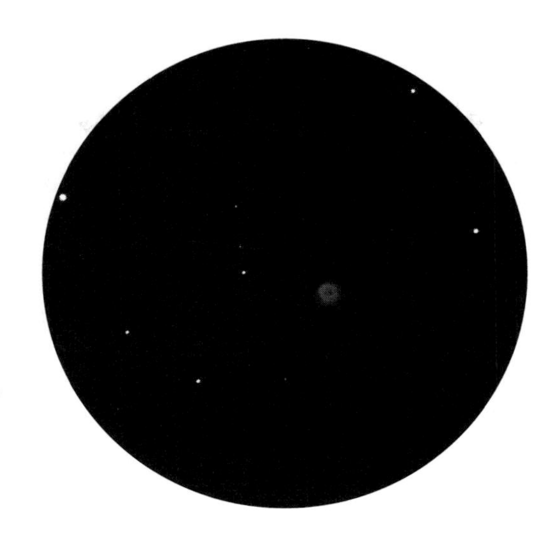

Eyepiece Impression 22. *NGC 6818, as seen through the author's 200-mm Newtonian and 7-mm Nagler eyepiece. South is up.*

(ρ^1) and Rho^2 (ρ^2) Sagittarii, then continuing northeastward for another 5° to a smaller arc of three 5th-magnitude stars. NGC 6818 lies a little over 1° to their north.

Most amateur instruments reveal NGC 6818 as a uniform disk of turquoise light. As suggested in Eyepiece Impression 22, telescopes of 200 mm and larger at higher magnifications add some personality to the Little Gem by hinting at the cloud's true ringlike structure. Arizona amateur Steve Coe describes its appearance as resembling the "CBS eye," the logo of the CBS television network. Because of the nebula's inherent brightness, trying to spy the eye's 14th-magnitude central star is best left to instruments of 305 mm and larger.

With a low-power eyepiece in place, look for a very faint patch of light about 40' south-southeast of NGC 6818. Do you see it? If so, congratulations — you've spotted NGC 6822, also known as Barnard's Galaxy (Figure 7.10). This is an irregular dwarf galaxy and a member of the Milky Way's Local Group. Just because it is nearby, however, doesn't mean it is that easy to spot. Its presence is masked both by its low altitude when viewed from mid-northern latitudes and by the dimming effects of intervening dust clouds. The huge area (10') of NGC 6822 makes a wide-field eyepiece a must. Even if you do glimpse this galaxy, you are likely to see little more than a barely discernible smudge. Through his 200-mm Schmidt-Cassegrain telescope, Arkansas amateur Dean Williams describes NGC 6822 as "a round glow of very low surface brightness. There are a number of stars dotting the field, with a chain of three or four stars trailing from one end."

M75 (NGC 6864) is a second globular cluster found in eastern Sagittarius. Like its neighbor M55, it can prove difficult to locate, especially the first time around. Begin at Tau (τ) Sagittarii, another of the stars in the Teapot's handle. Slide 11° due east to a kite-shaped pattern formed by A, b, c, and Omega (ω) Sagittarii. Extend a line toward the northeast from b through A Sagittarii for 4½° to a faint arc of three 6th- and 7th-magnitude suns and, beyond, M75.

While M55 is the least condensed globular in the Messier catalog, M75 holds the distinction of being the most concentrated. Resolution of its many faint stars is impossible in all but the largest backyard telescopes. The rest of us will have to settle for a round, nebulous smudge set in an empty field. Messier's own notes mention that he fleetingly saw a few individual stars within M75, though this would have been impossible in his telescope. Instead he probably glimpsed three unrelated 12th-magnitude stars superimposed in front of it. At a distance of 60,000 light-years, M75 is also the most remote of the Messier globulars.

Scorpius

Bordering Sagittarius to the west, Scorpius is another veritable smorgasbord for observers who like to stuff themselves with deep-sky treats. With the Milky Way skirting its eastern edge, the Scorpion is home to some of the summer sky's most impressive open and globular clusters, as well as some of the finest star fields for those who simply enjoy "star surfing" with binoculars. With the Scorpion low in the southern sky for most Northern-Hemisphere amateurs, it is important to schedule observations for when the constellation crosses the meridian. Scorpius reaches midnight culmination in early June, bringing it into the early-evening skies of July and early August.

Beta (β) Scorpii, one of the season's finest double stars, marks the tip of the Scorpion's northern claw. Even the smallest telescopes have little trouble resolving the system's 2.6- and 4.9-magnitude components separated by about 14". A variety of colors feature in descriptions of these two stars by some famous observers of the past. In the 19th century, the Rev. Thomas W. Webb described them as "pale yellow and greenish," while Mary Proctor saw them as "white and lilac." More recently, in his book, *Astronomical Objects for Southern Telescopes,* E. J. Hartung has interpreted them as a "splendid pale-yellow pair." Perhaps my imagination is not keen enough, for I have always seen them both as simply white.

Incidentally, these two components are designated the system's A and C stars — the missing B companion is held tightly to the primary. With ideal seeing the largest of amateur telescopes equipped with a high-power eyepiece might just be able to resolve Beta Scorpii B as a magnitude-9.5 dot a mere 0.5" to the primary's southeast. Both Beta Scorpii A and C are also spectroscopic binaries, making the system a quintuple star.

M80 (NGC 6093). "The richest and most condensed mass of stars which the firmament can offer." So wrote William Herschel when he first saw the globular cluster M80 in 1785, four years after Messier's discovery. M80 shines at 7th magnitude and is just about midway between Antares (Alpha [α] Scorpii) and Graffias (Beta [β] Scorpii). In ideal sky conditions a 200-mm will just show some of

Figure 7.11. *The globular cluster M4 in Scorpius is easily found just west of the brilliant star Antares. There are so many cluster stars recorded in this photo by Martin C. Germano that the "bar" of bright stars described in the text is lost. South is up.*

the 14th-magnitude points of light that make up this huge globe of stars. Fairly high magnification is a must, however, as the cluster's high stellar density hampers resolution at less than about 150×. M80 is thought to be about 27,000 light-years away, with a diameter of 40 light-years. *Burnham's Celestial Handbook,* volume 3, offers the thought-provoking fact that, if the Sun were viewed from the distance of M80, it would appear as a 20th-magnitude speck!

M4 (NGC 6121) is easily found a little less than 2° due west of fiery Antares (Figure 7.11). Under crystal-clear skies, the unaided eye may just be able to pick out this 6th-magnitude globular cluster as a faint glint against the gentle glow of the Milky Way. M4 is one of the sky's easiest globulars to resolve into individual stars. Even a 75-mm reveals a few of its brighter members around the edges. A 200-mm shows many stars across the cluster's face, with a unique "star bar" bisecting the cluster from north to south. William Herschel was the first to notice this curious feature, describing it as a "ridge of 8 to 10 pretty bright stars" forming a distinctive central spike. My own notes recall this feature as "quite obvious at 150×" through my 200-mm f/7 Newtonian.

NGC 6124 is a fine open cluster in the outskirts of southwestern Scorpius. To get to NGC 6124, follow a zigzag trail that begins at Zeta (ζ) Scorpii in the Scorpion's tail and wanders about 2° northwest to a close pair of 6th-magnitude stars, then southwest another

2° to a lone 5th-magnitude star, and finally 1½° northwest again to the cluster. Covering nearly ½°, NGC 6124 consists of about 100 stars of 9th magnitude and fainter that collectively create a 6th-magnitude object. About two dozen points are visible through binoculars, with a distinctive string of seven found along the cluster's southern edge. Telescopes and high-power binoculars also reveal a close-knit group of five suns toward the cluster's center.

NGC 6231. Scan ½° north of Zeta Scorpii to find this next outstanding open cluster. One glance at will immediately tell you that this object is something special. It is estimated that 120 Type-*O* and Type-*B* stars are crammed within the cluster's small 15' breadth. When viewed with a 200-mm telescope at low power, NGC 6231 displays about a quarter of its jewel-like stars shining between magnitudes 5 and 13, the remaining fainter cluster members creating a triangular wedge of celestial mist. Larger apertures and higher magnifications blow some of the cloudiness away, revealing even more stars.

The brilliance of NGC 6231 becomes all the more impressive when we consider its great distance from us, about 5,900 light-years. Imagine the spectacle if this cluster could be magically moved to just 400 light-years away, the same distance as the Pleiades. From there the brightest stars of NGC 6231 would rival even blazing Sirius.

NGC 6281. Located about 2½° due east of Mu^1 (μ^1) and Mu^2 (μ^2) Scorpii, NGC 6281 is an easy open cluster to glimpse with just about any optical assistance. Binoculars reveal 7th- to 9th-magnitude stars in a tight swarm whose shape has been described very differently by different observers. For instance, in *Touring the Universe Through Binoculars* I liken the cluster's shape to a crooked cross, while in the *Observing Handbook and Catalogue of Deep-Sky Objects* authors Luginbuhl and Skiff describe it as looking like "Christmas tree lights." Perhaps you will see something completely different!

As aperture and magnification increase, the structure of NGC 6281 seems to mutate, with many fainter stars flooding the field. A 200-mm telescope resolves about two dozen suns, while a 305-mm instrument adds an additional dozen or so points within the cluster's 8' span. When viewing NGC 6281 at low magnification, look for a lone 6th-magnitude star just to the cluster's northwest. This star, designated **HD 153919,** is a powerful x-ray source. Studies reveal that it is pouring out about 800 times as much x-radiation as our Sun. This strange object is also an eclipsing binary system that fluctuates between magnitudes 6.51 and 6.60 over a period of 3.4120 days. As a variable star, it has the alternative designation V884 Scorpii.

NGC 6322 lies a third of the way from Eta (η) to Theta (θ) Scorpii in the Scorpion's tail. Viewed through binoculars, this open cluster appears as a faint smudge in the Milky Way. Telescopes show a distinctly triangular object, with three equally spaced 8th-magnitude stars close to the cluster's edge. Poured into and around this stellar trio are 27 other stars ranging from 10th to 12th magnitude. NGC 6322 has an overall magnitude of 6.0, making it easy to spot

Figure 7.12. *The open cluster M7 in Scorpius is an outstanding sight through binoculars and finderscopes but is probably too large to fit into a telescope's restricted field of view. Sharp-eyed observers can also spot the globular cluster NGC 6453 (far right, above center) and the dark nebula Barnard 287 (bottom center). North is up in this photo by Martin C. Germano.*

even with modest equipment.

NGC 6388 is a 7th-magnitude globular cluster near the constellation's southern border, about 1° south of Theta (θ) Scorpii. Small amateur instruments show it as a fuzzy ball of unresolved light. A mottled, grainy texture may be detected in medium-size telescopes, but only the largest amateur instruments stand much chance of resolving any individual stars, since none shine brighter than about 15th magnitude.

M6 (NGC 6405) and **M7** (NGC 6475) form a pair of beautiful open clusters above the stinger of the Scorpion. Both are visible to the naked eye as faint patches of light.

M6, nicknamed the Butterfly Cluster for its remarkable resemblance to one of these graceful insects in flight, lies about 5° northeast of Lambda (λ) Scorpii. Most observers can immediately spot two "wings" of stars as they spread out from the group's more densely packed "body." Unfortunately, long-focal-length telescopes have fields of view that are too restricted to encompass the full glory of M6. Binoculars and rich-field instruments, on the other hand, display a wondrous sight. Of the 80 stars that make up the celestial butterfly, more than a third are bright enough to be seen through 7× binoculars. Observers viewing through giant glasses can count almost 50, while 150- to 200-mm rich-field telescopes reveal just about all of the cluster members. The brightest star within M6 is a blazing orange lighthouse known as **BM Scorpii,** an irregular variable. This star fluctuates between magnitudes 6.8 and 8.7 across an average period of 850 days.

M7 (Figure 7.12) was first recorded in the second century A.D., when Ptolemy identified it in his monumental work the *Almagest* as a nebulous patch just 4° east-northeast of Lambda Scorpii. Sixteen centuries later, Charles Messier included it in his now-famous catalog of deep-sky objects; it is the list's southernmost entry. Messier's description of M7 reads, in part, "a cluster considerably larger than the preceding [M6]." To me, this seems a bit sterile for such a magnificent family of stars. My 7 × 50 wide-angle binoculars create a three-dimensional effect, many of the brighter members appearing to float in front of a field strewn with fainter points of light. Colors abound in the stars of M7, several exhibiting subtle hues of yellow and blue. Brightest of all is a Type-*G* yellow star of 6th magnitude lying close to the group's center.

A careful scan of M7 with a telescope might reveal a faint globular cluster just northwest of the second-brightest cluster member. This is **NGC 6453** (Figure 7.12), first spotted by John Herschel in June 1837. NGC 6453 shines dimly at 10th magnitude, and is only 4' across — not exactly impressive figures, but it adds to the thrill of the hunt. Even with its southern declination, I have little trouble picking out the cluster with my 200-mm reflector from a dark-sky site on Long Island.

NGC 6496, a 9th-magnitude globular cluster, rides along the border with Corona Australis. Provided you have a good view toward the south, this object is easiest to find by starting at Theta (θ) Scorpii, then heading southeast for 3° to 5th-magnitude 96 Scorpii. NGC 6496 ought to fit into the same eyepiece field, less than ½° to the star's east. A 100-mm instrument shows only a faint nebulous glimmer with no trace of any stars. Stellar resolution is still difficult in a 254-mm telescope, due partly to the faintness of the cluster members and partly to the group's weak concentration.

Scutum

Edward Emerson Barnard once referred to Scutum as "the gem of the Milky Way." That is certainly the case, for though it is one of the smallest constellations it holds a greater abundance of fine deep-sky objects per square degree than just about any other. The few listed here represent just the tip of Scutum's deep-sky iceberg. In fact, the entire constellation might be thought of as one big naked-eye deep-sky object: the Scutum Star Cloud. This is a star-rich region of the Milky Way, which stands out prominently against its surroundings when viewed under a dark sky. Scutum passes midnight culmination in early July and is carried high into the early-evening skies of late August and early September.

M11 (NGC 6705). Possibly already on your list of favorite summer objects, M11 is one of the richest and brightest of the open clusters visible from our planet. Although the stars of Scutum are all quite faint, M11 is easy to find thanks to some nearby stars in Aquila. First, locate 4th-magnitude Lambda (λ) Aquilae, the brightest of the Eagle's tail feathers. Through your finder or binoculars, you'll see that Lambda is one of three stars set in an arc that curves toward the southwest. Follow the arc past its middle star, 12 Aquilae, and stop at its third, Eta (η) Scuti. From here M11 should be visible through your finder as a hazy "star" another 1½° farther west.

To its discoverer, Gottfried Kirch (1639–1710), who spied it in 1681, M11 was "a small, obscure spot with a star shining through." Sure enough, if you look at M11 through a small telescope it reveals only a lone 8th-magnitude sun set amid a nebulous glow. Larger instruments burst the cluster into a crowd of fainter stars shining between 11th and 14th magnitude. Although it appears symmetrical at first, most observers eventually notice that M11's stars actually lie in a blunt V pattern. To Adm. Smyth the cluster thus resembled "a flock of wild ducks," giving rise to its popular nickname, the Wild Duck Cluster.

Serpens

Depicting a giant serpent carried by Ophiuchus, Serpens holds the distinction of being the sky's only constellation split into two parts. The western half, the head of the Serpent (sometimes called Serpens Caput), contains a stunning globular cluster and many faint galaxies. The eastern, or tail, end (Serpens Cauda) dips near the Milky Way and includes several interesting open clusters and nebulae. Serpens Caput culminates at midnight in late May and is high in the south during the evening hours of July. Serpens Cauda reaches midnight culmination about a month later and is highest in the evening skies of August.

M5 (NGC 5904) is a stunning globular cluster that ranks with M22 in Sagittarius and M13 in Hercules. Finding it is made easy thanks to its proximity to several handy guidestars. Begin by identifying the naked-eye triangle of Alpha (α), Lambda (λ), and Epsilon (ε) Serpentis. Draw a line southwestward for 8° from Lambda through Alpha to a flattened diamond formed by the stars 4, 5, 6, and 10 Serpentis, all around 5th magnitude. M5 lies just northwest of 5 Serpentis, the northwest corner of the diamond. On a clear, dark night, M5 might just be seen with the naked eye as a 6th-magnitude glow.

A 100-mm instrument (perhaps even an exceptional 75-mm) is all that is needed to resolve some of the cluster's tiny stars. As shown in Eyepiece Impression 23, a 333-mm telescope resolves a multitude of stellar points across the entire disk. Some observers also comment that many of cluster members appear to form curious strings. In their *Observing Handbook and Catalogue of Deep-Sky Objects,* authors Luginbuhl and Skiff note "a string of four or five 12.5-magnitude stars [running] across the northern side of the core . . . giving the impression of a rich open cluster superimposed on a bright galaxy." According to current estimates, over half a million stars form M5.

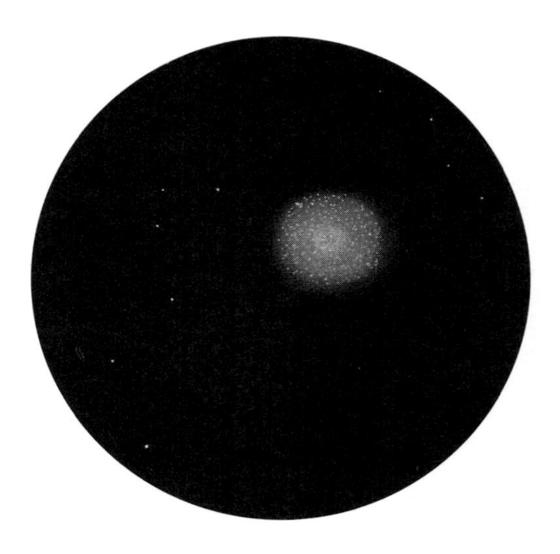

Eyepiece Impression 23. *M5, as seen through the author's 333-mm Newtonian and 12-mm Nagler eyepiece (125×). South is up.*

M16 (NGC 6611) lies adrift in the Milky Way and is bright enough to be seen through small field glasses on clear summer nights. Unfortunately, there are no nearby bright stars to aid us in finding it, so we'll have to get a little adventurous. With binoculars or a finder-scope, find 3rd-magnitude Delta (δ) Aquilae to the southwest of Altair. Move 9° southwest to 3rd-magnitude Lambda (λ) Aquilae in the Eagle's tail feathers, then leap 13° farther southwest to 5th-magnitude Gamma (γ) Scuti. With Gamma centered in your field of view, look for M16 as a smudge of light about 2½° to the west-northwest.

M16 might be considered a "Jekyll and Hyde" object. Over the years, visual observers have come to know it as a fine open cluster of about five dozen suns scattered across about ⅓° of sky. Long-exposure photographs, however, reveal a great complex of beautiful nebulosity threaded among the cluster stars. Until recently, the nebula (known popularly as the Eagle Nebula) was considered a difficult visual challenge. But thanks to the widespread use of contrast-enhancing nebula filters, the Eagle may now be seen to soar where it had never been seen before. Using my lowest-power, widest-field eyepiece teamed with a narrowband nebula filter, my 333-mm Newtonian reveals wonderful contrasts between bright and dark clouds, as well as Fred Hoyle's "Elephant Trunk" dark nebula formation. Hoyle contends that here the bright nebulosity is rapidly expanding into cooler gases.

IC 4756. For those with binoculars and rich-field telescopes, the open cluster IC 4756 in Serpens Cauda is sure to become an instant seasonal favorite. To find it, draw an imaginary line from Rasalgethi (Alpha [α] Herculis) to Rasalhague (Alpha [α] Ophiuchi). If you extend this line for 22°, passing 72 Ophiuchi along the way, you'll eventually come to the 4th-magnitude star Alya (Theta [θ] Serpentis). IC 4756 also lies on that line, about 4½° before Alya. Actually, this cluster is bright enough that if you just aim your finder in its di-

Eyepiece Impression 24. *STAR 22, the Mini-Coathanger, as seen through the author's 200-mm Newtonian and 24-mm Tele Vue Wide Field eyepiece (58×). South is up.*

rection and sweep back and forth slowly, you are bound to run into it. Just don't confuse IC 4756 with NGC 6633, a much denser swarm of stars about 3° to the northwest in Ophiuchus (described previously under its home constellation). Spanning almost 1°, IC 4756 is a coarse distribution of some 80 stars of 9th magnitude and fainter — a bright object set against a nice backdrop that is ideal for low-power viewing.

Ursa Minor

While just about every amateur astronomer has heard of Ursa Minor and its alter ego, the Little Dipper, few have taken the time to explore its few deep-sky offerings. While several galaxies are scattered among the stars of the Little Bear, all are too faint to be included in this survey. Instead, listed here are a well-known double star and a little-known asterism. Although Ursa Minor is visible throughout the year from most of the Northern Hemisphere, technically it reaches midnight culmination in early May and is therefore highest in the early-evening skies of June and July.

Polaris. Every good Boy or Girl Scout has learned to recognize Polaris, the North Star, shining at the end of the handle of the Little Dipper. But did you know that Polaris (Alpha [α] Ursae Minoris) is a double star? Even a 50-mm telescope reveals Polaris B, a 9th-magnitude orb set about 18" to the primary's southwest. Both show little evidence of color. Polaris is actually a quadruple star, with two other very faint companion stars detectable in instruments of 200 mm and larger. Polaris C at 13th-magnitude is about 45" to the primary's east-northeast, while 12th-magnitude Polaris D lies some 83" south-southeast of Polaris A.

STAR 22, nicknamed the Mini-Coathanger, is one of those celestial surprises that you will occasionally stumble upon in the course of surveying the night sky. I first heard of the Mini-Coathanger, so named for its resemblance to the Coathanger Cluster in Vulpecula

Figure 7.13. *Collinder 399, the Coathanger Cluster in Vulpecula, is a favorite binocular sight of summer stargazers. Note its unmistakable resemblance, with the hanger's four-star hook curving under the straight bar of six suns. North is up in this photo by Brian Kennedy.*

(see Collinder 399, described below), while attending the Starfest '93 astronomy convention in Kingsport, Tennessee. Tom Whiting of North East, Pennsylvania, told how he first bumped into this unique pattern of stars in August 1993. He found it hanging just to the north of a diamond-shaped asterism of four 6th- and 7th-magnitude suns about 2° south-southwest of Epsilon (ε) Ursae Minoris in the Little Dipper's handle.

The Mini-Coathanger, shown in Eyepiece Impression 24, is formed from 11 stars ranging in brightness from 9th to 11th magnitude and spanning about 30'. Their combined likeness to a coathanger is unmistakable, several of them delineating the hook and others making up the crosspiece. A low- or medium-power eyepiece gives the best views. Thanks to its circumpolar location, the Mini-Coathanger can be enjoyed year-round.

Vulpecula

Vulpecula, the Fox, appears as a dim, starless void wedged between the constellations Cygnus to the north and Aquila to the south. Although it lacks any bright naked-eye stars, Vulpecula holds one of the sky's premiere deep-sky treasures, the famous Dumbbell planetary nebula. Also within Vulpecula is a pair of unique open clusters that are both well worth the search. The constellation passes midnight culmination in late July and so is well placed for early-evening observation by the latter part of August and into September.

Collinder 399. It's easy to spot this obscure-sounding open cluster simply by turning your finderscope or binoculars 8° south of Albireo in Cygnus. Look for a smudge of grayish cosmic cotton speckled with

Figure 7.14. *M27, the famous Dumbbell Nebula in Vulpecula, is the brightest planetary nebula in the summer sky. Compare this photo with Eyepiece Impression 25. Photo courtesy the U.S. Naval Observatory. South is to the lower left.*

a few faint points of light. That patch is the cluster (Figure 7.13). We have the late Walter Scott Houston to thank for introducing Collinder 399 to untold thousands of amateurs through his Deep-Sky Wonders columns in *Sky & Telescope*. Although that catalog designation might not strike a familiar chord with you, the names Brocchi's Cluster or the Coathanger Cluster may. All three refer to the same neat collection of stars. Best of all, you don't need any optical aid to see it. On exceptional nights the naked eye can pick out the brightest stars in the cluster — 4, 5, and 7 Vulpeculae — though binoculars are needed for a good view. With 7× binoculars, six stars aligned in a row appear to form the Coathanger's crossbar, while four others curve away to mark the hook. In all, Collinder 399 contains some five dozen stars spanning a full 1° of sky.

M27 (NGC 6853). Of all the planetary nebulae in our sky, none are as easy to see as M27, the famous Dumbbell Nebula (Figure 7.14). Even 7× binoculars show it as a "not-quite-stellar" star. But since the stars in Vulpecula are all but invisible from most observing sites, I recommend using the stars of neighboring Sagitta instead. Aim your telescope toward Gamma (γ) Sagittae, the easternmost star in the Arrow. Head 3½° north to 6th-magnitude 14 Vulpeculae. M27 lies in the same field of view as this star, a little less than ½° to its southeast.

Through a 200-mm Schmidt-Cassegrain telescope, Glenn Bock from Newark, Delaware, sees M27 as "easy, large, and rectangular with rounded corners; the southern half is brighter than the northern half." With a narrowband light-pollution filter in place, this same size instrument will faintly reveal some of the tenuous filaments that reach out of the nebula's core. And the view of M27 through a large telescope is sure to leave you breathless!

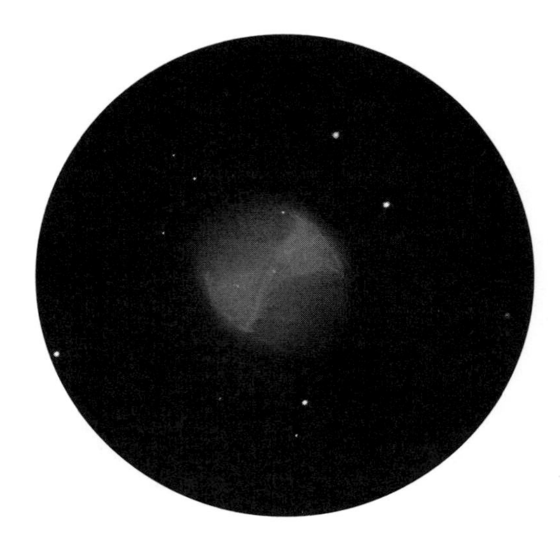

Eyepiece Impression 25. *M27, the Dumbbell Nebula, as seen through the author's 333-mm Newtonian and 12-mm Nagler eyepiece (125×). Compare the view of both the nebula's central star and some of the cloud's internal intricacies with Figure 7.14. South is to the lower left.*

Different observers have likened the distinctive shape of this planetary to a weight lifter's dumbbell, an hourglass, a bow tie, or even an apple core. Let your imagination run wild. As shown in Eyepiece Impression 25, several stars can be seen in front of M27, with the true central star shining at magnitude 13. Observers should note, however, that the central star is most easily found *without* a light-pollution filter in place; a filter lowers the contrast between the star and surrounding nebulosity just enough to make the star disappear.

NGC 6940 is a splendid open cluster lying just inside the northern border of Vulpecula. To get there slide southward from Epsilon (ε) Cygni, along the Swan's eastern wing, to 4th-magnitude 52 Cygni. After pausing to admire the Veil Nebula (a portion of which is superimposed over the star; see its description earlier in this chapter under Cygnus), place 52 in the northeast corner of your finder and look for 4th-magnitude 41 Cygni in the field's northwest corner. NGC 6940 resides about 3½° southwest of 52 and 2⅓° southeast of 41.

All five dozen stars within this rich collection shine like sapphires against a velvet black backdrop, save for one renegade. That maverick, **FG Vulpeculae,** also happens to be the group's brightest star, a ruby red semiregular variable. FG ranges in brightness between magnitudes 9.0 and 9.5 over a period that averages 80 days and offers a striking color contrast to its blue-white brethren.

On the next clear summer evening, leave your world behind and take a vacation in the Milky Way galaxy. Without ever leaving your backyard, you will see sights that are without equal anywhere on Earth. Bon voyage — and don't forget to write!

CHAPTER 8

Falling for galaxies

Even while the memory of warm summer nights is still fresh in our minds, there is no denying that the change in seasons is upon us. The summer sky, rich in clusters and nebulae, is now giving way to the galaxy-strewn heavens of autumn (Figure 8.1). Here we find objects aplenty to test our observational skills and techniques. The following menu offers just a few samples of what this season has cooked up for us. Try your luck with some of these morsels, and then branch out on your own to come up with your own list of autumn's tempting treats.

Andromeda

Riding high overhead on autumn nights is the constellation Andromeda, the Princess. Best known to deep-sky observers as the home of M31, the closest major galaxy to the Milky Way, the Princess also harbors several spectacular objects that lie within our own galaxy. Andromeda culminates at midnight around the first day of autumn and is excellently placed during the evening hours of November and December.

NGC 7662 is one of autumn's outstanding planetary nebulae. Beginning at Alpheratz (Alpha [α] Andromedae), recognizable as the northeastern star in the Great Square of Pegasus, move about 16° to

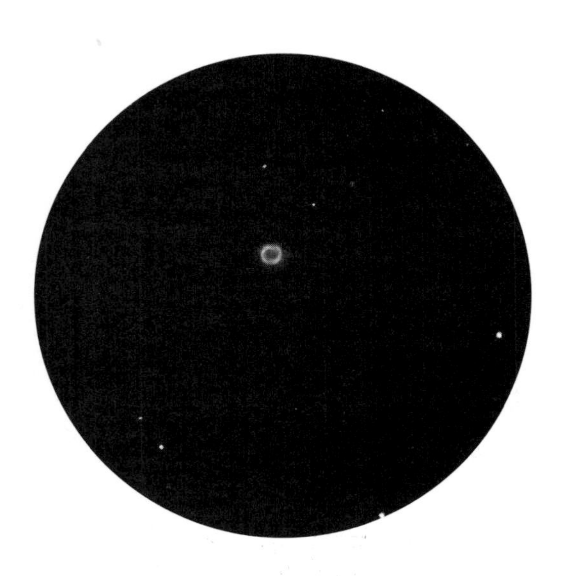

Eyepiece Impression 26. *NGC 7662 in Andromeda is frequently thought of as autumn's answer to summer's Ring Nebula. The similarity is clear in this south-up drawing made through the author's 200-mm Newtonian and 7-mm Nagler eyepiece (200×).*

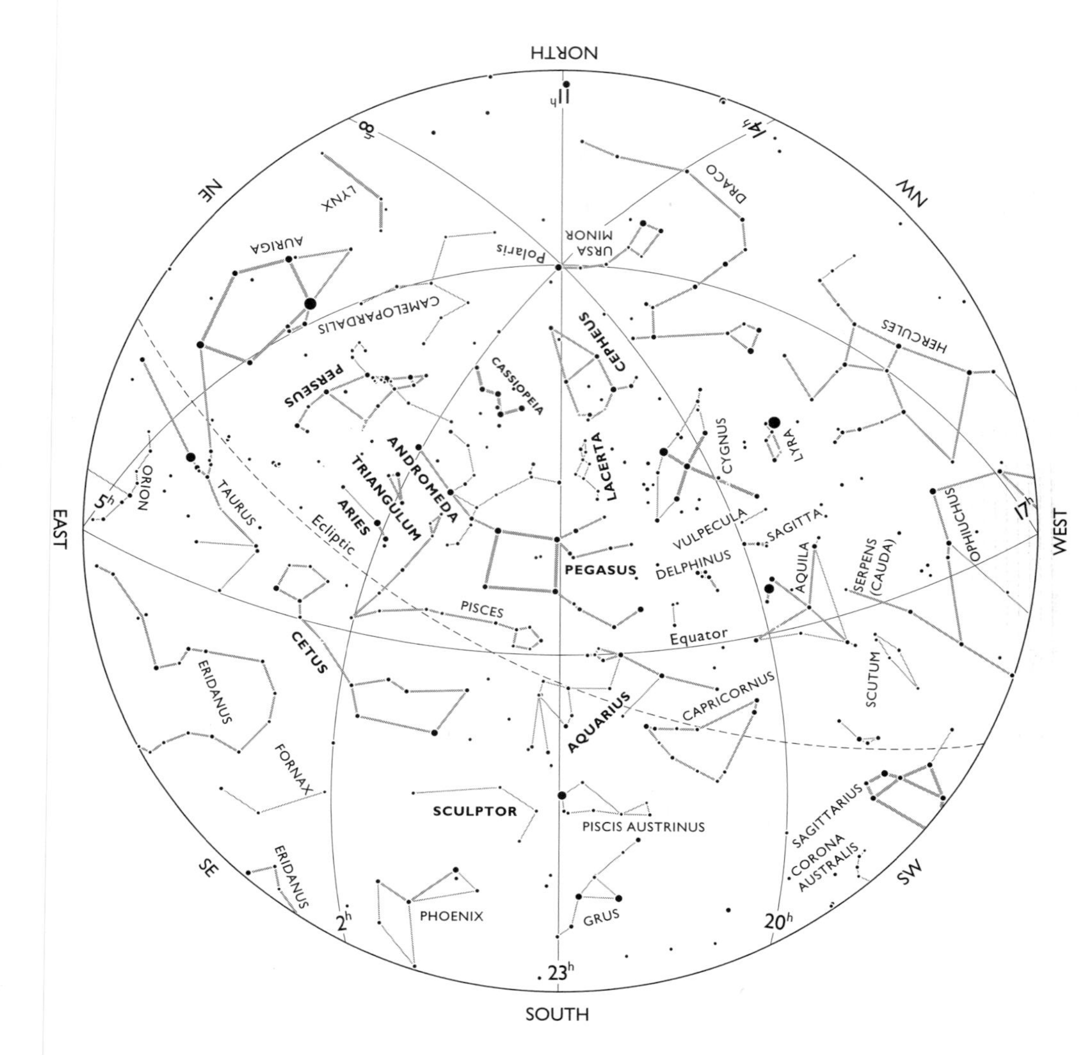

Figure 8.1. *The autumn naked-eye sky. Constellations discussed in this chapter are shown in boldface. The chart is plotted for 30° north latitude. For viewers north of that latitude, stars in the northern part of the sky will appear higher above the northern horizon and stars in the south lower (or possibly below the southern horizon).*

the northwest until you find Iota (ι), Kappa (κ), and Lambda (λ) Andromedae. All three of these 4th-magnitude stars are set in a distinctive north-south arc. From Iota, the arc's southernmost star, slide 2° west to NGC 7662.

Small telescopes show NGC 7662 as a circular 9th-magnitude disk of pale blue light; the intensity is enhanced in a 200-mm (8-inch) instrument. High magnifications reveal brighter rim sections toward the northeast and southwest edges of the planetary's 20" disk, as suggested in Eyepiece Impression 26. E. S. Barker, of Herne Bay in the

Figure 8.2. *M31, known popularly as the Andromeda Galaxy, is the closest major galaxy to our own Milky Way. Also captured in this photograph by Robert Bickel are two of M31's companion galaxies. M32 can be seen just above and to the left of M31's central core, while M110 lies below. South is to the upper left.*

United Kingdom, wrote in the *Webb Society Deep-Sky Observer's Handbook,* volume 2, that his 215-mm (8½-inch) telescope "traces a ring and dark centre" in this planetary. Observing through the light pollution of Lilburn, Georgia, with a 200-mm Schmidt-Cassegrain telescope, Eric Greene recalls NGC 7662 as a "very bright, very easy planetary with a dark center and faint ring structure around a bright outer rim visible at 400×; no sign of central star at any magnification." Indeed, NGC 7662's 13th-magnitude central star proves very difficult to see because of the low level of contrast between it and the nebula itself. An unrelated 13th-magnitude field star can be more readily seen near the eastern edge of the nebula.

M31 (NGC 224), an Sb spiral, holds the distinction of being the largest member of the so-called Local Group of galaxies to which the Milky Way belongs (Figure 8.2). And it shows. Even from moderately light-polluted sites, M31, better known as the Andromeda Galaxy, is seen with the eye alone as a subtle smudge of light.

To find M31 start from Mirach (Beta [β] Andromedae). If you are viewing through light-polluted skies, switch to binoculars or a finderscope; otherwise, with your eye alone, look just to the north to find the fainter star Mu (μ) Andromedae and, a little farther north, the even fainter Nu (ν). To the eye, or in binoculars or a finder, M31 ap-

Eyepiece Impression 27. *Both M31 and M32 (to the upper left) fit into this view drawn through the author's 200-mm Newtonian and 24-mm Tele Vue Wide Field eyepiece (58×). Note one of M31's dark dust lanes to the lower left of the galactic core. South is up.*

pears as a dim glow just northwest of Nu. Under dark, rural skies, the galaxy's nondescript appearance blossoms into a distinctly cigar-shaped apparition. Some sharp-eyed observers can trace the Andromeda Galaxy for an incredible 5°, a distance equaling ten full Moons stacked end to end! No other galaxy visible from the Northern Hemisphere can compare.

Binoculars and small telescopes resolve M31's pronounced core as a bright 10' sphere engulfed in the elongated glow of the galaxy's disk. A 100-mm (4-inch) telescope adds a distinctive dark lane girdling the plane of the galaxy to the northwest of the core. The vague outline of a second dark lane, parallel to the first but farther from the core, requires at least a 200-mm instrument to be seen. Both lanes are faintly shown in Eyepiece Impression 27. A third, still weaker dark lane runs along the other side of the core, but it is difficult to pick out in anything less than a 254-mm (10-inch).

From the core of M31, slide along the galactic plane about 1½° to the southwest until you arrive at **NGC 206,** a bright, sharply rectangular patch of light. Spanning close to 3,000 light-years, NGC 206 is an enormous cloud of stars within the Andromeda Galaxy. From our distant vantage point it shrinks to about 2' × 1', but remains bright enough to be seen in a 150- to 200-mm (6- to 8-inch) telescope on dark nights. Larger backyard telescopes begin to add a mottled texture to the surface of NGC 206; the bright regions mark pockets of increased stellar density. Medium- and high-power eyepieces give the best views.

Experienced deep-sky observers with large telescopes can spy several other faint deep-sky objects within M31. No fewer than seven of M31's family of globular clusters shine brighter than 15th magnitude, bringing them within the range of 305- to 355-mm (12- to 14-inch) telescopes. The brightest, at magnitude 14.2, is indexed as G280 (in the *Atlas of the Andromeda Galaxy*) and located east of the core.

Eyepiece Impression 28. *The faint glimmer of the galaxy NGC 404 lies just 6' northwest of 2nd-magnitude Beta Andromedae, seen to the upper right. This drawing was made through the author's 200-mm Newtonian and 12-mm Nagler eyepiece (116×). South is up.*

M32 (NGC 221), an E2 elliptical system, is the most prominent of four small companion galaxies swarming around M31. Even steadily held 7 × binoculars will reveal 8th-magnitude M32 as a not-quite-stellar point of light set about 24' south of the parent galaxy's bright core. With a small telescope, be on the watch for an oval patch spanning about 3' × 2' and highlighted by a bright, nonstellar core. As aperture grows, so grows M32, with soft extensions of the galactic halo coming into view. Photographs expand it to about 8' × 6', oriented northwest-southeast.

M110 (older observing references usually note this E6 elliptical only by its alter ego, NGC 205) is a second companion of the Andromeda Galaxy, about 36' northwest of M31's core. In 1968, in the original edition of his book, *Messier's Nebulae and Star Clusters*, Kenneth Glyn Jones pointed out that, although Messier did not comment on it in his catalog, he had indeed observed and recorded this second companion of M31, and Jones proposed that NGC 205 be added to the listing as M110. Thus, two centuries after it was first published, the Messier catalog received its final entry.

Although M110 has a slightly higher integrated magnitude than M32 (magnitude 8.0 versus 8.2), it appears noticeably fainter than its more flamboyant neighbor. Why the discrepancy? As mentioned above, M32 is concentrated over an area of 8' × 6'; M110, on the other hand, extends across 17' × 10'. Thus, while M110 is technically brighter than M32, its light is scattered over a much larger area, and so it has a lower surface brightness. Through most backyard telescopes, M110 appears as a faint glow stretching northwest-southeast.

The two remaining galaxies accompanying M31 seem to have wandered far from home. NGC 147 and NGC 185 are found within 1° of each other, but about 7° north of M31 in neighboring Cassiopeia. They are described in their home constellation's section, later in this chapter.

Figure 8.3. *NGC 891 in Andromeda is a classic example of an edge-on spiral galaxy. Note the dark lane that slices along the galaxy's edge. South is up. Photo by Dale E. Mais.*

NGC 404 has been a favorite galaxy of mine ever since I "discovered" it in 1974. You mean you didn't read about my discovery? I guess it wasn't exactly headline news, but I bumped into this small lenticular galaxy strictly by chance on a clear night nearly two decades ago while using the star Mirach (Beta [β] Andromedae) to align my finderscope.

As shown in Eyepiece Impression 28, NGC 404 lies just 6' northwest of that 2nd-magnitude star, making it at once extremely easy to locate but hard to see. The galaxy's proximity to Beta makes it simple enough to locate the field, and at 10th magnitude NGC 404 is bright enough to be spotted in a 150-mm instrument *provided* the star is moved out of the field of view. Through a medium-power eyepiece the star will be just outside the field, and the galaxy should appear as a small circular smudge of feeble light accented by a dim central nucleus. And my discovery? NGC 404 came as a surprise to me because the most popular star atlas in use at the time, the *Skalnate Pleso Atlas of the Heavens*, did not plot it. Today, most atlases show the galaxy, but I wonder how many other amateurs in the past also accidentally discovered "my" galaxy?

NGC 752. Found in eastern Andromeda just over the border from Triangulum, NGC 752 is one of the largest and brightest of the non-Messier open clusters north of the celestial equator. Sweep about 5° south-southwest of Gamma (γ) Andromedae (see the next entry) to a triangle of 6th- and 7th-magnitude stars and, a little farther west, NGC 752. (Incidentally, the brightest member of that triangle is **56 Andromedae**, an easy double star for small telescopes.)

With its five dozen members scattered across nearly a full 1°, this stellar troupe is made for giant binoculars and rich-field telescopes. A few of its brightest stars are visible in 7× binoculars, while 11× binoculars begin to reveal a multitude of fainter cluster members as well. The best view I can recall of NGC 752 was through my daughter's 108-mm (4¼-inch) f/4 rich-field Newtonian a few years back. Dozens of stellar denizens, many set in attractive pairings and long, intertwining threads, glistened across the field.

Looking through a pair of binoculars, try to picture NGC 752 as a ball. Then imagine a pattern of stars to its west as a golf putter. That's the basis for **STAR 14,** an idea advanced by John Davis of Amherst, Massachusetts. Davis suggests that the small isosceles triangle to the southwest of the open cluster, which we used to find NGC 752 in the first place, marks the head of a golf club, while a line of four 6th- and 7th-magnitude suns tilted toward the northwest forms the club's shaft. In all, the putter's head is about ½° long, while the shaft extends for about 1½°.

Gamma (γ) Andromedae, one of autumn's premier double stars, creates a wonderfully colorful scene through nearly all telescopes. Its 2.3-magnitude Type-*K* primary radiates a stunning yellowish gleam, while the 5.5-magnitude Type-*B* secondary shines a deep blue. Although Gamma Andromedae has been known to be double since the late 18th century, there has been no appreciable change in the stars' separation or position angle since discovery. The stars remain separated by about 10", the bluish Gamma B lying to the northeast of Gamma A.

In 1842, Otto Struve discovered that Gamma Andromedae B is itself a close binary star. Its 6th-magnitude companion, known as Gamma Andromedae C, orbits once every 61 years in a highly elliptical path. As seen from Earth, the two stars at their closest are separated by less than 0.05". At their greatest separation Gamma C is 0.55" away, when it should just be visible in a high-quality 200-mm telescope in extremely steady seeing. Just where is Gamma C now? It passed greatest separation in 1982 and is now on the way back toward the system's B star. With each passing year it draws nearer to its 2013 close encounter.

NGC 891. One of the most photogenic edge-on spiral galaxies, NGC 891 presents a challenge to backyard deep-sky hunters (Figure 8.3). To find it, cast off from the double star Gamma (γ) Andromedae, heading more or less due east toward Algol (Beta [β] Persei). Pause about a third of the way and look through your finderscope for a trapezoidal pattern of four 6th- and 7th-magnitude stars. NGC 891 lies inside the trapezoid, just northwest of its easternmost point.

Figure 8.4. *NGC 7009, the Saturn Nebula, in Aquarius. Notice how the two protrusions extending from the planetary nebula bear a remarkable likeness to Saturn. South is up. Photo by Martin C. Germano.*

NGC 891 is 13' long but a mere 3' wide — a cigar-shaped sliver of dim gray light. At 10th magnitude it sounds relatively bright, but NGC 891 is one of those deceptive objects with a low surface brightness that elude many an observer's eye. Use a low-power eyepiece to find it, and once you have it in view switch to a moderate magnification for a closer look.

A 200-mm telescope operating at 100× will show the galaxy's disk as a long, thin, almost ghostlike glow. The dark lane that bisects it so prominently in photographs may be just discernible but only under fairly dark skies. Sue French of Glenville, New York, sums it up nicely. Through her 254-mm reflector, she sees NGC 891 as a "very faint, very elongated hazy blur; fairly large, but because of its low surface brightness, the galaxy was barely visible."

Aquarius

Aquarius, the Water Bearer, dwells in a large, barren portion of the autumn sky below the more prominent constellation Pegasus. Although it is composed of 2nd-magnitude and fainter stars, Aquarius hosts many interesting telescopic targets. The constellation culminates at midnight toward the end of August and is conveniently placed for evening observation during October and November.

Planetary nebula **NGC 7009,** located in the extreme western part of the constellation, marks the first stop on our Aquarius tour (Figure 8.4). It is most easily found by beginning at the triangle of neighboring Capricornus, the Sea-Goat. Locate 4th-magnitude Theta (θ) Capricorni, along the north side of the triangle, then scan 6° north-northeast to 5th-magnitude Nu (ν) Aquarii, and finally turn westward another 1° to find your target.

On nights of good transparency, NGC 7009 is visible in steadily supported binoculars and small telescopes as a fuzzy greenish point of light. It stands out surprisingly well, due partly to its rather bleak surroundings but mostly to its remarkably high surface brightness

Eyepiece Impression 29. *The two unique extensions that give NGC 7009 its nickname, the Saturn Nebula, are apparent in this drawing made through the author's 333-mm Newtonian and 12-mm Nagler eyepiece (125×). South is up.*

and a greenish color produced by doubly ionized oxygen (O III). A telescope of 150 mm or larger will reveal why this strange little enigma has been called the Saturn Nebula. Looking carefully, can you detect the two faint lobes protruding from either side of the cloud's round disk, as shown in Eyepiece Impression 29? The overall effect at low power is strongly reminiscent of our solar system's sixth planet.

M2 (NGC 7089), one of autumn's best globular clusters, is easy to find thanks to a westward-aimed arrow created by the stars in the constellation's Water Jar asterism (Gamma [γ], Zeta [ζ], Eta [η], and Pi [π] Aquarii), and Alpha (α) Aquarii at the tip. The arrow points directly at M2, some 8½° west of Alpha. M2 was first spotted in 1746 by Giovanni Maraldi (1709–1788) as an ill-defined smudge. A modern 100-mm telescope betters Maraldi's view by resolving some of the stars around the edges of M2. Yet, as shown in Eyepiece Impression 30, it takes at least a 305-mm to detect stars across the entire disk. Like some other globulars, M2 appears slightly oval in shape. This effect is thought to be due to the group's collective rotation, causing the globular as a whole to flatten along its rotational axis and broaden along the "equator." M2 lies about 50,000 light-years away, and claims some 100,000 members.

While in the area, be sure not to pass up a second globular cluster, M15, to the north (see its description under its home constellation, Pegasus, later in this chapter).

NGC 7293, nicknamed the Helix Nebula, is the sky's largest planetary nebula. To pinpoint it, return to the Water Jar. Draw a short line from Pi (π) Aquarii to Zeta (ζ) Aquarii and extend it southeastward for 9½° to Lambda (λ) Aquarii. Turn due south and drop just over 8° to Delta (δ) Aquarii. Peering through your finderscope, slide 7° southwest, past a wide pair of 5th-magnitude stars, to end up at 5th-magnitude Upsilon (υ) Aquarii. The Helix lies just 1° to Upsilon's west.

As the brightest and largest planetary in the sky, NGC 7293

Eyepiece Impression 30. *One of the season's most spectacular globular clusters, M2 in Aquarius is depicted here through the author's 333-mm Newtonian and 12-mm Nagler eyepiece (125×). South is up.*

sounds as if it should be the easiest to find. Imagine a 6.5-magnitude planetary nebula half the size of the full Moon. It should practically tap you on the shoulder and say, "Looking for me?" But when you aim toward its location and gaze into the eyepiece, it's not there! Why? As with so many deep-sky objects, its listed magnitude is its integrated magnitude — a measure of the object's brightness if you were somehow able to condense its light into a starlike point. Since the nebula is not a point, but ¼° across, its surface brightness (or brightness per square area) is much lower than its magnitude suggests. That turns an apparently easy target into a difficult (but not impossible) test.

The Helix is best seen through giant binoculars and low-power rich-field telescopes. While higher magnification helps to reveal cloud structure and the 13th-magnitude central star, the overall view is less attractive in a more restrictive field. The most aesthetically pleasing encounter I have had with the Helix was through my daughter's 108-mm f/4 rich-field Newtonian at 17× with a narrowband light-pollution filter. The gray disk and central "hole" of the Helix were clearly visible against a starry field.

Aries

To the naked eye, the constellation Aries, the Ram, appears as little more than an arc of three unspectacular stars west of the much more prominent winter constellation Taurus, the Bull. While Aries is crowded with many galaxies, none is bright enough to be included in this short sky survey. But what Aries lacks in quantity of bright objects, it makes up for in quality by contributing one of the season's most spectacular binary stars. Aries culminates at midnight at the end of October and rides high in December's early-evening skies.

Gamma (γ) Arietis (Mesartim) is visible to the naked eye as the faintest and southernmost of the constellation's three main stars. Robert Hooke (1635–1703) was the first to detect the double nature of Gamma way back in 1664, making it one of the first telescopic doubles ever discovered. Even a small telescope at low power will show Gamma as two nearly identical, pearly white magnitude-4.8 stars. The two are separated by just under 8" and are positioned exactly north-south of each other. Like Gamma Andromedae mentioned earlier, Gamma Arietis A and B have shown no appreciable change in either separation or position angle since the stars were first measured by F. G. W. Struve in 1830. Still, they are believed to form a true binary system because of their shared proper motion in the sky.

Cassiopeia

Reigning over our autumn sky, Cassiopeia, the Queen, possesses some royal open clusters and nebulae. Many enjoyable hours may be spent at the eyepiece soaking in the beauty of this constellation and its many outstanding deep-sky objects. The Queen reaches midnight culmination in early October and is best seen during the early-evening hours of November and December. But keep in mind that since Cassiopeia is a circumpolar constellation, many of its treasures may be enjoyed year-round.

STAR 29 was independently discovered by Sam Bissette of Wilmington, North Carolina, and John Davis of Amherst, Massachusetts. The asterism covers an area of about 1½° × 2°, and all of its stars shine between 5th and 7th magnitude, making it an ideal target for observers using low-power, wide-field binoculars. The center of STAR 29 is found in Cassiopeia, though some of its stars lie across the border with Cepheus. To find it, scan about 8° to the west of Beta (β) Cassiopeiae, the westernmost star in Cassiopeia's familiar W pattern. (In most 7× binoculars, that's about two fields of view.)

The nine stars in STAR 29 form a nearly perfect, albeit sideways, number 7, leading to the group's nickname of Lucky 7. The two brightest stars in the pattern are the 5th-magnitude 1 and 2 Cassiopeiae, the latter marking the bend in the 7. (Davis adds that if a curved arc of stars to the north of the 7 is included as a handle, the pattern resembles a submachine gun.)

While just about all of the other asterisms listed in this book were discovered visually, Bissette caught the Lucky 7 photographically on a wide-field long-exposure picture of Cepheus. He is carrying out an unusual observing program in which he photographs wide expanses of the sky and then examines them with a microscope. If he finds an interesting portion of a slide, he then connects his camera to the microscope and takes a close-up photograph.

M52 (NGC 7654). Often ignored by observers because of its relatively remote location, the open cluster M52 nonetheless puts on a fine show for those who take the time to seek it out. To find it, trace a line from Alpha (α) to Beta (β) Cassiopeiae and extend it slightly more than the same distance again to the northwest of Beta. There you should find 5th-magnitude 4 Cassiopeiae. M52 lies less than 1° to the south.

M52 reminds different observers of different things. In his classic *Bedford Catalogue,* Adm. William Smyth described it as triangular with an 8th-magnitude orange star at the apex, the entire effect resembling "a bird with outspread wings." In *The Messier Album,* John Mallas and Evered Kreimer described the view of M52 through a 100-mm refractor as "a needle-shaped inner region inside a half circle." My 200-mm f/7 Newtonian displays about 20 stars within the cluster's 13' border, which is only about 20 percent of the group's estimated population. The remaining stars, all too faint to resolve individually, blend their light into a faint background glow.

Just northeast of M52 is **STAR 12,** an attractive binocular asterism of about 14 stars from 5th to 9th magnitude. I first came across it while compiling *Touring the Universe Through Binoculars,* describing it as "a wide triangular asterism . . . a very attractive group in low-power binoculars." The inventive eye of Massachusetts amateur John Davis sees it more as an airplane, with the asterism's brightest star (4 Cassiopeiae) as the plane's eastern wing. A curved north-south line of stars to its west makes up the plane's fuselage, while a lone 6th-magnitude star marks the tip of the western wing. In all, Davis's airplane spans 1 square degree.

The Airplane also makes a handy reference when trying to land NGC 7538 and NGC 7635, a pair of bright nebulae that I will leave you to discover on your own.

NGC 7789 is famous for being one of the densest open clusters north of the celestial equator. To locate this rich stellar family, start at Caph (Beta [β] Cassiopeiae), the westernmost star in Cassiopeia's W. Shift about 2½° southwest to 5th-magnitude Rho (ρ) Cassiopeiae, then another 1° to the south-southeast for NGC 7789.

Current estimates put the total population of NGC 7789 at almost 600 stars, all crammed into just 16'. This great density, coupled with the fact that none of the stars shine brighter than magnitude 10.7, makes resolution difficult in many smaller backyard telescopes. A 150-mm instrument reveals about five dozen cluster members sprinkled across a moderately bright background glow, while a 200-mm increases the star count to about 100. Resolution steadily increases as aperture grows, with close to 200 stars visible in 305- to 355-mm instruments.

My 333-mm (13.1-inch) f/4.5 Newtonian reveals that the cluster's few bright stars are set in a loose arc. Similarly, while observing with a 200-mm Schmidt-Cassegrain from Ouachita National Forest in Arkansas, Dean Williams notes what he describes as "some darker lanes creating spiraling patterns, reminiscent of the petals of a rose opening." Have others noticed this same effect?

NGC 147 and **NGC 185,** mentioned earlier as two of the Andromeda Galaxy's companion systems, lie 1° apart but about 7° north of M31 over the border in Cassiopeia. To locate this galactic pair, follow an imaginary line from Mirach (Beta [β] Andromedae) to Schedar (Alpha [α] Cassiopeiae). Just past the midway point

Figure 8.5. *The bright nebula NGC 281 in Cassiopeia is easy to see in this photo by Martin C. Germano, but don't let that lull you into a sense of false confidence! It's tough to pick out visually. South is up.*

along this line, watch for a crooked north-south row of three 5th-magnitude stars. Aim at the northernmost member of the trio, Omicron (ο) Cassiopeiae.

NGC 185, found about 1° due west of Omicron, is the easier of the pair to spot. An E0 elliptical galaxy, it can be glimpsed as a dim smudge through telescopes as small as 75 mm (3 inches) in aperture. My 200-mm Newtonian reveals a slightly oval disk and a bright core; the overall effect reminds many of an unresolved globular cluster. NGC 147, lying another 1° west of NGC 185, requires a more concentrated search because of its lower surface brightness. Most amateur telescopes show only a ghostly blur that lacks any central concentration.

Eta (η) Cassiopeiae. Here's a pleasant double star for telescopes large and small alike. Eta Cassiopeiae can be seen with the unaided eye as a 4th-magnitude star a third of the way from Schedar (Alpha [α] Cassiopeiae) to Gamma (γ) Cassiopeiae in the constellation's W formation. Telescopically, Eta splits into a pair of stars that are best viewed with 75× to 100×. The 7.5-magnitude secondary, 12" to the 3.5-magnitude primary's northwest, shines with a weak orange tint, while the primary itself appears somewhat yellow. The marked difference in brightness means that steady seeing conditions are helpful for resolving both stars in telescopes smaller than 75 mm. With a

Eyepiece Impression 31. *NGC 457, the Owl Cluster, is drawn flying upside down in this sketch made through the author's 333-mm Newtonian and 24-mm Tele Vue Wide Field eyepiece (63×). The two bright stars in the center mark the owl's eyes, and two wings may be imagined stretching to either side; the body and tail extend toward the upper part of the drawing. South is up.*

larger instrument you should have little trouble making out both, even under less than ideal seeing.

NGC 281 is a bright nebula (Figure 8.5), which may be found by first aiming your telescope at Schedar (Alpha [α] Cassiopeiae) and then letting the sky drift across the eyepiece field for 16 minutes, after which time it will be in view. Observing under dark skies through a 200-mm f/7 Newtonian, I noted NGC 281 as a large, faint, amorphous mist surrounding the multiple star Burnham 1 (sometimes written as β1). Four of the five stars in this system are visible in amateur telescopes, with an 8th-magnitude primary accompanied by 9th- and 10th-magnitude B, C, and D companions. My impression was of a triangular blur with a rounded tip and a brighter region just west of the 8th-magnitude star.

NGC 457 is a delightful open cluster for deep-sky observers, not just because of its inherent beauty but also because it is so easy to find! You should have little trouble spotting it through just about any optical instrument; it appears to surround Phi (φ) Cassiopeiae, a 5th-magnitude star about 2° southwest of Delta (δ) Cassiopeiae.

What makes NGC 457 especially attractive is its distinctive shape, as shown in Eyepiece Impression 31. To many it appears reminiscent of a big-eyed bird, leading to its popular nickname the Owl Cluster. Others see different patterns among its stars. For instance, Sue French of Glenville, New York, comments that it looks more like a dragonfly. Bug or bird, you decide. Regardless of which creature you see, the body may be traced out from about a dozen stars of magnitudes 9 to 11, while a pair of 10th-magnitude stars marks the tail feathers. Two arcs, each containing about a half dozen suns, form the wings. The east wing is highlighted by a distinctive 8th-magnitude orange star, one of the brightest members of the cluster. But what really draws your attention are the dazzling "eyes," marked by 5th-magnitude Phi Cassiopeiae and 7th-magnitude HD 7902. Phi is a cluster member, and at absolute magnitude –8.5 it is one of the most luminous stars known; HD 7902 does not belong to the cluster.

M103 (NGC 581). While in the general area of Delta (δ) Cassiopeiae, be sure to pay a visit to this fine galactic cluster, lying about 1° to Delta's northeast. M103 represents a footnote in astronomical history as the final object listed in Messier's original catalog (M104 through M110 are 20th-century additions made on the basis of Messier's unpublished notes.) In Messier's words, M103 is simply "a cluster of stars," but through modern telescopes it has a luster that he could never have imagined. Today's amateurs record it as a sparkling collection of stardust set in the pattern of an arrowhead measuring about 6' across.

Marking the tip of the arrowhead is the attractive multiple star Struve 131 (Σ131). Of the star's five components, amateur telescopes most easily reveal the 7.3-magnitude primary and its two companions of magnitudes 10.5 and 10.8. Most authorities conclude that the association between Struve 131 and the cluster is purely coincidental, the star lying between us and M103. Of the 100-plus true members of the cluster, most shine at 10th magnitude or fainter, though one of the brightest is an orange 9th-magnitude star lying just southeast of the cluster's center. Many of the fainter stars also display subtle hints of color, especially when their images are slightly defocused.

STAR 13. From John Davis of Amherst, Massachusetts, comes this asterism that is perfect for low-power binoculars. Dubbed the Queen's Kite, its distinctive diamond-shaped body can be found about 2° southeast of Delta (δ) Cassiopeiae. The top point of the kite is marked by 5th-magnitude Chi (χ) Cassiopeiae. Six other stars of 6th and 7th magnitude fill out the rest of the kite's diamond-shaped body, seen cocked toward the southeast, while five more stars in an arc winding toward the west-southwest make up the tail. Like me, you'll probably find that the kite's pattern is much more distinctive in light-polluted suburban skies than in darker rural skies, where a multitude of fainter suns obscure the design.

NGC 663, a 7th-magnitude open cluster, is bright enough to be seen in most finderscopes if you have a sharp eye. Look for a faint smudge about 1° southeast of a lopsided diamond of one 6th-magnitude and three 7th-magnitude stars about halfway between Delta (δ) and Epsilon (ε) Cassiopeiae. Despite its relatively high magnitude, NGC 663 reveals only a few points of light against an unresolved mass when viewed with small amateur telescopes. Through 200- to 254-mm instruments, many stars within the gravitational grip of this 7,000-light-year-distant swarm begin to shine through the warm glow of still fainter cluster members. My 333-mm Newtonian reveals several dozen stars from 8th to below 12th magnitude. Most are asymmetrically clumped into two densely populated regions inside the cluster's 16' diameter, with several set in close pairs.

Stock 2 is a fine, albeit little-known open cluster lying less than a binocular-field to the north of the famous Double Cluster in

Perseus. Close to one-third of the 50 stars that span the 1° width of Stock 2 shine between 8th and 10th magnitude and are visible through 7× glasses.

Many observers enjoy playing connect-the-dots among the stars of Stock 2 and have created a variety of shapes. For instance, some have likened it to a bug with antennae; some picture a pineapple, when a V-shaped stellar asterism to the east is added. Others see it as simply rectangular. But to me John Davis of Amherst, Massachusetts, came up with the best analogy when he likened Stock 2 to a weightlifter flexing his (or her) muscles. The aforementioned V-shaped asterism represents the weight lifter's legs; the body is marked by a thin line of stars extending to the west; and the arms are created by two curves of stars near the center of the cluster. Completing the picture, 7th-magnitude suns at the end of either arc mark the lifter's hands.

Iota (ι) Cassiopeiae, one of the finest triple stars that the sky has to offer, is faintly visible to the unaided eye against the backdrop of the Milky Way. To find Iota, extend an imaginary line from Delta (δ) to Epsilon (ε) Cassiopeiae (the two easternmost stars in the familiar Cassiopeia "W") and continue it an equal distance to the northeast. Iota will appear as a faint point of 4th-magnitude light to the naked eye, though it is easy to see through finderscopes.

Although Iota is actually a quadruple star, amateur telescopes can pick out only three of its components. The system's primary, Iota A, shines at magnitude 4.6. The secondary, 7th-magnitude Iota B, is the closer of the two faint companions and may be found only a couple of arcseconds to the primary's west. A 75-mm telescope will probably have difficulty in differentiating it from the brighter gleam of Iota A. Iota Cassiopeiae C is an easier catch in small instruments, as it lies nearly three times as far from the primary, toward the southeast.

Each component of Iota Cassiopeiae is of a different spectral class, yet all appear white, at least initially. Only after careful study will each star's subtle colors be detected. Although some observers see Iota A as yellow, it always looks white to me. Iota B and C strike many observers as being ashen blue. How do you perceive the colors?

STAR 15. From Father Lucian Kemble of Ontario, Canada, comes another kite-shaped pattern of stars tucked just inside the constellation's eastern border. Extend the line connecting Ruchbah (Delta [δ] Cassiopeiae) to Epsilon (ε) Cassiopeiae for 5° northeast to Iota (ι) Cassiopeiae. Shift another 8° to 5th-magnitude Gamma (γ) Camelopardalis; the Kite flies 2° farther northwest.

Binoculars and small telescopes easily show Kemble's Kite. Its frame is made up of six 6th- to 8th-magnitude stars in a diamond formation, trailed by four other stars that make up its tail. Father Kemble writes, "What first caught my attention was the seeming miniature Orion's belt [three equally spaced stars forming the Kite's eastern edge], but then the kite-and-tail shape became quite evident." Thanks to the lack of surrounding stars, Kemble's Kite flies boldly against the barren surroundings.

Cepheus

Cepheus, Cassiopeia's husband, holds little in store for the casual naked-eye stargazer. The constellation's five main stars form a pentagon, but as none of the five are brighter than 3rd magnitude the figure is an unfamiliar naked-eye sight from many urban and suburban observing sites. Even the Milky Way, passing through the southern tier of Cepheus, seems here to lack the brilliance that it adds to neighboring Cygnus and Cassiopeia. These factors can make finding Cepheus's deep-sky objects a challenge, but they can also add to the observer's satisfaction when one is finally spotted. Cepheus is a circumpolar constellation, so many of the objects to be described are visible throughout the year. The King, however, ascends to his highest at midnight in late August and is best seen in the early-evening hours of October and November.

NGC 6939, a bright though little-observed open cluster, is located just across the border from the summer constellation Cygnus. To find it, begin at 3rd-magnitude Alderamin (Alpha [α] Cephei), the southwestern star in the pentagon. Proceed about 4° west to 4th-magnitude Eta (η) Cephei. NGC 6939 lies just 2° farther southwest, at the southwestern point of an equilateral triangle formed by it, Eta, and Theta (θ) Cephei.

Turning a telescope its way, notice a closely knit family of 80 stars crammed into a tiny 8' circle. Although their light combines to an overall 8th magnitude, no individual shines brighter than 12th. With a 150- or 200-mm telescope, most observers will spot many dim cluster stars set in a diamond pattern and filled with the subtle glow of other, still fainter suns. In the *Webb Society Deep-Sky Observer's Handbook,* volume 3, Kenneth Glyn Jones recorded "five mag. 11 stars and many fainter" through a 200-mm telescope at 100×. Larger instruments are capable of displaying more and more points surrounding and within the diamond.

Less than 1° to the southeast of NGC 6939, straddling the Cygnus-Cepheus border, lies **NGC 6946,** a magnificent face-on Sc spiral galaxy. (Many references give its constellation as Cygnus, though it is now generally agreed that the galaxy belongs to Cepheus.) Like so many other face-on spirals, NGC 6946 photographs well but is hard to see. When turned its way, my 200-mm Newtonian unveils a faint, oval, grayish patch of light augmented by an ever so slightly brighter central core. Through her 254-mm Newtonian from Glenville, New York, Sue French sees a "large, low-surface-brightness nebulous patch with a hint of spiral structure imagined. Fits in the same eyepiece field (48×) as NGC 6939."

No two ways about it, NGC 6946 is a tough catch. Arizona amateur Steve Coe makes an important point about the galaxy's visibility. He writes, "This face-on spiral galaxy has a low surface brightness and therefore responds to the atmosphere more than edge-on systems. For that reason I have called this object "pretty faint" on a night I rated 5/10 and then called it 'pretty bright' on a night that was 8/10 in the mountains." In other words, it might be best to put

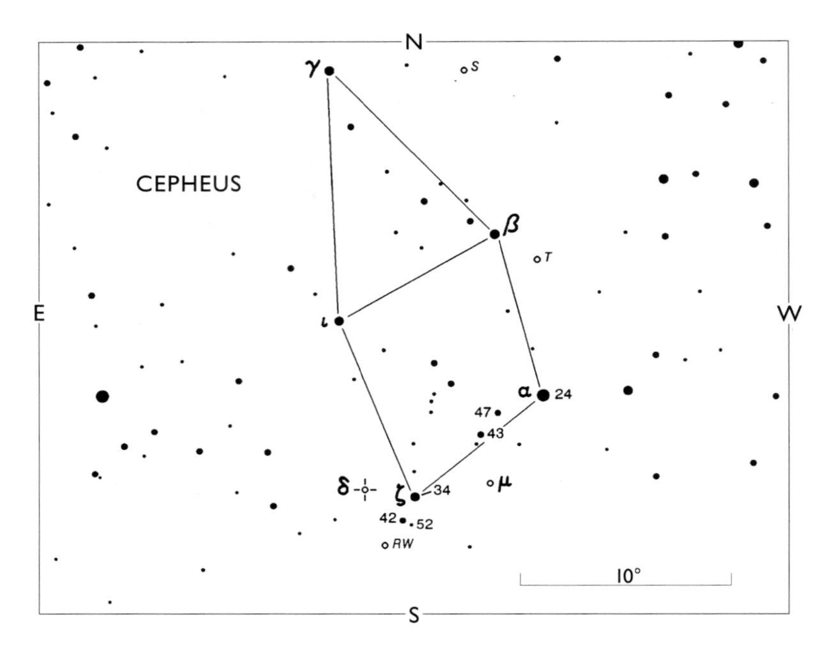

Figure 8.6. *Use this finder chart to try your luck at estimating the magnitude of the variable star Delta (δ) Cephei. Magnitudes of several comparison stars are given, with decimal points omitted. The objects* RW, S, *and* T *are other large amplitude variable stars in Cepheus. North is up. Courtesy the American Association of Variable Star Observers.*

NGC 6946 on hold until one of those special once-in-a-season nights when the air is clear and dry.

Delta (δ) Cephei, the easternmost of three stars set in a triangle marking the southeast corner of the Cepheus pentagon, was discovered to be a variable star in 1784 by English astronomer John Goodricke (1764–86). Its brightness fluctuates between magnitudes 3.5 and 4.4 in a period of 5.37 days. Use Figure 8.6 to try your luck at estimating the brightness of Delta Cephei by comparing it to the many surrounding fixed-brightness stars.

Delta Cephei holds a special place in the history of astronomy, as it was the first example of a Cepheid variable to be discovered. Cepheid variables are a special class of yellow-giant variable stars, which have contributed greatly to our current picture of the universe. Cepheids exhibit a tight relationship between their period and their inherent brightness (absolute magnitude). Once the period is known, the absolute magnitude can be derived. Then, by comparing the star's absolute magnitude with its apparent magnitude, its distance can be deduced. From observations of Cepheids in several nearby galaxies, astronomers have been able to calculate the distances to those faraway island universes. This fixes a crucial lower rung on the distance-scale ladder by which we attempt to estimate the size of the universe.

NGC 40. Not necessarily the easiest object in the sky to find, the

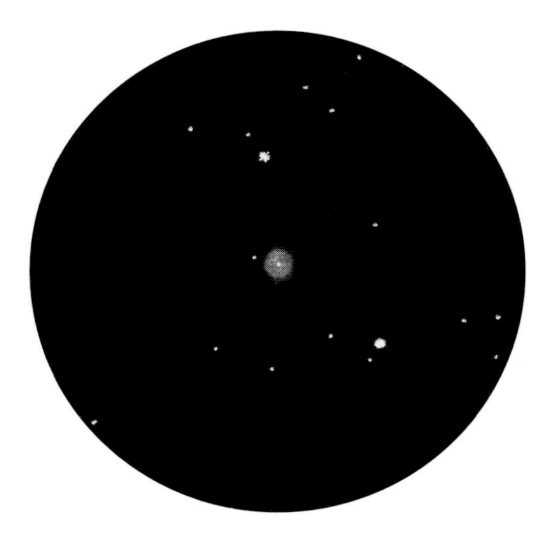

Eyepiece Impression 32. *The planetary nebula NGC 40 dares observers to tread the star-barren region of eastern Cepheus in order to find its grayish disk. Dave Kratz made this drawing through a 355-mm Schmidt-Cassegrain telescope at 140×. South is up.*

planetary nebula NGC 40 is nonetheless visible through telescopes as small as 150 mm if one knows exactly where to look. Draw an imaginary line between Gamma (γ) Cephei, at the top of the pentagon, and Kappa (κ) Cassiopeiae, just north of that constellation's W pattern. NGC 40 is located on that line, about a third of the way from Gamma to Kappa and framed between two 6th-magnitude stars.

A low-power eyepiece with a 150- or 200-mm telescope shows the characteristic round disk of this planetary with relative ease. At first glance the disk appears uniform in brightness and texture. A closer look, however, reveals irregular dark areas near the 11.5-magnitude central star. Eyepiece Impression 32 shows the view through a 355-mm Schmidt-Cassegrain telescope.

Cetus

Portrayed as either a whale or a sea-monster on ancient star maps, Cetus occupies a large, barren part of the autumn sky south of Pisces and Aries. Within its broad boundaries are one of the season's finest planetary nebulae, a classic long-period variable star, and many groupings of galaxies. Cetus reaches midnight culmination in mid-October, so it is highest in the southern sky during November and December evenings.

NGC 246 would be one of the sky's better-known planetary nebulae were it not so far from most deep-sky enthusiasts' regular haunts. Finding it really isn't that difficult if you begin at the Square of Pegasus. Trace the line that marks the Square's eastern side, from Alpheratz (Alpha [α] Andromedae) to Algenib (Gamma [γ] Pegasi). Continue southward for a little over 30° to the bright star Diphda (Beta [β] Ceti). Viewing through your finderscope, head 7° north to the wide pair of 5th-magnitude stars Phi^1 (ϕ^1) and Phi^2 (ϕ^2) Ceti. NGC 246 lies about 1° south of a point midway between these two 5th-magnitude suns, and forms an equilateral triangle with them. Be sure to use a low power when searching, then switch to a medium

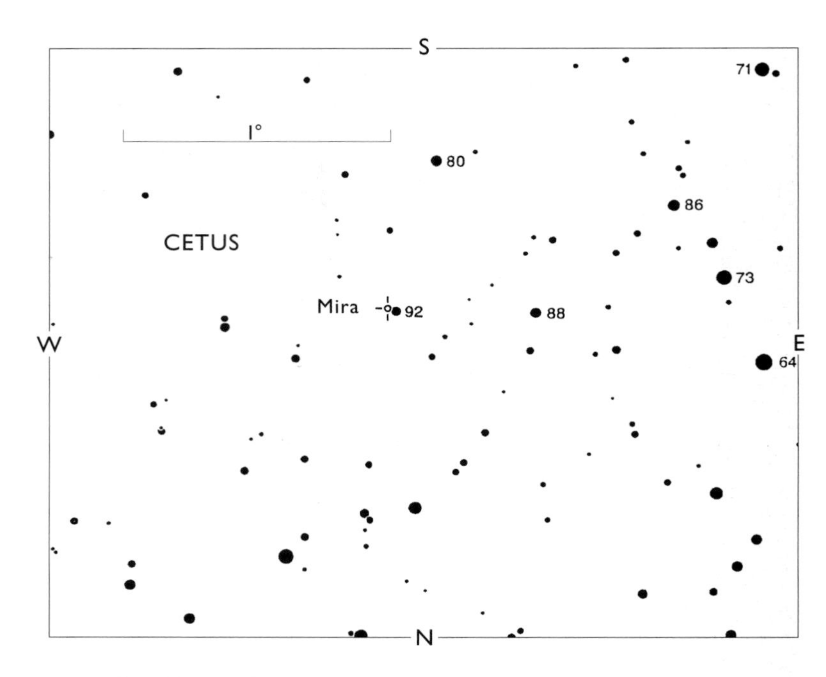

Figure 8.7. *Use this finder chart to estimate the magnitude of the variable star Mira [Omicron (o) Ceti]. The magnitudes of several comparison stars are given with decimal points omitted. Mira spends about 75 percent of its time below naked-eye visibility; you will need a clear night to catch sight of it. South is up. Courtesy the American Association of Variable Star Observers.*

power for closer inspection.

While zeroing in on NGC 246 is fairly simple, actually spotting its large, faint disk may require several minutes of concentrated searching before it becomes evident. Under superior sky conditions my 200-mm Newtonian displays a dim ring of grayish light that appears to encompass several faint stars. When viewed from my light-polluted backyard, the nebula's annular structure appears broken, while darker skies show a full ring with the brightest segment toward the northeast. The cloud's stellar parent, a 12th-magnitude centralized point of light, is an easy object in telescopes as small as 150 mm. The other stars surrounding NGC 246 are not actually associated with it; they are merely chance line-of-sight affiliates.

Mira (Omicron [o] Ceti) is the quintessential example of a long-period variable. You'll find it to the southwest of the Whale's pentagonal tail. The German astronomer David Fabricius (1564–1617) was the first to notice the star's peculiar flickering, though the depth of Mira's variability could not be confirmed until it was studied through a telescope. It normally oscillates between 3rd and 9th magnitudes, with an average period of 332 days, though on rare occasions it has brightened to magnitude 2.0 and faded to magnitude 10.1. According to one 1779 notation, it rose to 1st magnitude. But don't be deceived; Mira spends about 75 percent of its time below

Figure 8.8. *The Seyfert galaxy M77 dominates the top edge of this photo by Martin C. Germano. Although Messier himself missed it, the edge-on spiral galaxy NGC 1055 (lower left) is visible through telescopes as small as 150 mm in aperture. South is up.*

naked-eye visibility. Use the chart in Figure 8.7 to estimate Mira's brightness on the next clear night.

M77 (NGC 1068). Due to its sparse surroundings, this out-of-the-way galaxy (Figure 8.8) is often bypassed by observers in favor of other seasonal treats. Yet for those who take the time to search it out, M77 rewards their diligence with a high surface brightness that makes it easy to spot. In fact, M77 is bright enough to be seen on good nights through steadily held 10× binoculars.

Finding M77 is not as difficult as it might seem at first. Begin by locating the naked-eye pentagon marking the tail end of Cetus. Aim your finderscope at the star Delta (δ) Ceti, directly to the south, then look for an isosceles triangle of 8th-magnitude stars about 1½° to its

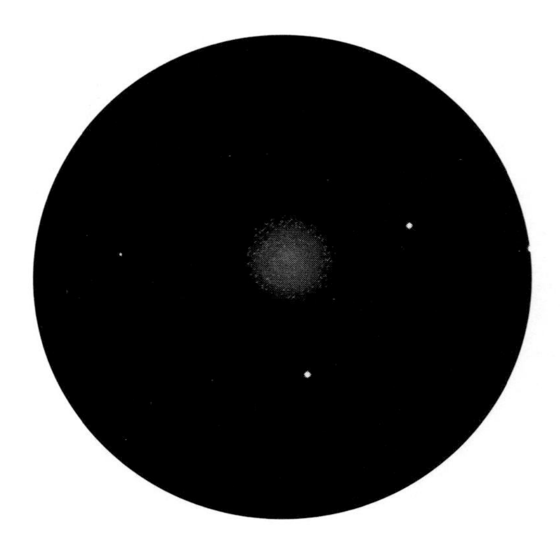

Eyepiece Impression 33. *M15 in Pegasus is one of the season's finest globular clusters. This drawing was made through the author's 200-mm Newtonian and 12-mm Nagler eyepiece (116×). South is up.*

southeast. Ninth-magnitude M77 lies about halfway between Delta and the triangle, just 1½' west-northwest of a 10th-magnitude field star. For those viewing through a 100- or 150-mm telescope, M77 will show a bright core surrounded by the subtle glow of its spiral arms. Larger telescopes help increase the galaxy's apparent size and brightness, and will reveal some structural details. Photographs bring out the tightly wrapped spiral arms and expand M77 to 7' × 6' across.

Although classified as a type-Sb spiral, M77 is also distinguished as a *Seyfert galaxy*. Seyfert galaxies — named for their discoverer, American astronomer Carl Seyfert (1911–1960) — exhibit unusually bright nuclei and may be related to quasars. M77 was also one of the first galaxies found to have a large redshift. This discovery in 1913 by Vesto Slipher (1875–1969) was an important steppingstone to the theory of the expanding universe.

Of the other galaxies that lie near Delta Ceti, 9th-magnitude **NGC 1055,** to the northwest of M77, is the brightest (Figure 8.8). Through amateur telescopes, this Sb spiral galaxy appears as a dim, ghostly cigar. A faint field star lies just off its the northern edge, while two brighter stars lie one on either side. Arizona amateur Steve Coe notes that by using a 445-mm (17½-inch) telescope and averted vision, he is able to make out a dark lane running along the galaxy's edge.

Other galaxies adjacent to M77 and NGC 1055 include NGC 1072, 1073, 1087, 1090, and 1094 — an ample menu of offerings good for at least one evening's entertainment.

Lacerta

Lacerta, the Lizard, occupies an empty region of the early autumn sky between Cygnus and Andromeda. Though it contains no bright naked-eye stars, the rich Milky Way star fields in this region hold several open clusters, the finest of which is featured here. Lacerta culminates at midnight toward the end of August and is best seen on October evenings.

Few amateurs are familiar with the glittering treasure chest of **NGC 7209,** probably due to its remote location. Begin your quest for these stellar jewels at Deneb in neighboring Cygnus and scan about 9° east to 4th-magnitude Rho (ρ) Cygni. Continue the eastward trek another 6° until you come to a tight isosceles triangle of 5th- and 6th-magnitude stars. NGC 7209 lies about 1° to their north, near a lone 6th-magnitude star.

Nearly all amateur telescopes will show NGC 7209 well. In a 150-mm telescope the cluster appears as a warm glow peppered with many stars of 9th magnitude and fainter. E. S. Barker remarked in the *Webb Society Deep-Sky Observer's Handbook,* volume 3, that, through a 215-mm telescope, it is "difficult to define the borders; sparse star distribution to the south, ending in a starless lane." Larger instruments increase the count to almost 100 suns scattered loosely across 25'. Several faint pairs are found in the cluster's northeastern quadrant, while a lone yellowish point highlights the opposite border.

Pegasus

Marked by the famous "Great Square," Pegasus, the Flying Horse, hosts a lone globular cluster and several interesting galaxies. You'll find Pegasus flying past midnight culmination in late August, and prominent in the evening skies of October, November, and December.

M15 (NGC 7078) is a magnificent globular cluster lying in far-western Pegasus. It may be found by tracing a line from Theta (θ) Pegasi to Enif (Epsilon [ε] Pegasi) and continuing half again as far to the northwest. There you will see a 6th-magnitude star and, just to its west, a dim glow. That dim glow is M15. With a visual magnitude of 6.4, it should be visible through finderscopes without much trouble. A 150-mm telescope will display a few individual stars around the edges, even though the cluster is tightly structured. Larger telescopes, such as the 200-mm reflector used for Eyepiece Impression 33, will resolve many more stars across the cluster's 12' diameter.

Within M15 is a small, faint planetary nebula. Cataloged as **Pease 1,** it was discovered on photographic plates taken with the 100-inch (2.54-meter) telescope at Mount Wilson. Studies conclude that the planetary is actually within M15, not simply superimposed on it. *Burnham's Celestial Handbook,* volume 3, records the planetary as 1" in diameter, and shining faintly at photographic magnitude 13.8. This would certainly prove a challenge for amateurs with very large telescopes. Has Pease 1 ever been detected visually?

NGC 7217, a 10th-magnitude face-on spiral galaxy, lies in the barren regions of northwestern Pegasus. To find it, start off from Scheat (Beta [β] Pegasi), marking the Square's northwestern corner. Shift 12½° farther northwest to 6th-magnitude Pi^1 (π^1) and 4th-magnitude Pi^2 (π^2) Pegasi. NGC 7217 lies a little less than 2° to their south-southwest, about halfway between the pair of Pis and 6th-magnitude 23 Pegasi, and less than 1° northeast of an 8th-magnitude field star.

Figure 8.9. *Stephan's Quintet, a paradoxical family of five galaxies in Pegasus, lie about a half degree south-southwest of much brighter NGC 7331 (not seen in this photo). Kim Zussman captured this view of the Quintet nestled amid a field of faint stars. The irregular clump just left of center looks like one object but is actually two interacting galaxies. South is up.*

A 150- or 200-mm telescope will reveal this as a relatively large, circular glow highlighted by a bright central core. Even instruments three times this aperture fail to resolve the spiral arms, though some subtle mottling is evident within the disk region. A faint star is also seen superimposed on the galaxy's northern edge, while a magnitude-10.5 star lies about 3' to the galaxy's southeast.

NGC 7331 is the brightest of the dozens of galaxies scattered throughout and around the Great Square of Pegasus. Begin your quest at Eta (η) Pegasi, located just beyond the Square's northwestern star. Scan about 4° north-northwest of Eta for a pair of 6th-magnitude stars. NGC 7331 lies just over 1° to their south.

NGC 7331 strikes most amateurs as a tiny version of the Andromeda Galaxy, M31. In reality it is every bit as large as M31 but lies about 50 million light-years away (M31 is comparatively close, at 2.2 million light-years). Small telescopes show NGC 7331 as a sharply oval, 9th-magnitude smudge of grayish light that fades quickly away from a brighter core. A telescope of 200 mm or larger will expand the disk, but the arms themselves elude detection since

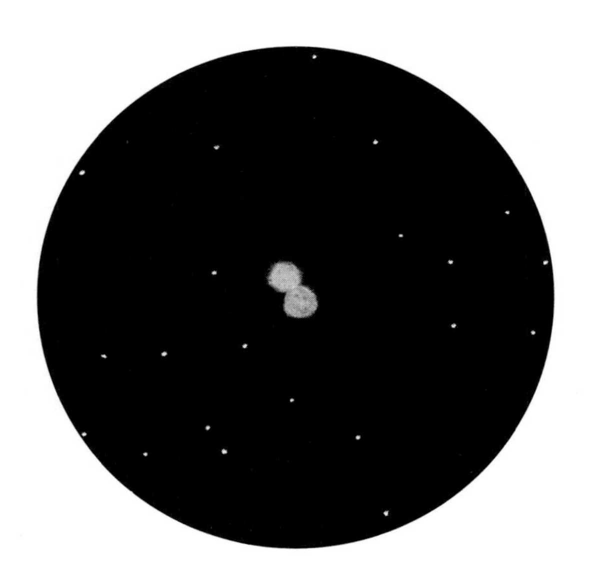

Eyepiece Impression 34. *M76, nicknamed the Little Dumbbell for its miniaturized resemblance to M27, lies in the constellation Perseus. Note the twin shells of nebulosity expanding from the object's center. This drawing was made by Dave Kratz through a 355-mm Schmidt-Cassegrain telescope at 140×. South is up.*

the plane of the galaxy is tilted only slightly to our line of sight.

Surrounding NGC 7331 are many fainter galaxies to challenge deep-sky observers. The brightest, 13th-magnitude NGC 7335, is located about 3.5' to the east-northeast. About 30' south-southwest of NGC 7331 is the famous Stephan's Quintet clan (Figure 8.9). This group, consisting of NGC 7317, 7318A, 7318B, 7319, and 7320, is visible in an instrument of 254 mm or larger as small fuzzy patches nestled among a collection of stars. (For a review of the many galaxies in the area, consult *Sky & Telescope*'s September 1988 and September 1991 issues.)

Perseus

Perseus, the Warrior, lies along the Milky Way just east of Cassiopeia. Within its borders are several of the season's finest open clusters as well as a wedge of bright nebulosity that can be notoriously difficult to find. Perseus passes midnight culmination in early November and is high in the early-evening skies of December and January.

M76 (NGC 650–51), nicknamed the Little Dumbbell for its resemblance to the Dumbbell Nebula (M27) in Vulpecula, is famous for having the faintest magnitude of all the Messier objects. This has led many to conclude, erroneously, that it is difficult to find and see. To find M76, trace the western arc that outlines Andromeda's body. At the end of this arc lies the 4th-magnitude variable star Phi (ϕ) Persei. Center Phi in your finderscope, then shift 1° north to a reddish 7th-magnitude field star. M76 lies just to that star's west.

Telescopes as small as 75 mm will show M76 as a tiny, faint point of fuzzy light, which, at first glance, may well look like an ordinary star. But upon closer examination, at 100× and up, most observers notice that this "star" just doesn't "look right." Its appearance in a 100- or 150-mm telescope is described by many observers as rectangular, oriented approximately north-south. To Dave Kratz of Poquoson, Virginia, who drew Eyepiece Impression 34, it looks like a

Figure 8.10. *NGC 884 (left) and NGC 869 (right) comprise the famous Double Cluster in Perseus. Although Messier apparently missed it, the Double Cluster is clearly visible to the naked eye on clear, dark nights and is striking through binoculars. This photo by Martin C. Germano is oriented with north up.*

celestial peanut! Larger instruments display M76 as two oval spheres of grayish light that appear to touch one another (hence the double designation of NGC 650 and NGC 651). For Arkansas amateur Dean Williams, viewing through his 200-mm f/10 Schmidt-Cassegrain telescope, "the best image of the nebula was obtained at 161×. The dumbbell shape was easily observed. M76 shows a more definite separation of its two halves than M27, and is more elongated in appearance."

So, *never* be intimidated by an object's listed magnitude. In some cases, such as the spiral galaxy M33 in Triangulum (described later in this chapter), the listed magnitude creates an expectation that an object will be brighter, and therefore more accessible, than it actually appears. With others, such as M76, it misleads observers into thinking that an object will be fainter and hence more elusive than it appears. This is not to say that M76 is bright — just that it appears brighter than you might expect for a 12th-magnitude planetary nebula.

NGC 869 and **NGC 884.** No discussion of the finest deep-sky objects would be complete without paying homage to NGC 869 and 884, the Double Cluster — two open clusters less than 1° apart (Figure 8.10). Few autumn clusters compare to either NGC 869 or 884, but taken together they provide us with a view that overflows with splendor. They are most easily found by tracing a line from Gamma (γ) Cassiopeiae to Ruchbah (Delta [δ] Cassiopeiae), respectively the center and southeastern stars in the Cassiopeia W, and ex-

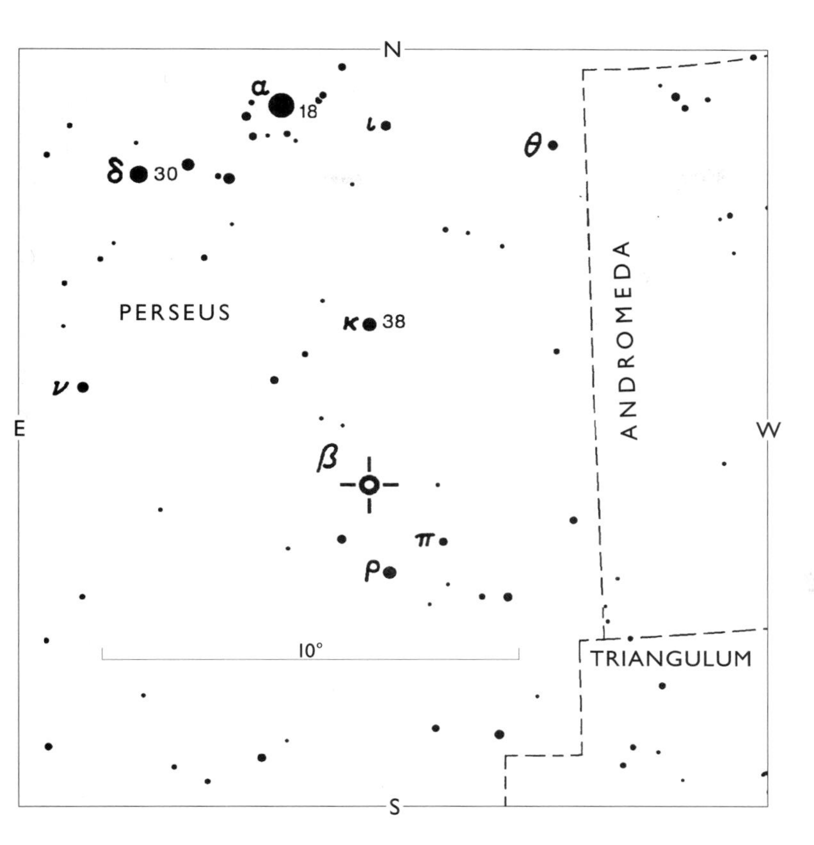

Figure 8.11. *This chart, from the* AAVSO Variable Star Atlas, *shows the position of the eclipsing binary star Algol [Beta (β) Persei]. The surrounding stars' magnitudes are given with decimal points omitted. North is up. Courtesy the American Association of Variable Star Observers.*

tending the line toward the northern tip of Perseus. You'll find the clusters about halfway between the constellations. Under even moderately dark skies, their combined glow is visible without optical aid as a faint patch in the Milky Way.

Messier never recorded the Double Cluster (if he really did miss it, I will never understand how), but its discovery dates to antiquity. Records show that Hipparchus was first to record its existence as far back as the second century B.C. Today, even the smallest of instruments begins to unleash the beauty that is the Double Cluster. The finest view I have ever had of them was not too many years ago at the Stellafane amateur telescope makers' convention in Springfield, Vermont. Through a pair of 11 × 80 binoculars, the clusters took on a fiery radiance against the star-filled backdrop. A few stars shining with subtle hues of yellow and red could be seen scattered throughout a field dominated by blue-white suns.

NGC 869 and 884 do not lie right next to each other in real space, though they are close. The former is estimated to be 7,200

Figure 8.12. *NGC 1499, nicknamed the California Nebula for its likeness to the Golden State, is an astronomical oxymoron. While the deep-red cloud is clear in photographs, visual observers will find it difficult to view on even the darkest nights. South is to the right in this photo by Chuck Vaughn.*

light-years from Earth, while the latter is estimated at 7,500 light-years away. More surprising is their disparity in age, with NGC 869 thought to be 5.6 million years old and NGC 884 a comparative infant at 3.2 million years.

M34 (NGC 1039) is a wonderful cluster to view through telescopes and binoculars alike. Visible to the naked eye on exceptional nights as a faint smudge near the Perseus-Andromeda border, M34 explodes into stardust with only the slightest optical aid. Small finderscopes will show M34 as a misty glow a little less than halfway from Algol (Beta [β] Persei) to Almach (Gamma [γ] Andromedae). In fact, 8 × 50 and larger finders will reveal a number of individual stars within the cluster set against a nebulous backdrop of fainter, unresolved suns. To me, the group as a whole looks decidedly rectangular, while some observers comment about strings of stars that run across the object. In his book, *The Universe from Your Backyard,*

David Eicher notes three "arms" of stars extending away from the cluster's center. Can you spot this unusual feature?

Two binary stars highlight the view through most amateur instruments. Almost centrally located within M34 is the double **h1123,** a pair of 8th-magnitude whitish stars separated by 20". The second stellar duet, designated **OΣ 44,** is a closer pair of white stars set only 1.4" apart. Look for it a little to the southeast of center.

Algol (Beta [β] Persei) was believed to be possessed by evil powers in ancient times, as early astronomers watched in fear and wonder while the star changed in brightness. Today we realize that Algol is an eclipsing binary, with a smaller star orbiting a larger one. As seen from Earth, the orbit is tilted almost edge on, causing the secondary star to pass alternately in front of and behind the primary. Although the stars are too close to one another to be resolved, we see these passages as fluctuations in the system's total brightness. Over a period of 2.87 days, Algol's brightness fluctuates between a maximum magnitude of 2.1 and a minimum of 3.4. The greatest dip in magnitude happens when the fainter secondary passes in front of the brighter primary, temporarily blocking part of its light. Each of these eclipses lasts about ten hours, of which about two hours are spent at minimum brightness. A secondary dimming amounting to only about 0.1 magnitude occurs as the companion star is eclipsed by the primary.

Figure 8.11 shows a chart that may be used to estimate the magnitude of Algol in comparison to some nearby stars of fixed magnitude. For readers interested in monitoring the changing face of Algol, *Sky & Telescope*'s Web site, SKY Online, lists the dates and times of minimum and maximum magnitudes; the URL address is http://www.skypub.com/whatsup/algol.html. The information is also available in the Royal Astronomical Society of Canada's annual *Observer's Handbook.*

Melotte 20 is better known to most amateurs as either the Alpha (α) Persei Moving Cluster or the Alpha Persei Association. Call it what you will, this huge collection of about 70 loosely bound stars is scattered across 3° of central Perseus. Because of its wide girth, this stellar congregation is best viewed through a finderscope or low-power binoculars. The stars in Melotte 20 shine at 10th magnitude or greater, so nearly all are visible through 7× binoculars on a clear, dark night. Many are set in close-knit duets, trios, and quartets and add greatly to the group's visual impact. Brightest of all is 3rd-magnitude Mirfak (Alpha Persei), the most brilliant gem in this celestial diamond mine. Studies indicate that the age of Melotte 20 is only about 51 million years, younger than many diamonds found here on Earth.

NGC 1499, the California Nebula, is a paradoxical object (Figure 8.12). Many catalogs note it as 6th magnitude, yet it is one of the faintest bright nebulae in the sky — an astronomical oxymoron. It frequently eludes observers using 305- or 355-mm telescopes, yet under dark rural skies it can be glimpsed with binoculars! Although hard to see, the California Nebula is easy to locate as it lies just north of 4th-magnitude Xi (ξ) Persei, its illuminating star. Our difficulty in distinguishing the California Nebula stems from its large extent of

Figure 8.13. *NGC 253, a spiral galaxy belonging to the Sculptor Group of galaxies, resides in the far southern autumn sky. The mottled appearance of the galactic disk is prominent in this photo by Kim Zussmann. South is to the upper left.*

145' × 40'. While its combined light gives it an integrated magnitude of 6, the surface brightness is a mere magnitude 14 per square arcminute. Still, in ideal conditions several keen-eyed observers have spotted it with 10× binoculars and nebula filters.

One especially clear January night a couple of winters ago, my daughter's 108-mm f/4 rich-field telescope equipped with a 26-mm Plössl eyepiece and a hydrogen-beta filter successfully revealed it. The shape of the entire "state," while extremely subtle, could be seen against an ocean of stars. As my eyes became more accustomed to the scene, some of the cloud's delicate texture could also be discerned. Indeed, the California Nebula is a truly memorable sight for those persistent enough to find it.

Sculptor

Sculptor, one of the sky's least conspicuous constellations, scrapes the southern horizon of mid-northern latitudes at this time of year. With none of its stars shining brighter than 4th magnitude, just finding the constellation is a tough enough test for most. Yet within this seemingly empty region are some of the season's most exciting objects. Bear in mind that, since the visibility of these heavenly sights will be adversely affected by atmospheric extinction because of their low altitude in Northern-Hemisphere skies, it is best to plan your observing session when they are at or near culmination. Sculptor culminates at midnight in late September and is ideally placed for early-evening observation during November and early December.

NGC 55, one of the showpiece galaxies of the autumn sky, goes ignored and unappreciated by most northerners because of its extreme southerly declination and great distance from bright reference stars. If viewed from a southern vantage point, however, it is sure to become a seasonal favorite. NGC 55 is most easily located by starting at Fomalhaut (Alpha [α] Piscis Austrini). Viewing through your finder, slide 5° southeast to 4th-magnitude Gamma (γ) Sculptoris, then 6° farther southeast to 4th-magnitude Beta (β) Sculptoris. Shift 7° east-southeast to a close pair of 7th-magnitude stars, then another 1½° farther east to NGC 55.

When the galaxy is highest in the sky, even an 8 × 50 finderscope will reveal its long, needle-thin disk. In fact, NGC 55 is so elongated that most telescopes need a low-power, wide-field eyepiece to squeeze its full ½° extent into the field of view. Once you have spotted it, switch to a medium-power eyepiece for a better look at the spiral arm halo. Medium- and large-aperture instruments reveal NGC 55 to be an amazingly intricate object riddled with many bright and dark patches.

NGC 253 is an exciting galaxy to view through just about any instrument (Figure 8.13). A member of the Sculptor Group of galaxies (as is NGC 55), NGC 253 has been described by many as the Southern Hemisphere's answer to the Andromeda Galaxy. To find NGC 253 first locate the bright star Diphda (Beta [β] Ceti), which lies in a barren part of the southern autumn sky well below the Great Square of Pegasus and northeast of the brighter star Fomalhaut in Piscis Austrinus. From Diphda, move 7° due south, watching through your finder for a triangle of three 5th-magnitude stars and, farther south, a diamond of 6th-magnitude stars. Just west of the diamond's southern point, look for a half dozen 7th- and 8th-magnitude stars set in a bow-tie pattern. NGC 253 lies within that bow tie.

Even from northern observing sites, NGC 253 stages an impressive performance. Its oval nucleus and long, slender galactic disk shine brightly in a 150- or 200-mm instrument, while 254 mm or larger will disclose the entire 25' × 7' span of this magnificent object. As the galaxy reaches culmination, a mottled surface texture reminiscent of NGC 55 may be seen through many backyard instruments. Sprinkled across the galactic disk are many foreground stars, including the two stars that form the bow tie's knot.

NGC 288. Time for a break. In the midst of all the galaxies strewn across the autumn sky is this globular cluster, a refreshing deep-sky oasis. Be forewarned, however; NGC 288 is not as easy to spy as

Eyepiece Impression 35. *M33, the great spiral galaxy in Triangulum, is the nemesis of many first-time deep-sky observers. This drawing, made through the author's 333-mm Newtonian and 24-mm Tele Vue Wide Field eyepiece (63×), shows subtle hints of the spiral arms and other latent features. The galaxy's largest bright nebula, NGC 604, is depicted here as a detached smudge just east (right) of the galaxy.*

many other globulars. Its southerly declination, coupled with its large apparent diameter and inherent dimness, can challenge even seasoned observers. As with NGC 253, mentioned above, begin your hunt at Diphda (Beta [β] Ceti). Tumble southward, passing the stellar signposts for NGC 253 as well as the galaxy itself, and continue until you strike 4th-magnitude Alpha (α) Sculptoris. Turn back northward and, peering through your finder, hop to a 6th-magnitude star about 1½° north-northeast of Alpha. Hop an equal distance farther north-northwest and take a look through your telescope. NGC 288 ought to lie in, or at least near, the field of view. In fact, given dark skies, it may even be visible in a finder.

A 200-mm telescope will begin to resolve some of NGC 288's individual stars, but use averted vision for the best view. Not surprisingly, larger scopes will resolve more cluster stars, but your observing location can make an even bigger difference. For instance, Geoff Chester notes that, observing from Virginia with a 200-mm f/6

Figure 8.14. *Use this photograph to identify some of the myriad star clouds and nebulae that populate M33. As you can see from the labels, many carry their own NGC and IC designations. The easiest to spot visually is NGC 604, located around 5 o'clock (below right) of the galaxy's nucleus. South is up. Photo by Alfred Linge.*

Newtonian, NGC 288 "resolves nicely" through a 15-mm eyepiece. Contrast this with Sue French's impression, who saw NGC 288 as a "large, round, faint patch; some mottling visible at higher power" with her 254-mm telescope in Glenville, New York.

Triangulum

Triangulum, the Triangle, occupies a small wedge of the autumn sky and goes unnoticed by most casual skywatchers. For the deep-sky aficionado, however, this tiny constellation is big-game territory, for within its limited reaches lies one of the best-known — and potentially most frustrating — galaxies in the sky. Triangulum reaches midnight culmination toward the end of October and is best placed for early-evening observation between November and January.

M33 (NGC 598). Few of the Messier objects cause as much consternation among new deep-sky observers as does this face-on Sc spiral. An integrated magnitude of 5.7 implies that M33 should be not just visible, but easily visible in just about all binoculars and telescopes. But is it? To find out for yourself, center your finderscope on Alpha (α) Trianguli, then shift about 2½° west until you come to a 6th-magnitude star. Heading 3° farther west you'll come to a 7th-magnitude star. M33 lies a little north and about two-thirds of the way from the 6th-magnitude star to the 7th.

This sounds simple enough, yet when observers turn their instruments its way, the galaxy is nowhere to be found. Why? Like many other greatly extended deep-sky objects, M33 has a very low surface brightness. Its apparent diameter is huge — a full degree. As a result, the small field of view of most telescopes causes many amateurs to pass right by without noticing it. In cases like this, binoculars and rich-field telescopes enjoy a great advantage over other instruments. Instead of seeing only a small portion of a large target, as is the case with most telescopes with focal ratios of f/6 or greater, these instruments squeeze the entire object into a single field with room to spare. Even then, and using averted vision, most observers see M33 as little more than an ill-defined blob at first.

As the observer's eye becomes more accustomed to the galaxy's appearance, M33 begins to yield riches unsurpassed by any other galaxy north of the celestial equator. In a high-quality 200-mm instrument fitted with a low-power eyepiece under dark skies the two delicate spiral arms can be seen unwinding from the core; Eyepiece Impression 35 shows the view through a 333-mm scope. (Note, however, that averted vision will probably be necessary to see the arms through telescopes smaller than about 305 mm.) A careful search of the disk region will turn up numerous bright pockets of nebulosity, as shown in Figure 8.14. Many of these carry separate NGC and IC designations and are either large regions of bright nebulosity or star clouds. The brightest of the two-dozen-plus such patches visible through a 200- or 254-mm telescope is separately cataloged as **NGC 604.**

I find it ironic that NGC 604 is one of brightest nebulae in the midautumn sky, yet it is not even in our galaxy, lying as it does an estimated 2.7 million light-years away. And while M33 is infamous for its low surface brightness, NGC 604 stands out surprisingly well. In fact, under slightly hazy skies, the nebula may actually be visible while the galaxy is nowhere to be found! Over a thousand times the size of the Orion Nebula, NGC 604 is visible in 8 × and larger binoculars and finderscopes as a small, slightly fuzzy "star." Telescopes expand it into a soft blur set to the northeast of the galaxy's core. An instrument of 254 mm or larger will reveal this mammoth H II region to be noticeably oval, with bright concentrations. The largest amateur telescopes may just be able to detect the brightest members of the nebula's associated cluster, shining between 15th and 18th magnitude. Try a high power for the best view.

Although the autumn sky may lack the naked-eye "pizzazz" of summer and winter evenings, it is well endowed with a wide variety of telescopic treats. Sure, some may take a little longer to find, but each will pay off richly in the end.

CHAPTER 9

A winter deep-sky wonderland

As most of the Northern Hemisphere prepares for the first snow to blanket the land, amateur astronomers are readying their equipment to greet the cold, dark nights of winter (Figure 9.1). Gone are the haze and smog that hang over us during the warmer months. In their place clear, frigid air lets starlight cascade toward our waiting eyes.

As the Milky Way blazes overhead, many observers take the opportunity to revisit some old stellar friends scattered across the sky. Other amateurs prefer to stretch both their telescopes and their skills to the limit by probing the depths of space for challenging targets. Still others find that the true challenge of winter observing is simply to stay outside for more than a few minutes without freezing solid! Regardless of your personal observing agenda, the celestial wonders offered here are sure to warm the soul of any deep-sky observer on even the coldest nights.

Auriga

Marked by the brilliant star Capella, the constellation Auriga, the Charioteer, rides in the soft glow of the winter Milky Way. It is littered with many fine open star clusters that offer hours of fascinating viewing for deep-sky observers. Auriga reaches midnight culmination in late December and is high in the sky for evening observation during January and February.

STAR 4. The first stop on our tour of Auriga is an asterism created from 16, 17, 18, 19, and IQ Aurigae that I referred to as Harrington 4 in *Touring the Universe Through Binoculars*. Massachusetts amateur Jack Megas was first to point out this pattern to me, at the 1987 Astronomer's Conjunction annual convention in Northfield, Massachusetts. From a relatively dark observing site, the collective light from these stars merges into a naked-eye blur seen a little to the lower right of the constellation's center. Binoculars resolve a dozen or more stars in an area 75' × 30' across. In his Backyard Astronomy column in *Sky & Telescope,* associate editor Alan M. MacRobert once likened the stars' overall pattern to a "flying minnow." What shape can you see among these stars?

M38 (NGC 1912), the westernmost of Auriga's three Messier open clusters, is easy to spy through even the smallest telescope. It

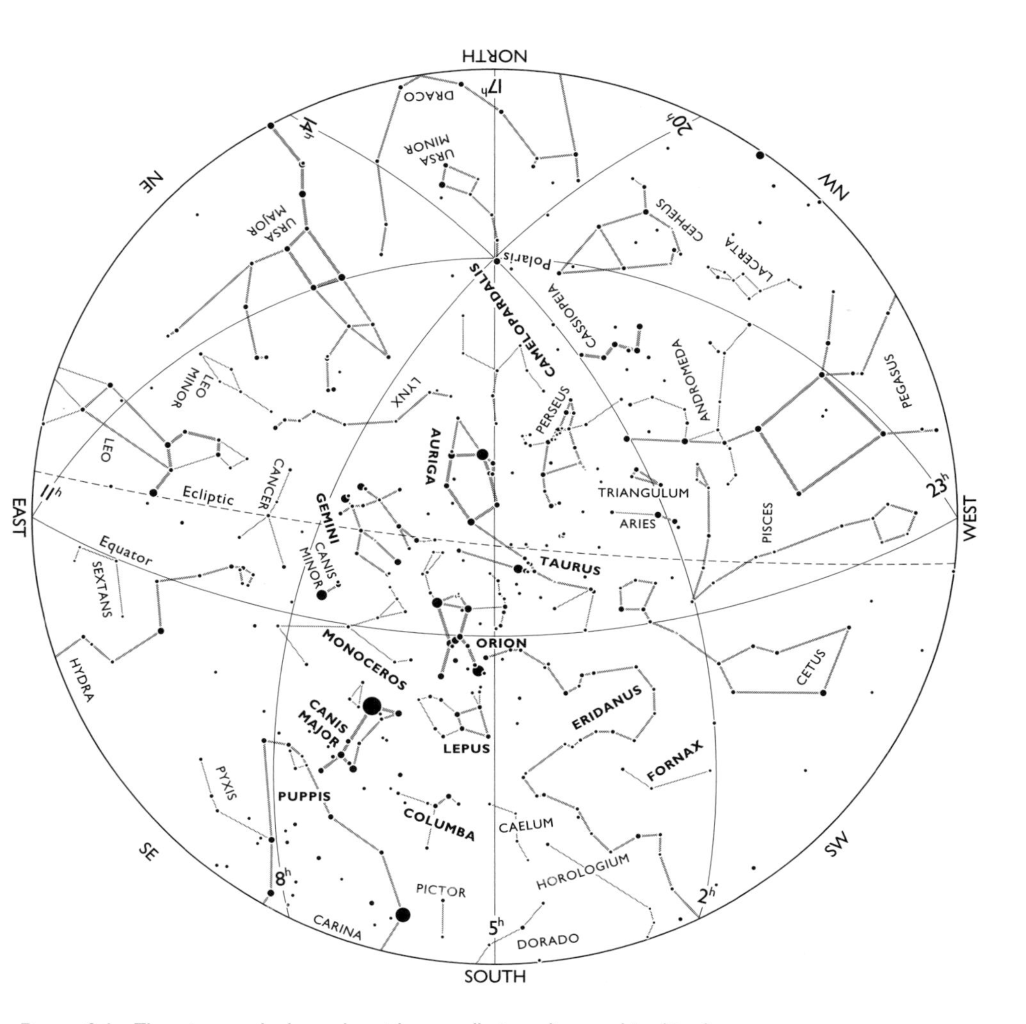

Figure 9.1. *The winter naked-eye sky, with constellations discussed in this chapter shown in boldface. The chart is plotted for 30° north latitude. For viewers north of that latitude, stars in the northern part of the sky will be higher and stars in the south lower (or possibly below the horizon).*

can be found about a third of the way from Phi (ϕ) to Sigma (σ) Aurigae, a pair of 5th-magnitude stars near the center of the constellation's pentagonal form. There, look for a 6th-magnitude star and, just to its northwest, the glow of M38. Through most finderscopes and binoculars, M38 appears as an ill-defined nebulous patch. At higher aperture and magnification the nebulous effect disperses, leaving a pleasant open cluster in its wake.

Over the years, the 100 stars that form M38 have been likened to several different shapes. Perhaps the one that is most often alluded to is an oblique cross, highlighted by a bright star at its center. Other observers see it more as the Greek letter pi (π). Can you see a

Figure 9.2. *M37 is the richest of the three Messier open clusters that call the constellation Auriga home. Telescopes with 100-mm and larger apertures resolve the group into a swarm of tiny points highlighted by a 9th-magnitude orange-red sun. This photo taken by Martin C. Germano is oriented with south up.*

definite pattern among the stars of M38? While in the area, be sure to stop by **NGC 1907,** another nice open cluster about 1° south of M38.

NGC 1931. Let's pause momentarily from our open-cluster tour to take in this surprisingly bright patch of nebulosity set about halfway between Phi (ϕ) Aurigae and a 6th-magnitude star to its east. The high surface brightness of NGC 1931 allows easy capture through most backyard telescopes. At first glance it appears as a perfectly circular misty patch of light surrounding a lone magnitude-11.5 star. At high magnification a slight ellipticity may be detected, with the major axis running from northeast to southwest. Four central stars can then be faintly spotted as well. David Allen reported in the *Webb Society Deep-Sky Observer's Handbook,* volume 2, that these points of light reminded him of a miniature version of the famous Trapezium multiple-star system in the Orion Nebula.

Open cluster **M36** (NGC 1960) is a fine grouping of stellar fireflies that is easy to spot with just about any optical aid. From Beta (β) Tauri, at the tip of the Bull's northwestern horn, scan 4° north-northwest to Chi (χ) Aurigae (itself an easy optical double star for binoculars and finderscopes). M36 lies another 2° north-northwest, nestled between a pair of 6th-magnitude field stars. (Note that since it is shared by both Taurus and Auriga, Beta Tauri is also designated Gamma [γ] Aurigae on some star charts.)

Although a little smaller than M38, M36 outshines its neighbor by about a magnitude, and stands out well against its star-rich surroundings. A 200-mm (8-inch) telescope reveals more than three dozen stars spanning about 20' set in a crooked Y pattern. To Kenneth Glyn Jones, author of *Messier's Nebulae and Star Clusters,* M36 appears similar to "the constellation Perseus in miniature." Take a look south of the cluster's center for the double star Struve 737 (Σ737). See if you agree with Adm. Smyth, who in the 19th century described it as "a neat double star . . . both white." Just over 10" separates its two 9th-magnitude components, making them resolvable in just about all amateur telescopes.

M37 may be found by drawing an imaginary line between Beta (β) Tauri and Theta (θ) Aurigae to its north. Move to the halfway point between the two, then drift due east, where you'll come upon M37 just north of a 6th-magnitude field star. This exceedingly rich open cluster is comprised of about 150 stars scattered across nearly ½°, with most shining between about 9th and 13th magnitude (Figure 9.2). Although small instruments show M37 as an unresolved swarm of stardust, the cluster puts on a dazzling show through telescopes of 100 mm (4 inches) and larger. These reveal many fainter points crammed tightly together and surrounded by a striking backdrop of field stars. The group's most distinguished member, a central 9th-magnitude stellar ember, beams a radiant orange-red. Subtle tinges of color are also suggested by some of the fainter cluster stars. Perhaps the Rev. Webb summed up this cluster best in his *Celestial Objects for Common Telescopes:* "even in small instruments, [M37 is] extremely beautiful; one of the finest in its class. Gaze at it well and long!"

Camelopardalis

One of the most difficult constellations to discern with the unaided eye, Camelopardalis, the Giraffe, contains several interesting albeit infrequently visited deep-sky objects that reward those who seek them out. The constellation reaches midnight culmination in early February. It is best placed for evening observation toward the end of winter, though for most northern observers much of the constellation is circumpolar.

STAR 3 is an outstanding pattern of stars hidden in the vast emptiness of Camelopardalis. Draw a line from Gamma (γ) Cassiopeiae to Epsilon (ε) Cassiopeiae and continue for 12° to their east. Peering through binoculars or a finderscope, you should spot a stream of about two dozen stars from 5th to 9th magnitude pouring over about 2½° of the northern winter sky. According to Walter Scott Houston, this long string of stars was also first noticed by Father Lucian Kemble back in 1980. In his honor, Houston proposed calling it Kemble's Cascade, an apt name indeed. Sharp-eyed binocularists may notice a hazy smudge toward the southeastern end of Kemble's Cascade. This smudge is the open cluster NGC 1502, the next stop on our tour of Camelopardalis.

NGC 1502 hosts some 45 stars in an 8' area. Smaller backyard telescopes are capable of resolving only about a quarter of the cluster

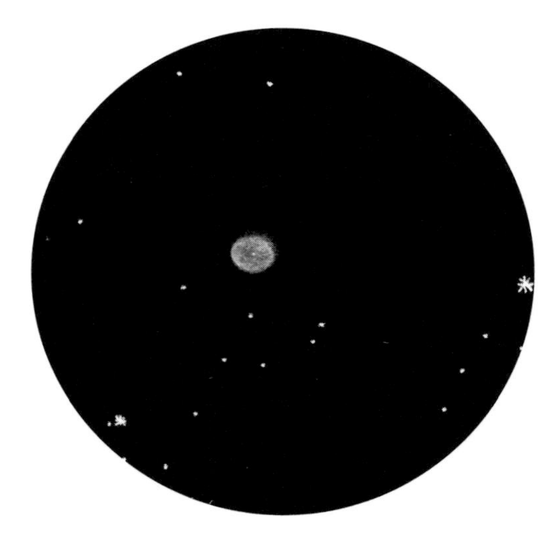

Eyepiece Impression 36. *NGC 1501, noted for its beautiful blue color, is a victim of circumstance. Even though it is one of the season's brightest planetary nebulae, few amateurs ever take the time to hunt it down in the dim constellation Camelopardalis. Drawing by Dave Kratz through a 355-mm Schmidt-Cassegrain telescope at 140×. South is up.*

members into individual points of light, while 254-mm (10-inch) and larger apertures increase the count to more than 40. All telescopes reveal a pair of 7th-magnitude stars (cataloged as ADS 2984) near the cluster's center. The system's B star, also known as **SZ Camelopardalis**, is an eclipsing binary that fluctuates between magnitudes 7.0 and 7.3 with a period of 2.7 days. These changes are probably too subtle to be noticed visually except by experienced variable-star observers.

NGC 1501 is one of the brightest planetary nebulae of the season, though finding it is not necessarily an easy task. First locate the open cluster NGC 1502 as outlined above. Move southward from NGC 1502 along a line of three 7th-magnitude stars, then hop about 1° to a lone 7th-magnitude point. NGC 1501 lies just to this final star's northwest.

A 150-mm (6-inch) telescope reveals NGC 1501 as a perfect spot of dim bluish light about 1' across. My 200-mm f/7 Newtonian adds a small amount of detail to the planetary's disk, with just a hint of a brighter outside edge seen at high powers. The nebula's central star, shown faintly in Eyepiece Impression 36, glows weakly at 13th magnitude and requires at least a 254-mm telescope to be seen unless sky conditions are extraordinary.

NGC 2403 is one of the galactic gems of the late winter sky. Situated in southeastern Camelopardalis far from any bright naked-eye stars, NGC 2403 proves challenging to hunt down. Your best chance for sniffing it out is to begin at Omicron (o) Ursae Majoris, the star that marks the nose of the Great Bear. Wind northwestward along a crooked 8° line of three 6th-magnitude stars and pause at the end star, 51 Camelopardalis. Even through suburban skies, the galaxy itself may be visible in an 8 × 50 or larger finderscope as a faint smudge about a degree west of 51. Telescopes reveal that it is just east of a close pair of 9th-magnitude stars.

Eyepiece Impression 37. *The spiral galaxy NGC 2403 is an oft-neglected jewel of the late-winter sky yet is bright enough to be seen through finderscopes. The author made this drawing through his 200-mm Newtonian and 12-mm Nagler eyepiece (116×).*

Through most backyard instruments NGC 2403 appears as shown in Eyepiece Impression 37 — a large oval disk oriented northwest-southeast and accented by a bright core. Two stars, each around 11th magnitude, appear to stand in attendance on either side of the galaxy. In long-exposure photographs NGC 2403 covers an area of 18' × 10', though it will probably appear smaller to visual observers. A 254-mm instrument will begin to reveal some of this type-Sc galaxy's delicate spiral structure, the most prominent arm extending toward the southwest. Some observers also report sighting a dust lane crossing just north of the galactic core.

Canis Major

Silhouetted in front of the Milky Way and standing obediently by the side of Orion, its master, is the familiar winter constellation of Canis Major, the Big Dog. This portion of the southern winter sky hosts several fine open clusters and other galactic objects. And while you may already be familiar with the first target listed, several lesser-known deep-sky wonders also await your arrival. Canis Major, spotlighted by the brilliant star Sirius (Alpha [α] Canis Majoris), reaches midnight culmination early in January, and is ideally placed for early-evening observation in February and March.

M41 (NGC 2287) was one of the first deep-sky objects ever recorded. According to *Burnham's Celestial Handbook,* volume 1, it was known as early as 325 B.C., when Aristotle described "a cloudy spot" to the south of Sirius. We may make a similar naked-eye observation on any winter night when interfering lights and clouds are absent. Look about 4° due south of brilliant Sirius. M41 is one of the most dazzling open clusters in the sky regardless of what instrument is used (Figure 9.3). Even through suburban skies, low-power binoculars can pick up about two dozen of the cluster's stars, while through any telescopes the cluster bursts with splendor.

On their first encounter with M41, many observers immediately

Figure 9.3. *M41 in Canis Major is one of the brightest and most colorful open clusters that the winter sky has to offer. Awareness of its existence predates the telescope by more than a millennium: Aristotle first noted it as a "cloudy spot" with his unaided eye. South is to the upper left. Photo by Martin C. Germano.*

notice that some of the brightest members create a central figure that is strongly reminiscent of the "keystone" in Hercules. In addition, many of the cluster stars are arranged in close pairs and triples. When the image is defocused slightly, many are seen to shine with hints of yellow, orange, and blue-white. Most of M41's members are crammed into 30', though some stragglers expand the diameter to nearly 40'. I wonder what Aristotle's reaction would have been had he known just how beautiful his "cloudy spot" actually is?

Herschel 3945. Provided your southern horizon is not blocked by some earthly obstruction, Herschel 3945 (sometimes abbreviated h3945) will prove to be an easily resolvable double star that is ideal for small telescopes. Center your aim on 2nd-magnitude Wezen (Delta [δ] Canis Majoris), the apex of the triangle of bright stars that marks the dog's hindquarters. Scan 3° to the northwest to 3rd-magnitude Omicron^2 (o^2) Canis Majoris, then turn 3° east-northeastward to our target star. Herschel 3945 marks the right angle of a small triangle formed with two unrelated 6th-magnitude stars. Through my 100-mm refractor, the system's 5th-magnitude primary appears orangish, while the 7th-magnitude companion shines with a subtle bluish tint. With the pair separated by nearly 30", even tripod-mounted 10 × binoculars can resolve them.

NGC 2359, an emission nebula tucked away in a corner in northeastern Canis Major, is overlooked by many amateurs in favor of other attractions that are easier to find. This is a pity; if you take the time to search for it, I promise you will not be disappointed. NGC 2359 is most easily approached by casting off from Sirius (Alpha [α] Canis Majoris), hopping eastward to Iota (ι) Canis Majoris, then northeastward to Gamma (γ) Canis Majoris. Scan 3° due east to a 5th-magnitude star, then veer northward through 2½° to a group of four 8th-magnitude stars. NGC 2359 lies less than 1° to their north.

My 200-mm f/7 reflector shows the nebula to be trapezoidal in shape; a bright, misty cloud spangled with many stars. Some refer to this as Thor's Helmet, while others, for reasons that elude me, simply call it the "Duck." Whatever your preferred nickname, be sure to search within the cloud for the 11th-magnitude blue-white Wolf-Rayet star that is responsible for exciting the nebula to shine. Larger telescopes and photographs also bring out faint filaments extending across the cloud.

Collinder 140. Although it is known to few observers, this bright, large open cluster is an ideal target for binocularists. Collinder 140 is set just inside the extreme southeast corner of the constellation, meaning that a good view to the south is required to spy it. Scan about half a binocular field (about 4°) due south of the star Aludra (Eta [η] Canis Majoris) until you spot a right triangle of 5th-magnitude stars set among a smattering of fainter points of light — all in all, 30 stars of 9th magnitude and brighter. Low-power glasses give the best view.

Columba

Another low-riding winter constellation is the faint pattern of Columba, the Dove. Columba flies just above the horizon in mid-northern latitudes, south of the prominent constellation Orion and the fainter Lepus. The Dove carries with it one of the few globular clusters visible during this season. Many galaxies are also scattered through the constellation, though they are all too faint to be included in this brief survey. Columba culminates at midnight in late December and is best suited for evening observation in January and February.

NGC 1851 is a challenging globular cluster that is sure to test your skills as an observer. To try your luck with NGC 1851, first find the constellation! Three of Columba's brightest stars form a slender triangle that scrapes along the southern horizon from midnorthern latitudes. From Epsilon (ε) Columbae, the triangle's southwestern star, shift southward to another triangle formed from three 6th-magnitude suns. From this triangle's southwestern star, slide about 1° to the southwest to a smaller triangle of three 7th-magnitude stars, then, finally, another degree in the same direction to NGC 1851. Keep in mind that, while NGC 1851 shines at 7th magnitude, it never reaches an altitude greater than 10° above the southern horizon as seen from the middle latitudes of the United States and southern Europe. Unless atmospheric conditions are ideal, the cluster free of haze and clouds, it will likely remain lost from view.

Astronomers describe NGC 1851 as a Class II globular, indicating

Figure 9.4. *NGC 1300 in Eridanus is one of the sky's finest examples of a face-on barred spiral galaxy. Detecting the spiral arms visually, however, requires a large telescope and keen eyesight. South is up. Photo by Chuck Vaughn.*

that its stars are very densely packed. A 200-mm telescope will reveal a very bright central core, fading rapidly toward the group's edges, but with no hint of stellar resolution. A larger instrument is needed to resolve any of the cluster's individual stars, which are tightly packed and faint.

Eridanus

Eridanus, the River, is one of the sky's longest constellations, flowing from the western edge of Orion southward below most of our horizons. Only readers who are south of approximately 30°N latitude can spot the River's lone bright star, Achernar [(Alpha (α) Eridani]. Several deep-sky islands lie in Eridanus, including the galaxy, planetary nebula, and pair of double stars that are detailed below. Eridanus reaches midnight culmination in mid-November and is best seen in the early evening during December and January.

Theta (θ) Eridani. Even though it doesn't stand out from the crowd of faint naked-eye stars that form this great celestial river, 3rd-magnitude Theta Eridani (also known as Acamar) is an impressive double star when viewed through amateur telescopes. Here we find a pair of white-hot Type-*A* suns, with the 3.4-magnitude primary separated from the 4.5-magnitude secondary by a little more than 8". While these two stars have changed little in separation or position angle since they were first measured by John Herschel in 1835, there seems little doubt that they form a true binary system.

NGC 1300, with its pair of majestic spiral arms pinwheeling off the end of a prominent bar of starlight, is often presented in astronomical texts as the quintessential face-on barred spiral galaxy (Figure 9.4). While its outstanding form is well recorded on long-exposure photographs, finding it is a formidable task for many observers. Unfortunately there's just no easy way to zero in on its barren corner of the winter sky, but here's one suggestion. Begin at Lepus, the Hare, to the south of Orion. Viewing through your finderscope, draw a line from Beta (β) Leporis to Epsilon (ε) Leporis and continue it to the west for 17° until you arrive at Tau^6 (τ^6), Tau^7 (τ^7), Tau^8 (τ^8), and Tau^9 (τ^9) Eridani, a distinctive foursome of 5th-magnitude stars. Follow the curve of the celestial river to the northwest, passing Tau^5 (τ^5), to pause at Tau^4 (τ^4) Eridani. Turn north for about 3° to a close pair of 6th- and 7th-magnitude stars. NGC 1300 lies just 1° to their south.

NGC 1300 is visible through most backyard instruments as an oval patch of dim, gray light. Only when viewed through large amateur telescopes will the galaxy's true nature be apparent. A 455-mm (18-inch) instrument reveals the faintest hint of spiral arms curving away to the north and south of the galaxy's core. Keen-eyed observers may also spy a second, fainter galaxy just northwest of NGC 1300. This is **NGC 1297,** an elliptical system that shines about two magnitudes fainter than NGC 1300. It displays a 2' disk, and a nucleus can be made out.

32 Eridani. Although also made difficult to home in on by the lack of any nearby bright stars, the beauty of 32 Eridani makes it well worth the search. Move due west of brilliant Rigel (Beta [β] Orionis) for about 15° to Omicron^1 (o^1) and Omicron^2 (o^2) Eridani, a pair of 4th-magnitude suns. Draw a line from Omicron^2 to Omicron^1 and continue it to the northwest for about 6° to reach the 4th-magnitude 32 Eridani. With its 4.7-magnitude primary and 6.2-magnitude secondary separated by 7", 32 Eridani is easily resolved with a power of 50 × or so. The primary is a Type-*G* star, the same as our Sun, while the secondary is a Type-*A* star. Some observers, however, describe the secondary as slightly greenish, clearly an illusion as there are no such things as green stars! Perhaps the perception of green is an artifact of the color contrast with the yellow primary.

NGC 1535. This 9th-magnitude planetary nebula distinguishes itself as one of the season's finest even though its home constellation is nearly indistinguishable. With your finderscope, scan about 1½° west-southwest of Rigel to a short arc of three stars aligned east-west. The brightest star in the arc, Lambda (λ) Eridani, shines at 4th magnitude. Continue westward past the arc for 8½°, passing lone 6th- and 7th-magnitude stars, until you come to a slender north-south triangle of 5th- and 6th-magnitude stars. Continue another 4° farther west to Omicron^1 (o^1) and Omicron^2 (o^2) Eridani. From Omicron^2, drop 2½° due south to a 5th-magnitude star, then another 2½° to NGC 1535.

Through a 200-mm telescope, NGC 1535 appears surprisingly large for a planetary and glows with a characteristic bluish tint. The best

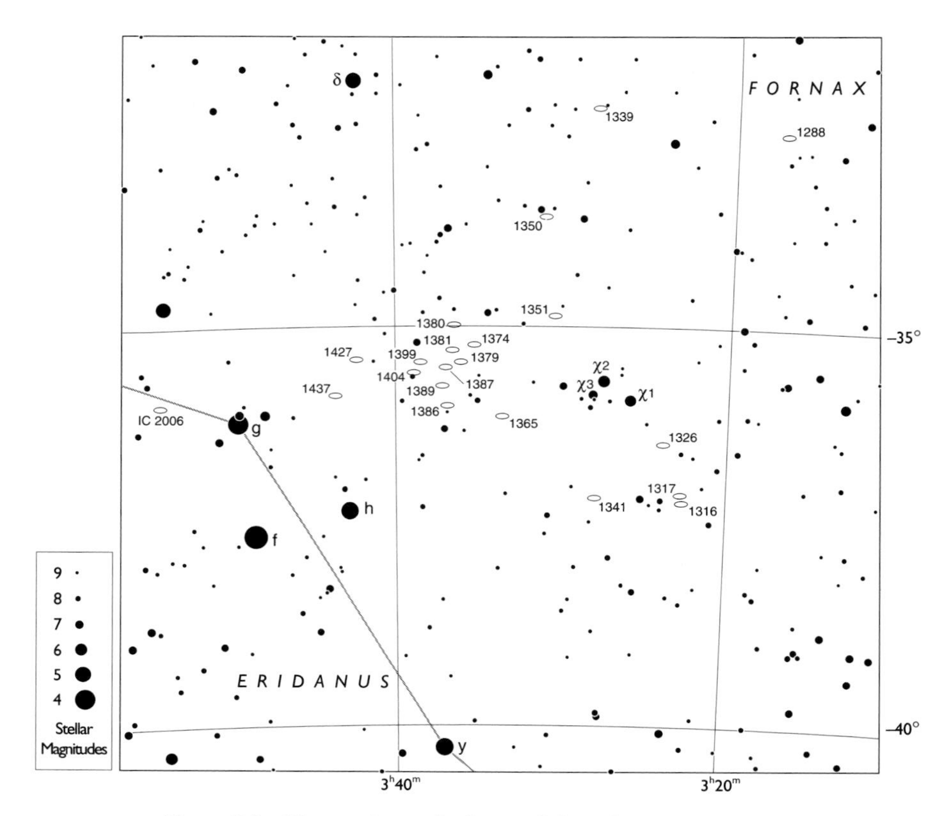

Figure 9.5. *After you locate the Fornax Galaxy Cluster using the star atlas at the end of this book, use this detailed map to identify which galaxy is which.*

views come at high magnifications, when averted vision enables the 12th-magnitude central star to be discerned. Many experienced deep-sky observers also note a dark, mottled center surrounding the central star as well as a faint outer halo encircling the brighter, oval disk.

Fornax

Fornax, the Furnace, is considered by many stargazers to be a constellation of the far south, a misconception since it has about the same declination as the summer constellation Scorpius. Given a good view in that direction, the Furnace will reveal its treasures to most of North America. Fornax culminates at midnight in early November. Look for its faint stars due south during the early-evening hours of December and January.

The **Fornax Galaxy Cluster** contains 18 close-set galaxies that are all within range of most backyard telescopes, yet few amateurs take the time to find them. While it is true that the Fornax Cluster is far from any bright stars, it can be spotted if you are patient. Begin at the constellation Lepus, directly below Orion. Draw an imaginary line from Alpha (α) to Epsilon (ε) Leporis and extend it about the same distance toward the southwest. There you will find three

4th-magnitude stars — Upsilon^2 (υ^2), Upsilon^3 (υ^3), and Upsilon^4 (υ^4) Eridani — forming a northeastward-pointing triangle. Binoculars and finderscopes add a fourth star (the 6th-magnitude Upsilon^1 [υ^1] Eridani) to create a trapezoid. A line from Upsilon^2 through Upsilon^4 extending farther southwest comes to an equilateral triangle made up of the 4th- and 5th-magnitude stars f, g, and h Eridani. You're almost there! From these scan westward to Chi^1 (χ^1), Chi^2 (χ^2), and Chi^3 (χ^3) Fornacis, a small triangle of three 6th-magnitude stars. Centering your view about 1° to the southwest of the Chi triangle, just west of a pair of 7th- and 8th-magnitude suns, should find you **NGC 1316.**

Classified as S0, the 9th-magnitude galaxy NGC 1316 proves to be an amazing sight when viewed through a medium or large backyard telescope and is one of the true unsung marvels of the early winter sky. Backyard instruments show it flaunting a bright, nearly circular disk measuring about 3' × 2'. The galaxy reaches a visual peak at the bright core, then fades off rapidly toward the edges. However, there is more going on here than meets the eye. Long-exposure photographs taken at the Cerro Tololo Inter-American Observatory in Chile uncover faint loops of nebulous material extending away from NGC 1316 and engulfing **NGC 1317,** an 11th-magnitude galaxy about 6' to the north. According to one theory, NGC 1317 will eventually be absorbed into the larger system.

Once NGC 1316 is spotted, the 9th-magnitude galaxy **NGC 1365** is easy. Return to the Chi triangle. With a wide-field eyepiece in place, move 1½° to the east-southeast, where our quarry lies in wait. NGC 1365 is one of the finest examples of a barred spiral in our sky. Its distinctive bar, seen so prominently in photographs, extends an estimated 45,000 light-years across much of the galaxy's apparent area of 10' × 6'. Through the telescope, the bar can be made out protruding to the north and south of the galaxy's central bright core.

NGC 1316 and NGC 1365 are but two of the many galaxies that populate this star-barren region of the southern winter sky. If you own a 200-mm or larger telescope, try aiming at the cluster's center at R.A. 03h 35m, Dec. –35° 40'. Nine other galaxies, all NGC objects but 11th-magnitude or fainter, lie within 1° of that point and are plotted in Figure 9.5. (For a complete discussion of the many other galaxies in the area, see *Sky & Telescope,* January 1988, page 109.)

NGC 1360. This surprisingly bright planetary nebula seems out of place in the midst of the area's many galaxies. To find it, begin in Lepus as you did to find the galaxies mentioned above. Extend a line from Delta (δ) to Epsilon (ε) Leporis and follow it toward the west. Through your finderscope or binoculars, look for a quadrilateral formed by Tau^6 (τ^6), Tau^7 (τ^7), Tau^8 (τ^8), and Tau^9 (τ^9) Eridani. Now, extend a line from Tau^9 through Tau^8, and continue for about 4° westward to a close pair of 6th-magnitude stars. NGC 1360 lies just south of a point halfway between the two stars. In fact, all three may just squeeze into a low-power eyepiece's field.

Figure 9.6. *M35 (right center) is Gemini's brightest and largest open cluster, reaching naked-eye visibility on crystalline winter nights. NGC 2158 (upper left) is a richer but more distant open cluster that happens to lie along the same line of sight. While NGC 2158 may be seen through giant binoculars, a 150-mm telescope is needed to see any of its myriad stars. South is up. Photo by Martin C. Germano.*

NGC 1360 is known to very few amateurs, even though it puts on a fine show in telescopes as small as 150 mm. Through my 200-mm reflector, it is a bright oval disk of grayish light highlighted by an 11th-magnitude, slightly off-center central star. In the past, the nature of NGC 1360 was often questioned — was it a true planetary nebula or an unusual bright nebula? Recent studies, however, seem to show that NGC 1360 is a bona fide planetary.

Gemini

Marked by the prominent stars Castor and Pollux, the twin brothers of Gemini appear to be standing on the faint lane of the Milky Way near the zenith on winter evenings. The Twins bring with them many fine deep-sky objects, including several open clusters, an interesting double star, and a unique planetary nebula. Gemini reaches midnight culmination in early January and is seen highest during the early-evening hours of February and March.

M35 (NGC 2168). Let's set off on the right foot — the foot of

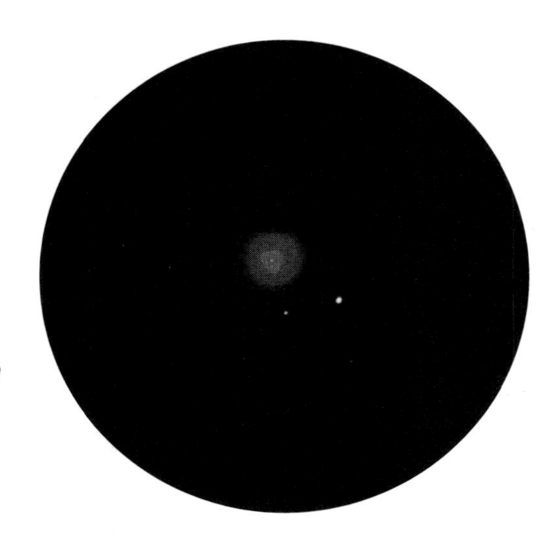

Eyepiece Impression 38. *NGC 2392, as seen through the author's 333-mm Newtonian, 7-mm Nagler eyepiece (214×), and 2× Barlow lens. South is up.*

Castor. The open cluster M35 is found in Gemini about 2½° northwest of Eta (η) Geminorum, the star marking the westerly twin's foot. Given a clear, dark night and a keen eye, many observers can see M35 without any optical aid as a misty patch of light along the main stream of the Milky Way (Figure 9.6).

Seven-power binoculars begin to show M35 as a dense swarm of stardust peppered with a half dozen stellar pinpoints. A 100- or 150-mm telescope will crack the star-filled vault of M35, while a larger instrument will resolve just about all of the 200 stars that call this group home. Be sure to use your lowest-power eyepiece for the best views. Most of the brightest cluster stars are blue-white, though a few are yellow and orange giants. Many appear to form graceful arcs and curves threading their way throughout the cluster, though there is a curious absence of stars near the center of the group.

NGC 2158. Take a look about ½° southwest of M35. There, you should spot the gentle glow of this second, more distant open cluster (Figure 9.6). NGC 2158 is believed to be 16,000 light-years away, while M35 is much closer, only 2,800 light-years. Appearing much fainter than its spectacular neighbor, NGC 2158 consists of hundreds of stars all densely packed into 5'. Collectively, they create an 8.6-magnitude object, bringing NGC 2158 into the range of telescopes as small as 75 mm (3 inches). Since none of the individual stars shine brighter than 12th magnitude, however, a 150-mm telescope is likely to be the smallest instrument capable of partial resolution. A 305-mm (12-inch) telescope resolves NGC 2158 into a multitude of dim specks set against the glow of other, unseen suns, while my 455-mm Newtonian extends stellar resolution to the cluster's core.

NGC 2392 is one of the finest planetary nebulae in the winter sky. Scan through your finder to the southeast of Delta (δ) Geminorum for three 5th- and 6th-magnitude stars that form a right triangle. Set your aim on the star marking the triangle's northern corner. With a

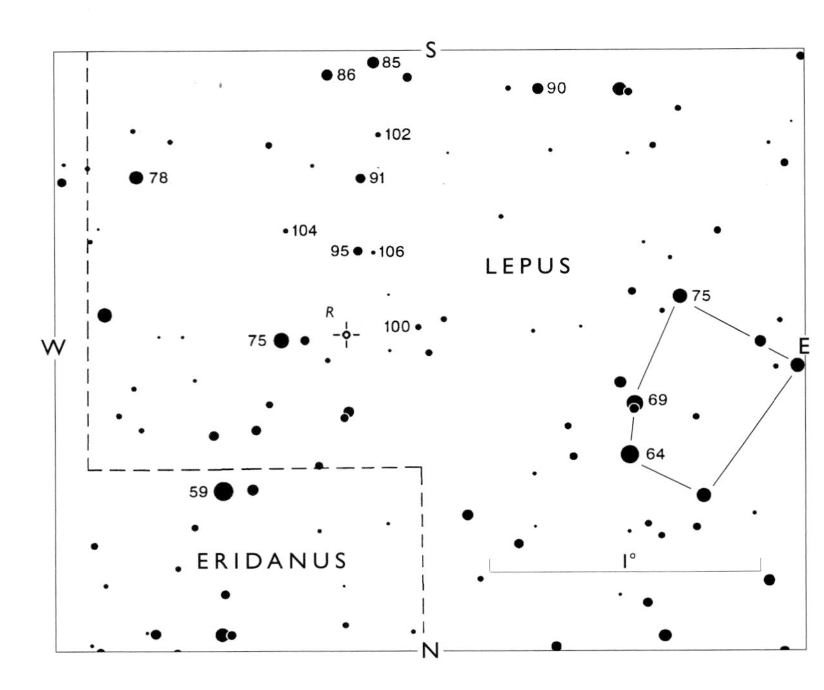

Figure 9.7. *Use this finder chart to locate the variable star R Leporis. Decimals have been omitted: the star labeled "75" is actually magnitude 7.5, and so on. The chart is oriented with south up to approximate the view through an inverting telescope. Courtesy the American Association of Variable Star Observers.*

low-power eyepiece in place, shift about ½° farther southeast, to where NGC 2392 lies.

NGC 2392 reveals its 9th-magnitude starlike disk in the smallest telescopes, while a 150-mm instrument captures both the nebula's distinctive bluish color as well as its 10th-magnitude central star. Years ago it was nicknamed the Eskimo or Clown-Face Nebula for its unique appearance in larger telescopes and photographs. At high magnification, NGC 2392 bares a complex structure of overlapping bright rings and dark patches. Eyepiece Impression 38 shows that the planetary's outermost ring has a fuzzy outer edge and a comparatively distinct inner border, while the innermost disk is shaped like a triangle with rounded corners. The combined effect has been likened to a human face, with the central star marking the nose, surrounded by a fur-lined hood or perhaps a clownlike ruffled collar.

Lepus

Take a careful look due south of Orion. Can you make out four comparatively faint stars set in a lopsided rectangle? If so, you have spotted the four main stars of the constellation Lepus, the Hare. Lepus contributes little to the naked-eye majesty of the winter sky, but it does hold the season's only bright globular cluster as well as a magnificent crimson star. The Hare bounds toward midnight culmination in mid-December and will appear low in the south

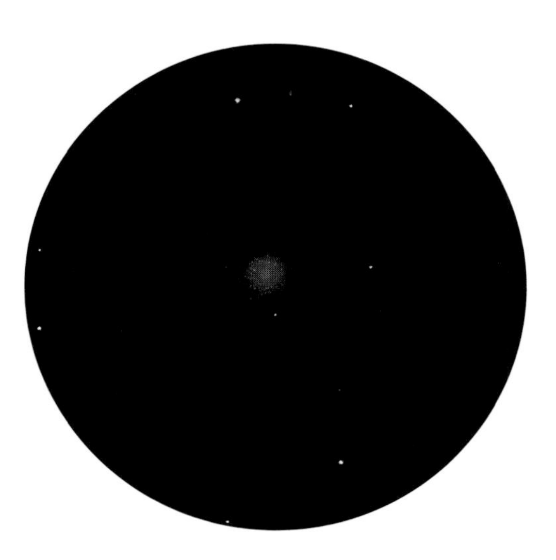

Eyepiece Impression 39. *Globular cluster M79 resides in the constellation Lepus, the Hare, hopping its way across the sky just south of Orion. A tightly packed swarm of perhaps 100,000 stars, M79 requires at least a 200-mm telescope for partial stellar resolution. The author made this drawing through his 200-mm Newtonian and 12-mm Nagler eyepiece (116×). South is up.*

during the evening hours of January and February.

R Leporis is one of the sky's most colorful and well-known variable stars. To find it, scan 3° northwest of Mu (μ) Leporis to a trapezoid formed by five 6th- and 7th-magnitude stars (a portion of it is shown along the eastern border of Figure 9.7). Slide 1° to the trapezoid's west, to find R just east of a fixed-brightness 7th-magnitude sun.

R Leporis is well known as a long-period variable star. Over a period that averages 432 days, it cycles from magnitude 5.5 to 11.7 and back again. What makes it so interesting is not so much its change in magnitude, but its deep red color. Discovered in 1845 by the English astronomer John Russell Hind (1823–95), R gleams with the sparkle of a brilliant stellar ruby against a velvet black backdrop of stardust. For its discoverer and its vivid ruddy tinge, which becomes especially intense at or near minimum brightness, R Leporis is known as Hind's Crimson Star.

M79 (NGC 1904), winter's lone bright globular cluster, resides in the southwest corner of Lepus, far from the season's many brilliant stars. To spot M79 begin by finding the quadrilateral of stars that is the Hare's body. Draw a line between the two western stars, Arneb (Alpha [α] Leporis) and Nihal (Beta [β] Leporis), and extend it an equal distance to the south-southwest, where you ought to spot a 5th-magnitude star through your finderscope. M79 is set just over ½° to that star's northeast.

This tight swarm, first spotted by Pierre Méchain in 1780, is listed as a Class V globular, indicating a moderately high degree of concentration. A 150-mm telescope will show a mottled surface to M79, as if on the verge of resolution. My 200-mm Newtonian, at which Eyepiece Impression 39 was drawn, just begins to resolve a few of the individual 14th-magnitude stars. Still larger telescopes turn M79 into a magnificent sight, with many stars scattered across its 9' disk.

Before you leave Lepus, return to that 5th-magnitude star close to M79. Cataloged as **h3752,** this is a fine double, which puts on a nice show through 75-mm and larger telescopes. You'll find a magnitude-5.4 Type-*G* yellowish primary separated from a magnitude-6.6 Type-*A* blue companion by about 3". Another, 9th-magnitude star about 1' to the primary's southeast makes this an optical triple system.

Monoceros

In stark contrast to its stellar surroundings, the area east of Orion looks like a starless hole in the sky. Much of that barren region belongs to the constellation Monoceros, the Unicorn. Even though the Milky Way passes centrally through Monoceros, this constellation contains no stars brighter than 4th magnitude. Yet, for all it lacks in naked-eye beauty, the Unicorn holds some of the season's finest deep-sky treasures. Monoceros culminates at midnight in early January and is well placed for early-evening observation in February and March.

NGC 2170 is the westernmost part of a small area of nebulosity in western Monoceros. Follow the line of the Belt of Orion toward the southeast, in the direction of Sirius (Alpha [α] Canis Majoris). Looking through your finderscope, move about 9° from the end of the Belt to 4th-magnitude Gamma (γ) Monocerotis (for scale, Sirius is about 20° from the end of the Belt). Gamma is at the eastern end of a crooked line formed with two 5th-magnitude stars to its west. Aim just a little north of the halfway point between these latter two for NGC 2170. A 150- or 200-mm instrument will uncover NGC 2170 as a fairly bright, amorphous glow surrounding a 10th-magnitude star. Larger telescopes expose the cloud's elliptical shape but offer little additional surface detail.

NGC 2182, NGC 2183, and **NGC 2185,** a trio of bright nebulae lying just east of NGC 2170, are also fairly easy to observe. NGC 2182 is illuminated by a hot Type-*B*4 star of 9th magnitude. Through my 333-mm (13.1-inch) reflector I have recorded this nebula as circular; other observers see it as more elliptical in shape. NGC 2183 and 2185 also fit into the same low-power eyepiece field. The instrument reveals NGC 2185 as a bright, round cloud engulfing three faint stars, while NGC 2183 encompasses a lone 13th-magnitude star and appears significantly fainter. Not surprisingly, a narrow-band nebula filter aids greatly in their detection.

All of the nebulae in this part of the sky are thought to belong to a large H II region that is obscured from our view by an opaque dust cloud. If that is the case, then what we see here is merely the tip of a celestial iceberg.

Beta (β) Monocerotis. This is one of winter's truly spectacular multiple stars. Beta Monocerotis is visible to the naked eye as a 4th-magnitude point of light set 10½° east-northeast of Saiph (Kappa [κ] Orionis) and an equal distance north-northwest of Sirius (Alpha [α] Canis Majoris). A small telescope quickly expands that single point into a brilliantly white stellar triple play. The 4.7-magnitude primary is separated from the 5.2-magnitude B star by about 7", while 10" separates it from the 6.1-magnitude C star. In addition,

Figure 9.8. *Like so many emission nebulae, NGC 2237 (the Rosette Nebula) in Monoceros is striking in photographs but only faintly visible through telescopes. The cloud's accompanying open cluster, NGC 2244 (centered in the nebula's "ring"), however, is easily detected through binoculars as a crooked rectangle of six bright stars and many fainter points. South is up. Photo by Brad Wallis and Robert Provin.*

sharp-eyed observers may also spot a fourth member of the system: a 12.2-magnitude point of light some 26" to the primary's northeast. When he discovered the split personality of Beta Monocerotis in 1781, William Herschel proclaimed it as "one of the most beautiful sights in the heavens."

The open cluster **NGC 2244** (Figure 9.8) is best known to deep-sky observers as the power source of the magnificent Rosette Nebula. To find it, trace a line from Orion's triangular "head" to Betelgeuse (Alpha [α] Orionis) and extend it to the southeast. About 8° beyond Betelgeuse, you will come to 5th-magnitude Epsilon (ε) Monocerotis and, about 2° farther east, NGC 2244.

Although the clouds of the Rosette are difficult to glimpse visually, the open cluster NGC 2244 is bright enough to be seen in finderscopes and even with the naked eye on exceptional nights. Binoculars immediately reveal the group's half-dozen brightest stars held in a distinctive rectangular pattern. Most brilliant of all is 6th-magnitude 12 Monocerotis, a yellowish sun, followed close behind by a nearby blue-white star. Scattered within and around the stellar rectangle are more than 90 fainter stars. Together they create a magnitude-4.8 splash across nearly ¼° of sky.

The Rosette Nebula, **NGC 2237,** is known to astrophotographers as a bright target that is fairly easy to capture with relatively short exposures given the proper conditions (Figure 9.8). Visual observers, however, may spend hours searching in vain for its dim glow. This is because, like all emission nebulae, the Rosette shines primarily at the red end of the spectrum, to which the human eye is much less sensitive than are photographic emulsions.

In 1888 J. L. E. Dreyer listed the Rosette in his *New General Catalogue* as four separate entries: NGC 2237, 2238, 2239, and 2246. These entries correspond to four bright portions of a huge, wreathlike cloud over 1° in diameter encircling the stars of NGC 2244. Under dark skies a 100- or 150-mm rich-field telescope will give a beautiful view of the Rosette complex. Three bright lobes interconnected by faintly perceptible wisps stand out against a star-filled backdrop. While in this region, take some time to casually scan the adjacent Milky Way star fields, for they are striking.

NGC 2261. Once you find the open cluster NGC 2244, use it as a guidepost to find other deep-sky treasures in the Unicorn. For instance, only 4° northeast of NGC 2244 is NGC 2261, the famed, if enigmatic, Hubble's Variable Nebula. Most amateur telescopes have little trouble revealing this strange celestial apparition thanks to its high surface brightness. However, since it measures only 2' × 1' across, quite a high magnification is needed for the best view. Like M78 in neighboring Orion, NGC 2261 exhibits a strong resemblance to a comet, with a "tail" extending to the north of a bright "coma."

Discovered by William Herschel in 1783, NGC 2261 surrounds the variable star **R Monocerotis**, an irregular variable that fluctuates between 10th and 12th magnitudes. While the star's variability was noted as far back as 1861, the chameleon-like behavior of NGC 2261 went unnoticed until 1916. While comparing photographs of the nebula, Edwin Hubble found that the nebula itself was changing in both brightness and structure over the course of several weeks.

As far back as 1861 observers also noticed that NGC 2261 actually changes shape. In 1966 two astronomers, Low and Smith, suggested that R Monocerotis is not a true star but a "protoplanetary system." Investigations at Kitt Peak and Mauna Kea observatories in 1983 seem to confirm this theory. If so, then R Monocerotis is the central condensation of a solar system in the making, surrounded by a halo of slow-moving, gravitationally bound matter.

NGC 2264. This spectacular open cluster, a bright smattering of 40 stars, lies just 5½° north-northeast of NGC 2244 and about 3° south-southwest of 3rd-magnitude Xi (ξ) Geminorum, and is easily visible with only the slightest optical aid. And what a distinctive shape! One glance and you will immediately know why Leland

Copeland nicknamed this group the Christmas Tree Cluster. Ten stars form the tree's main profile. The cluster's brightest star, S Monocerotis (an irregular variable), marks the tree's trunk, while the other nine appear like lights on imaginary branches.

Extending from the star at the top of the Christmas Tree is a large, triangular wedge of dark nebulosity nicknamed the Cone Nebula for its appearance on long-exposure photographs. Unfortunately, the Cone as well as most of the interstellar clouds in this region are extremely difficult to pick out visually. Only NGC 2261 (described above) can be spotted through smaller telescopes on good nights.

STAR 17 is a neat little arrowhead-shaped asterism that I chanced upon one night while star-surfing the winter sky with my daughter's 108-mm (4¼-inch) f/4 rich-field Newtonian. To find it, travel about 7° north of Sirius (Alpha [α] Canis Majoris), past a small three-star arc, to a triangle of three 5th-magnitude suns. Center your telescope's view on the triangle's westernmost star, then shift a little northwest to find STAR 17. Small telescopes show a distinctive V-shaped pattern of six 9th- and 10th-magnitude white stars spanning about ⅓°; a low power offers the best view. Given its home in Monoceros, it seems only fitting to christen this asterism the Unicorn's Horn. If you are looking at STAR 17, then you are only a few degrees away from another asterism, STAR 18, described below after NGC 2301.

NGC 2301. Here's a striking open cluster that is often overlooked by deep-sky observers. To find it, scan 9° southwest from Procyon (Alpha [α] Canis Minoris) to the binary star Delta (δ) Monocerotis; the cluster is 5° farther west-northwest. NGC 2301 is an amazing sight in all telescope apertures. With your lowest-power eyepiece in place, look for a distinctive string of 8th- and 9th-magnitude stars with a rich triangular wedge of fainter stars midway along it. I am always reminded of a bird in flight whenever I view this cluster, so I like to refer to it as the "Great Bird of the Galaxy" cluster. The bird's "wings," formed by the string of stars mentioned above, stretch for about 12', while the three-sided "body" is half that size. In all, 80 stars of 8th magnitude and fainter populate NGC 2301, producing an overall magnitude of 6.0.

STAR 18 lies to the southwest of the bright open cluster M50 (described later), along the southern tier of Monoceros. First spotted by Randy Pakan of Edmonton, Canada, this unique string of stars is easy to see in just about any backyard telescope and even through binoculars on dark, clear nights. To find it, begin at Theta (θ) Canis Majoris to the northeast of brilliant Sirius (Alpha [α] Canis Majoris). Move about 2° to the north-northwest, passing two pairs of 8th-magnitude stars, until you come to a close 7th- and 8th-magnitude duo. STAR 18 lies just to their east, and is visible through finderscopes as a wiggly line of faint stars.

As Eyepiece Impression 40 shows, STAR 18 is more than just another line of stars. When all 15 of its 9th- and 10th-magnitude stars

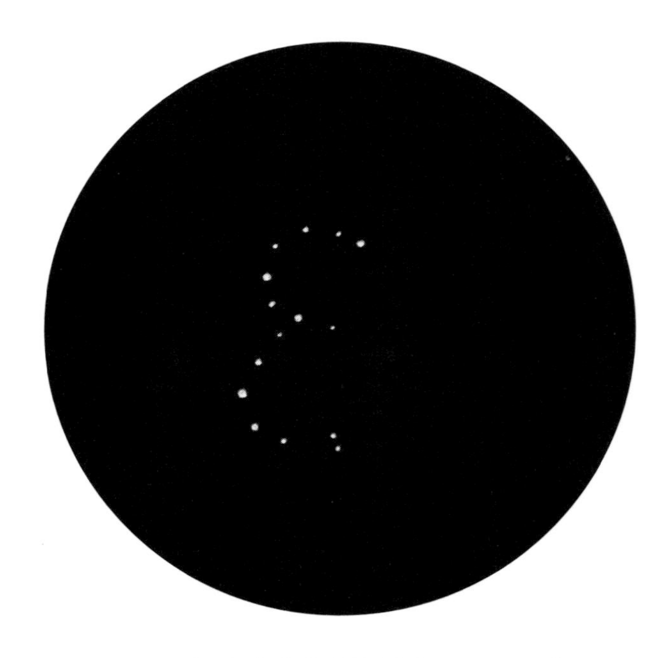

Eyepiece Impression 40. *STAR 18, nicknamed Pakan's 3, is clearly shown in this drawing by Randy Pakan through a 200-mm Newtonian and 40-mm (30×) eyepiece. South is up.*

are included, STAR 18 bears more than a passing resemblance to a reversed "3." Pakan's 3 is about 40' long and can just fit into a low-power eyepiece field. Adding a little extra interest to the pattern at the northern tip of the figure is an easy double star.

Nestled in a rich star field less than 2° off the galactic plane is our next target, **NGC 2316.** Although this combination emission-reflection nebula is far from any bright nearby stars, star-hoppers should still find it easily by first centering on open cluster M50 (our next target) and then jogging their telescopes about 1° northwest. NGC 2316 is readily visible in a 150-mm telescope as a tiny, evenly illuminated glow just north of three faint field stars. Powering the nebula is a dim binary that requires an aperture of 305 mm or more to be seen clearly.

M50 (NGC 2323), the only Messier object in Monoceros, is a magnificent open cluster that sadly seems to be ignored by many amateur astronomers. But for those who take the time to find it, it quickly becomes a favorite target to be revisited night after night. To locate this lovely flock, draw an imaginary line between Sirius (Alpha [α] Canis Majoris) and Theta (θ) Canis Majoris, a 3rd-magnitude star about 5° to the northeast. Using your finder, extend this line 4° farther northeast, to a triangle of faint stars. M50 lies within that triangle.

The cluster spans 16' of sky and is made up of 80 stars, of which about two dozen are visible in small telescopes. A 200-mm instrument easily quadruples the star count. K. Sturdy, of Helmsley in the United Kingdom, describes M50 seen through a 150-mm instrument as "slightly nebulous in the centre; stars in a rough Y-shape." Most of the cluster's members shine with the characteristic blue-white hue of young stars, though a lone ruddy beacon punctuates its center.

Figure 9.9. *M42, the Orion Nebula, is perhaps the single most amazing sight described in this book. A detached segment of the nebula, cataloged separately as M43, can be seen at the lower right. Compare this photograph taken by Martin C. Germano with the drawing on the next page. South is up.*

Orion

Widely acknowledged as the heaven's single most beautiful constellation, Orion, the Hunter, is a veritable playground for the deep-sky observer. Within its borders lie the magnificent Orion Nebula, the elusive yet captivating Horsehead Nebula, and several other deep-sky treats. Orion culminates at midnight in mid-December and stands high in the southern sky on January and February evenings.

NGC 1788, a bright nebula, resides in the constellation's southwest corner, about 2° north of Beta (β) Eridani. Look for a cigar-shaped object extending southward from a faint double star. A 150-mm is probably the smallest telescope that will show this tiny reflection nebula well, though it can be seen with averted vision through instruments as small as 75 mm. Instruments larger than 150 mm will also show several other dim stars superimposed on the cloud. Whatever telescope is used, however, the nebula's small size of 8' × 5' will necessitate a high power.

M42 (NGC 1942), the Great or Orion Nebula, is unquestionably the grandest deep-sky object visible from the northern hemisphere (Figure 9.9). Visible to the naked eye as a hazy patch of light

Eyepiece Impression 41. *M42, the Orion Nebula, as seen through the author's 200-mm Newtonian, 24-mm Tele Vue Wide Field eyepiece (58×), and DayStar narrowband LPR filter. South is up.*

surrounding **Theta (θ) Orionis**, the middle star in the Sword of Orion, this region quickly explodes into a glowing cloud of great intricacy when viewed with any optical aid. On especially transparent winter nights a medium- to large-aperture telescope displays a tremendous greenish cloud engulfing many stars. As the low-power view in Eyepiece Impression 41 demonstrates, the nebula fills the field. It looks to me like a cupped hand with tenuous, glowing fingers extending from the main body of the nebula toward the myriad of field stars.

In the 19th century, Adm. Smyth was the first to call attention to what looks like a fish's mouth, a dark protrusion at the cloud's northeast corner. Near its tip lie the stars of the **Trapezium**, also known as Theta^1 (θ^1) Orionis, one of the season's finest examples of a multiple star. Small telescopes show the system's four brightest members, while larger instruments reveal half a dozen stars surrounded by an array of dim points of light. Their combined energy excites the Orion Nebula into luminescence, resulting in the reddish color familiar from photographs. Although difficult to discern visually, hints of red can be seen along the nebula's misty fringes through large telescopes on exceptional nights.

Traditionally, the four brightest Trapezium members have been labeled A, B, C, and D in order of increasing right ascension rather than brightness, unlike other binaries. These stars, rated magnitudes 6.7, 7.9, 5.1, and 6.7, respectively, shine with a fiery blue-white intensity against the soft, gossamer clouds of the nebula itself. In 1826 F. G. W. Struve discovered an 11th-magnitude star only 4" north of A, while John Herschel found a second 11th-magnitude companion about 4" southeast of C. Now known as Theta^1 Orionis E and F, respectively, both may be spotted in high-quality 150-mm instruments when seeing is steady. Another Trapezium member, christened G, shines at 16th magnitude and therefore can be seen only in the largest amateur telescopes. It lies about 6' due west of D. In 1975 Theta[1] Orionis A, the Trapezium's westernmost star, was discovered to be an eclipsing binary that fades a full magnitude for 20 hours every 65.4 days. Theta[1] Orionis B, also known as BM Orionis, is another eclipsing binary, one with a 6.5-day period.

M43 (NGC 1982) is about 7' to the north of M42. Frequently, observers fail to distinguish this detached portion of the Great Nebula as a separate object. In fact, while M42 was telescopically discovered as early as 1610, M43 was not recognized as an individual entity for another 121 years. Mere words cannot possibly do justice to this magnificent pair. Even the finest drawings and photographs cannot capture the thrill of viewing M42–43 visually. All eyepiece powers work well, each offering a different perspective. Low powers are best for spotting faint, outlying wisps usually overlooked by the casual observer. Medium magnification reveals the nebula's complex structure and its varying colors and contrasts, while the area in and around the Trapezium is best served with high magnification. What a view!

NGC 1973, NGC 1975, and **NGC 1977** are the three brightest regions of a large emission-reflection nebula network just north of M42 and surrounding the 5th-magnitude 42 Orionis. Of the three, NGC 1977 stands out the best. Telescopes of 200 mm show it as a large, bright, bluish arc of nebulosity stretching between 42 Orionis and two dimmer stars. NGC 1973 is a comparatively small, faint object surrounding the variable KX Orionis, while NGC 1975, the dimmest of the three, has been recorded as greenish when viewed in larger telescopes. All require very dark skies for best visibility.

NGC 2022. Although Orion overflows with vast regions of bright and dark nebulosity, it can claim only a single planetary nebula that is bright enough to be seen in most backyard telescopes. That nebula is NGC 2022, found just east of Orion's triangular head. From Phi^2 (ϕ^2) Orionis, move 1° east to a pair of 8th-magnitude stars. Center the eastern star in the field of a medium-power eyepiece and shift the view ½° southeast to spot the starlike disk of NGC 2022.

At low power a high-quality 75-mm telescope is all that is needed to show NGC 2022 as a tiny 12th-magnitude point of light nestled in a starry field. While some amateurs comment on the nebula's greenish tint, most (including me) have difficulty detecting any color in its 18" disk. To see any detail requires a medium- or large-aperture telescope and high magnification. Only then, and by using averted vision, will the cloud's ringlike structure be seen. For instance, Kenneth Glyn Jones, observing in Winkfield, United Kingdom, remarks that through a 200-mm telescope at 200×, NGC 2022 appears as "an oval patch showing a slight impression of a ring; slightly brighter condensation on the northeastern edge." If you are observing from light-polluted surroundings a narrowband nebula filter might help, though from dark-sky sites it seems to make little difference whether you use one or not. And you can just about forget trying to spot the cloud's

Figure 9.10. *Barnard 33, the Horsehead Nebula, is the sky's best-known example of a dark nebula. Trying to spot it through a telescope, however, remains one of the great challenges for deep-sky observers. The problem is caused by the low contrast between it and IC 434, the emission nebula that silhouettes the Horsehead. Observers usually have better luck spotting NGC 2024, a patch of nebulosity set just to the east of the bright star Zeta (ζ) Orionis (the bright star to the left of center). West is up. Photo by Chuck Vaughn.*

central star. It shines at a bleak magnitude 15.2, restricting visibility to only the very largest amateur telescopes.

NGC 2023, NGC 2024, and **IC 434.** Photographs reveal that much of Orion is blanketed with faint wisps of nebulosity. The region surrounding Alnitak (Zeta [ζ] Orionis) is especially noteworthy, no fewer than six nebulae within 2° of the star having been accorded separate catalog entries. The brightest of them is NGC 2024, which appears visually as a complex, mottled gray patch of light with a couple of faint stars in the foreground (Figure 9.10). The source of excitation for this ½°-diameter object is hidden behind a dark, roughly rectangular absorption region, making it impossible to detect.

Although not as conspicuous as its larger neighbor, NGC 2023 is still relatively easy to find. Look for a fairly bright but small patch of mist diffusing outward from a central 8th-magnitude star. High magnification enhances the nebula's appearance.

IC 434, stretching 1° due south of Zeta, is famous as the backdrop for **Barnard 33**, the Horsehead Nebula (Figure 9.10). Both are notoriously difficult objects. Yet, would you believe that once,

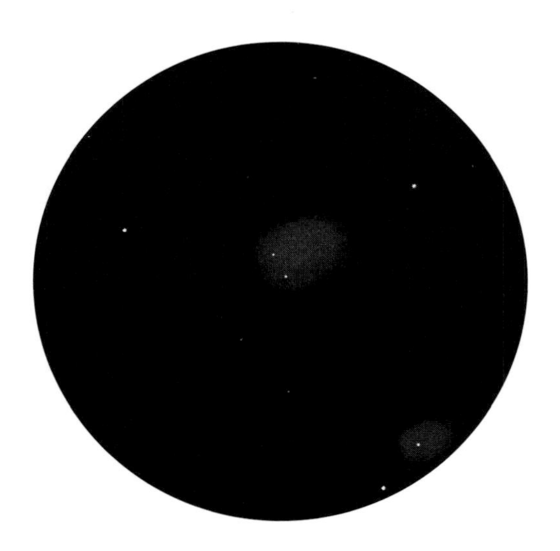

Eyepiece Impression 42. *M78 (center) and NGC 2071 (lower right), as seen through the author's 200-mm Newtonian and 12-mm Nagler eyepiece. South is up.*

under the dark skies of the Florida Keys, observer *extraordinaire* Tom Lorenzin actually showed it to me through his nebula-filtered 10 × 70 binoculars? Hard to believe, but true. From my suburban backyard, with its 5th-magnitude naked-eye limit, the Horsehead is a *little* more difficult to spy, though I have seen it faintly through my 333-mm Newtonian fitted with a 12-mm Nagler eyepiece and a hydrogen-beta nebula filter. In both instances, IC 434 could be seen as a faint foggy "wall" with a perfectly straight eastern edge; the surprisingly large Horsehead requires absolute concentration to be glimpsed.

M78 (NGC 2068). A little less than 4° east of Mintaka (Delta [δ] Orionis) is a clump of nebulosity, the most prominent region of which is designated M78. You might find it easiest to find M78 by centering on Mintaka, turning your telescope's clock drive off, and taking a 15-minute break. When you return, the Earth will have aimed your telescope toward it for you. M78, a reflection nebula discovered by Pierre Méchain in 1780, stands out well in small telescopes. Eyepiece Impression 42 shows its famous cometlike form, with a broad 6' × 4' fanlike "tail" extending to the south. Two nearly identical 10th-magnitude stars pose as "nuclei" to complete the comet illusion.

Several other patches surround M78. The brightest is **NGC 2071,** a fairly obvious glow that appears a little smaller than M78. Look for a round misty patch enveloping a 9.5-magnitude star north-northeast of M78. While clearly visible in a 200-mm telescope under dark, rural skies, it will probably require averted vision and/or a nebula filter to be seen with smaller instruments or from suburban locations. **NGC 2067** is a wispy target, which will prove difficult in telescopes smaller than 305 mm. It shines by the light of the same stars that illuminate M78 and is positioned about 4' to M78's northwest. Sharp-eyed observers might also be able to make out **NGC 2064,** just

Figure 9.11. *M47 is a bright, coarse open cluster in Puppis. The grouping consists of some 30 stars ranging between 6th and 12th magnitude, making this an easy object to spot with nearly all telescopes. South is up. Photo by Lee C. Coombs.*

southwest of M78. Although it has always eluded my probing eye, you may have better luck.

Puppis

Puppis is one of the four modern constellations into which the huge, ancient constellation Argo Navis was carved. Puppis represents the stern of the once-mighty mythical ship that carried Jason in search of the Golden Fleece. The other constellations formed from Argo include Carina the Keel; Pyxis the Compass; and Vela the Sails — all in the spring sky. Puppis culminates at midnight in early January and is highest in the early evening sky during February and March.

M47 (NGC 2422) is the western member of a close-set pair of open clusters residing in northern Puppis (Figure 9.11). Find it by extending a line from Sirius (Alpha [α] Canis Majoris) to Gamma (γ) Canis Majoris, then continuing eastward for about twice the distance to the ruddy variable star KQ Puppis and M47, which lies just

Figure 9.12. *Open cluster M46 is a densely packed collection of about 100 stars that lies about 1° east of M47. Together, M46 and M47 offer testimony to how different from each other two open clusters can be. Also note the tiny planetary nebula NGC 2438 superimposed on the center of M46. South is up in this photo by Preston Scott Justis.*

to the variable's east. Lying about 1,500 light-years from Earth, this coarse grouping is comprised of about 30 individual stars from 6th to 12th magnitude. Thus a 150-mm telescope can resolve just about all of these stellar sapphires as they burn against a velvet black backdrop. The beauty of the scene is further enhanced by surroundings that overflow with stars.

M46 (NGC 2437). Moving 1° east from M47, pause at the second member of this galactic team, M46 (Figure 9.12). While M47 is a coarse group, M46 is a densely packed throng bursting at the seams with 100 stars of 9th to 13th magnitude. All appear evenly distributed across the full 20' breadth of this family, which is 5,400 light-years distant.

Of special interest to observers with a 150-mm or larger telescope is a tiny planetary nebula (also visible in Figure 9.12) superimposed on the northern part of M46. **NGC 2438,** shining at photographic magnitude 10.1, is seen as a tiny, gray disk among the stars. Once thought to be actual companions in space, M46 and NGC 2438 are now believed to be in no more than a chance alignment as seen from our earthly vantage point, with NGC 2438 about 3,000 light-years away.

NGC 2440. Although Messier missed it, the planetary nebula NGC 2440 — an 11th-magnitude "star" adrift in the winter Milky Way — is visible in telescopes as small as 75 mm. While, as planetaries go, NGC 2440 is bright, pinpointing its exact location is tough since it is far from any bright reference stars. Your best bet is to find M46 first, then drop southward about 3½°. A 100- or 150-mm telescope will show it as a 20" circular patch of turquoise light. With high magnification and large aperture, a faint, strangely rectangular halo can also be seen engulfing the bright center. The nebula's progenitor central star shines weakly at 14th magnitude, limiting access to only the largest backyard telescopes. Many faint stars surround NGC 2440, including a yellowish one to its east. An instrument of 150 mm or larger will unveil many tiny stellar triangles around NGC 2440, adding to the area's interest.

M93 (NGC 2447) lies just 1½° northwest of 4th-magnitude Xi (ξ) Puppis, which in turn lies about 9° east-northeast of Wezen (Delta [δ] Canis Majoris). M93 is a fine example of a rich open cluster as it has some 80 members crammed into a tight 22' area. About a half dozen of the brighter members form a line that zigzags close to the cluster's center, while the remaining stars give the group an overall triangular shape. A 200-mm telescope resolves about two-thirds of the cluster stars, while a 305-mm reveals the rest as faint stellar fireflies seemingly flitting about a dim flame.

NGC 2451 is a large, bright open cluster easily seen in binoculars and finderscopes. Draw an imaginary line between Sirius (Alpha [α] Canis Majoris) and Wezen (Delta [δ] Canis Majoris) along the "back" of the Big Dog. Extend the line an equal distance southeast of Wezen, where you should find NGC 2451. If you come to 3rd-magnitude Zeta (ζ) Puppis, you have overshot the cluster by about 4°. Of the 40 stars in NGC 2451, 30 are resolvable in 11 × binoculars. The group's brightest member, the brilliant orange-red c Puppis, stands out from the crowd of white and blue-white stars. Collectively, the cluster forms a colorful welcome mat to the southern sky. Your best view of NGC 2451 will be through either binoculars or a rich-field telescope, as the cluster spans a full 45'.

Use the directions for locating NGC 2451 given above to discover another open cluster, **NGC 2477.** Lying about 2½° northwest of Zeta (ζ) Puppis and just north of 4th-magnitude b Puppis, NGC 2477 looks just like a ball of celestial cotton spanning about one Moon diameter when viewed through low-power binoculars. Through a rich-field telescope at low power, the cluster's fuzziness begins to dissolve into a myriad of faint points of light. In all, 160 stars dwell within NGC 2477, though none shine brighter than 10th magnitude.

NGC 2467 is a magnificent complex of an open star cluster intermingled with clouds of nebulosity about 2° southeast of Xi (ξ) Puppis. It is easily captured through binoculars and finderscopes as a circular glow sprinkled with stardust, while telescopes reveal a bright, homogeneous cloud containing many faint stars concentrated to the north of the cloud's brightest region.

NGC 2546, an open cluster about 3° northeast of Zeta (ζ) Puppis, is a loose grouping of about 40 stars scattered over nearly ¾°. Rich-field telescopes reveal an isosceles triangle of 7th- and 8th-magnitude stars framing about 20 stars of 9th magnitude and fainter. Use your widest-field eyepiece or, better still, binoculars for the best view.

Taurus

Standing at attention, almost daring Orion into action, is the constellation Taurus, the Bull. Taurus presents deep-sky observers with many fine targets within its boundaries, including a pair of easy open clusters, several difficult patches of nebulosity, and the expanding remnant of a once-powerful star. Taurus reaches midnight culmination in early December and is seen highest in the early-evening sky during January and February.

M45. Better known as the Pleiades or Seven Sisters, M45 must have been the first open cluster to be noticed in the northern hemisphere. Riding on the back of Taurus, the Seven Sisters appear to most naked-eye skywatchers as six or seven stars set in the shape of a tiny dipper. The number of stars may grow to eight, nine, or even ten on cold, crystal-clear winter nights. Some eagle-eyed observers can surpass even these numbers, with Walter Scott Houston once claiming 18. How many Pleiades can you see with just your eyes? Because of their wide span — almost 2° — the Pleiades demand a low power and a wide field to be seen at their best. Binoculars and rich-field telescopes cause a population explosion in the family of the Seven Sisters, revealing dozens upon dozens of fainter stellar siblings. Modern counts place perhaps 3,000 stars within the Pleiades.

Photographs reveal an amazingly intricate network of bluish reflection nebulosity intertwined around the stars of M45. Although much of this cloud is invisible to the visual observer, a few tufts faintly reveal themselves on clear, dark nights. **NGC 1432** and **1435** are the designations for the brightest tufts of the Pleiades' nebulosity. NGC 1435 surrounds 4th-magnitude Merope, the southeastern star of the dipper-shape's "bowl." With excellent sky conditions some observers can make out its form in binoculars. Through a 150- or 200-mm telescope NGC 1435 appears as a comet-shaped glow extending southward from the star; it becomes more apparent at higher apertures. NGC 1432 is a faintly shining circular patch of light around Maia, marking the northwestern corner of the bowl. Wait for that

Eyepiece Impression 43. *M1, the Crab Nebula, as seen through the author's 200-mm Newtonian, 12-mm Nagler eyepiece (116×), and DayStar narrowband LPR filter. South is up.*

special night when humidity is low and the air is calm for the best view. Even then, it will present a real challenge in anything less than a 150-mm instrument.

Incidentally, the stars of the Pleiades appear so bright through telescopes that many amateurs mistake the glow of starlight scattered by the telescopes' optics for the nebulosity. While internal light scatter can never be completely eliminated, it can be greatly reduced if all optical surfaces are clean and free of contamination. Otherwise, with dirty lenses and mirrors, you will see a cloud around every star in the sky!

STAR 16. Just north of the Hyades (the V-shaped cluster that marks the face of Taurus — see below) is a bright but loose asterism of 3rd-, 4th-, and 5th-magnitude stars that stands out well in low-power binoculars. From west to east the pattern consists of Omega (ω), 51, 53, 56, Kappa (κ), 67, Upsilon (υ), 70, and 72 Tauri, along with a scattering of fainter stars. John Davis of Amherst, Massachusetts, was first to bring this group to my attention, noting its resemblance to a "cute little dog whose nose got caught in a pencil sharpener!" Omega marks the tip of the dog's nose, 53 represents one of its eyes, and 51 and 56 Tauri lie at the tips of the dog's pointy ears. The dog's tail is outlined by the arc of Kappa, 67, Upsilon, and 72 Tauri, while is body and legs are formed from a pair of fainter triangular patterns to the south. In all, **Davis's Dog** extends across an area of 3° × 1°.

Hyades. The Hyades open cluster, also cataloged as Melotte 25, is famous for marking the face of Taurus. To the unaided eye six stars outline the Bull's head, with brilliant Aldebaran portraying its angry red eye. Aldebaran is not a member of the Hyades, just a foreground star in chance alignment. Estimates show that the Hyades lie about 130 light-years from Earth, while Aldebaran is almost exactly halfway between us and the cluster.

The Hyades ranks as one of the finest open clusters for low-power binoculars and finderscopes. All told, the Hyades are scattered across 5½° of sky, much too large an area for conventional telescopes to take in. Of the 380 cluster stars, about 130 are brighter than 9th magnitude and therefore visible in 7 × glasses. Brightest of all, at magnitude 3.3, is Theta² (θ^2) Tauri, which teams with the magnitude-3.9 Theta¹ (θ^1) Tauri to form a wide naked-eye double star. Several other group members also pair up into attractive double and multiple systems.

M1 can be spotted by sharp-eyed observers through telescopes and large binoculars as a faint glimmer near Zeta (ζ) Tauri. Look for a pair of faint stars to the north-northwest of Zeta and follow an imaginary line connecting them for about ½° to the west, where M1 awaits.

Since its discovery in 1731 by London physician John Bevis, M1 has attracted more attention than just about any other object in our survey. Charles Messier mistook it at first for a comet. His independent discovery in 1758 ultimately led to the compilation of his now-famous list of deep-sky objects. The advent of astrophotography renewed interest in M1. Film revealed detail that was never suspected by visual observations. Instead of an ill-defined smudge of gray light, M1 became known for its finely intertwined cloudy knots and wisps. The many spiny nebulous projections first revealed in these early deep-sky photographs led to M1 being nicknamed the Crab Nebula.

In the 1960s radio astronomers found that M1 contains a pulsar — the Crab's rapidly beating "heart." This incredibly dense stellar oddity is a neutron star rotating 33 times a second and sweeping a beam of energy across the Earth with every pass. The Crab Pulsar remains one of the fastest of its kind known today. Its discovery confirmed prior speculation that the Crab Nebula is the remnant of a tremendous supernova observed by Chinese astronomers and Native American skywatchers in A.D. 1054.

Small- to medium-aperture telescopes will show no more than the view reproduced in Eyepiece Impression 43 — an 8th-magnitude grayish smudge with little or no detail. Indeed, many first-time visitors to M1 often comment on the nebula's unimpressiveness, so don't feel bad if it leaves you wanting more. In a way that's good, for the next time you pause at M1 you may notice one or two subtle details that you missed before. The intricate irregularities, so prominent in photographs, begin to appear in 254-mm and larger instruments. Large apertures increase the mottled look of M1, though the famous crablike appearance is difficult to detect visually.

While the cold of winter may deter many from going outdoors, the die-hard deep-sky observer relishes the crisp, clear skies that the season often brings. By dressing wisely and planning ahead, you too can enjoy the beauty of the winter skies that is missed by fair-weather astronomers.

The Constellations

Those printed in bold have a section devoted to them in chapters 6 through 9. The last column lists the season when the constellation is best visible from the Northern Hemisphere in the evening. For spring constellations, see chapter 6; summer, chapter 7; autumn, chapter 8; and winter, chapter 9.

Constellation Name	Abbreviation	Genitive	Meaning	Season
Andromeda	And	Andromedae	Princess	Autumn
Antlia	Ant	Antliae	Air Pump	Spring
Apus	Aps	Apodis	Bird of Paradise	Summer
Aquarius	Aqr	Aquarii	Water-bearer	Autumn
Aquila	Aql	Aquilae	Eagle	Summer
Ara	Ara	Arae	Altar	Summer
Aries	Ari	Arietis	Ram	Autumn
Auriga	Aur	Aurigae	Charioteer	Winter
Boötes	Boo	Boötis	Herdsman	Spring
Caelum	Cae	Caeli	Chisel	Winter
Camelopardalis	Cam	Camelopardalis	Giraffe	Winter
Cancer	Cnc	Cancri	Crab	Spring
Canes Venatici	CVn	Canum Venaticorum	Hunting Dogs	Spring
Canis Major	CMa	Canis Majoris	Big Dog	Winter
Canis Minor	CMi	Canis Minoris	Little Dog	Winter
Capricornus	Cap	Capricorni	Sea-goat	Autumn
Carina	Car	Carinae	Keel (of the ship Argo)	Spring
Cassiopeia	Cas	Cassiopeiae	Queen	Autumn
Centaurus	Cen	Centauri	Centaur	Spring
Cepheus	Cep	Cephei	King	Autumn
Cetus	Cet	Ceti	Whale	Autumn
Chamaeleon	Cha	Chamaeleontis	Chameleon	Spring
Circinus	Cir	Circini	Compasses	Summer
Columba	Col	Columbae	Dove	Winter
Coma Berenices	Com	Comae Berenices	Berenice's Hair	Spring
Corona Australis	CrA	Coronae Australis	Southern Crown	Summer
Corona Borealis	CrB	Coronae Borealis	Northern Crown	Spring
Corvus	Crv	Corvi	Crow	Spring
Crater	Crt	Crateris	Cup	Spring
Crux	Cru	Crucis	Cross	Spring
Cygnus	Cyg	Cygni	Swan	Summer
Delphinus	Del	Delphini	Dolphin	Summer
Dorado	Dor	Doradus	Swordfish	Winter
Draco	Dra	Draconis	Dragon	Summer
Equuleus	Equ	Equulei	Little Horse	Autumn
Eridanus	Eri	Eridani	River	Winter
Fornax	For	Fornacis	Furnace	Winter
Gemini	Gem	Geminorum	Twins	Winter
Grus	Gru	Gruis	Crane	Autumn
Hercules	Her	Herculis	Giant	Summer
Horologium	Hor	Horologii	Clock	Winter
Hydra	Hya	Hydrae	Water Snake (male)	Spring
Hydrus	Hyi	Hydri	Water Snake (female)	Autumn

Constellation Name	Abbreviation	Genitive	Meaning	Season
Indus	Ind	Indi	Indian	Autumn
Lacerta	Lac	Lacertae	Lizard	Autumn
Leo	Leo	Leonis	Lion	Spring
Leo Minor	LMi	Leonis Minoris	Little Lion	Spring
Lepus	Lep	Leporis	Hare	Winter
Libra	Lib	Librae	Scales of Justice	Spring
Lupus	Lup	Lupi	Wolf	Summer
Lynx	Lyn	Lyncis	Lynx	Spring
Lyra	Lyr	Lyrae	Lyre	Summer
Mensa	Men	Mensae	Table	Winter
Microscopium	Mic	Microscopii	Microscope	Autumn
Monoceros	Mon	Monocerotis	Unicorn	Winter
Musca	Mus	Muscae	Fly	Spring
Norma	Nor	Normae	Square	Summer
Octans	Oct	Octantis	Octant	Autumn
Ophiuchus	Oph	Ophiuchi	Serpent-bearer	Summer
Orion	Ori	Orionis	Hunter	Winter
Pavo	Pav	Pavonis	Peacock	Summer
Pegasus	Peg	Pegasi	Flying Horse	Autumn
Perseus	Per	Persei	Warrior	Autumn
Phoenix	Phe	Phoenicis	Phoenix	Autumn
Pictor	Pic	Pictoris	Painter	Winter
Pisces	Psc	Piscium	Fishes	Autumn
Piscis Austrinus	PsA	Piscis Austrini	Southern Fish	Autumn
Puppis	Pup	Puppis	Stern (of the ship Argo)	Winter
Pyxis	Pyx	Pyxidis	Compass	Spring
Reticulum	Ret	Reticuli	Reticle	Winter
Sagitta	Sge	Sagittae	Arrow	Summer
Sagittarius	Sgr	Sagittarii	Archer	Summer
Scorpius	Sco	Scorpii	Scorpion	Summer
Sculptor	Scl	Sculptoris	Sculptor	Autumn
Scutum	Sct	Scuti	Shield	Summer
Serpens	Ser	Serpentis	Serpent	Summer
Sextans	Sex	Sextantis	Sextant	Spring
Taurus	Tau	Tauri	Bull	Winter
Telescopium	Tel	Telescopii	Telescope	Summer
Triangulum	Tri	Trianguli	Triangle	Autumn
Triangulum Australe	TrA	Trianguli Australis	Southern Triangle	Summer
Tucana	Tuc	Tucanae	Toucan	Autumn
Ursa Major	UMa	Ursae Majoris	Great Bear	Spring
Ursa Minor	UMi	Ursae Minoris	Little Bear	Summer
Vela	Vel	Velorum	Sails (of the ship Argo)	Spring
Virgo	Vir	Virginis	Maiden	Spring
Volans	Vol	Volantis	Flying Fish	Spring
Vulpecula	Vul	Vulpeculae	Fox	Summer

The Greek Alphabet

α	alpha
β	beta
γ	gamma
δ	delta
ε	epsilon
ζ	zeta
η	eta
θ	theta
ι	iota
κ	kappa
λ	lambda
μ	mu
ν	nu
ξ	xi
ο	omicron
π	pi
ρ	rho
σ	sigma
τ	tau
υ	upsilon
φ	phi
χ	chi
ψ	psi
ω	omega

Deep-Sky Objects

This table lists the deep-sky objects described in chapters 6 through 9, shown in the order in which they are described, and sorted by season. Types of object: **As** asterism, **BN** bright nebula, **DN** dark nebula, **GC** globular cluster, **Gx** galaxy, **OC** open cluster, **PN** planetary nebula, **Q** quasar, **Vr** variable star, ** double star. In the **Mag.** column, **p** denotes photographic magnitude. Constellation is abbreviated **Const. Size** is from *Sky Catalogue 2000.0;* objects (especially galaxies) may appear smaller in amateur telescopes. **Sep.** is a double star's apparent separation. **Per.** is a variable star's period of variability.

Object M	NGC	Other	Type	Const.	R.A. h m	Dec. ° '	Mag.	Size, Sep., or Per.	Remarks
SPRING									
		Xi	**	Boo	14 51.4	+19 06	4.7, 7.0	6"	ADS 9413
		Zeta	**	Cnc	08 12.2	+17 39	5.6, 6.0	1"	ADS 6650
							6.2	6"	
44	2632		OC	Cnc	08 40.1	+19 46	3.1	95'	Beehive or Praesepe
67	2682		OC	Cnc	08 51.4	+11 49	6.9	30'	
106	4258		Gx	CVn	12 19.0	+47 18	8.3	18'×8'	
94	4736		Gx	CVn	12 50.9	+41 07	8.2	11'×9'	
63	5055		Gx	CVn	13 15.8	+42 02	8.6	12'×8'	
51	5194		Gx	CVn	13 29.9	+47 12	8.4	11'×8'	
	5195		Gx	CVn	13 30.0	+47 16	9.6	5'×4'	companion of M51
3	5272		GC	CVn	13 42.2	+28 23	6.4	16'	
	2516		OC	Car	07 58.3	–60 52	3.8	30'	
	3114		OC	Car	10 02.7	–60 07	4.2	35'	
	3293		OC	Car	10 35.8	–58 14	4.7	6'	
	3372		BN	Car	10 43.8	–59 52	5	120'	Eta Carinae Nebula
	3532		OC	Car	11 06.4	–58 40	3.0	55'	
	3766		OC	Cen	11 36.1	–61 37	5.3	12'	
	5128		Gx	Cen	13 25.5	–43 01	7	18'×14'	Centaurus A
	5139		GC	Cen	13 26.8	–47 29	3.7	36'	Omega Centauri
	5460		OC	Cen	14 07.6	–48 19	5.6	25'	
		Alpha	**	Cen	14 39.6	–60 50	0.0, 1.2	22"	
							11.0	131'	Proxima Centauri
98	4192		Gx	Com	12 13.8	+14 54	10.1	10'×3'	
99	4254		Gx	Com	12 18.8	+14 25	9.8	5'	
100	4321		Gx	Com	12 22.9	+15 49	9.4	7'×6'	
	4322		Gx	Com	12 23.0	+15 54	13.9	1'	
	4328		Gx	Com	12 23.3	+15 48	13.5	1'	
		Mel 111	OC	Com	12 25	+26	1.8	275'	Coma Star Cluster
		17	**	Com	12 28.9	+25 55	5.3, 6.6	145"	ADS 8568
		24	**	Com	12 35.1	+18 23	5.2, 6.8	20"	ADS 8600
	4565		Gx	Com	12 36.3	+25 59	9.6	16'×3'	
64	4826		Gx	Com	12 56.7	+21 41	8.5	9'×5'	
53	5024		GC	Com	13 12.9	+18 10	7.7	13'	
	5053		GC	Com	13 16.4	+17 42	9.8	11'	
		U	Vr	CrB	15 18.2	+31 39	7.7–8.8	3.45d	
		S	Vr	CrB	15 21.4	+31 22	5.8–14.1	360d	
		R	Vr	CrB	15 48.6	+28 09	5.7–14.8		R CrB prototype
	4361		PN	Crv	12 24.5	–18 48	10.3p	45"	
		Delta	**	Crv	12 29.9	–16 31	3.0, 9.2	24"	ADS 8572
		STAR 20	As	Crv	12 35.7	–12 02		15'	Stargate
48	2548		OC	Hya	08 13.8	–05 48	5.8	55'	

Object			Type	Const.	R.A.	Dec.	Mag.	Size, Sep.,	Remarks
M	NGC	Other			h m	° '		or Per.	
	3242		PN	Hya	10 24.8	–18 38	8.6p	16"	Ghost of Jupiter
68	4590		GC	Hya	12 39.5	–26 45	8.2	12'	
83	5236		Gx	Hya	13 37.0	–29 52	7.6	11'×10'	
	2903		Gx	Leo	09 32.2	+21 30	9.0	13'×7'	
	2905		BN	Leo	09 32.2	+21 31			in NGC 2903
		Gamma	**	Leo	10 19.9	+19 52	2.2, 3.5	4"	ADS 7724
95	3351		Gx	Leo	10 44.0	+11 42	9.7	7'×5'	
96	3368		Gx	Leo	10 46.8	+11 49	9.2	7'×5'	
105	3379		Gx	Leo	10 47.8	+12 35	9.3	5'×4'	
	3384		Gx	Leo	10 48.3	+12 38	10.0	6'×3'	
65	3623		Gx	Leo	11 18.9	+13 05	9.3	10'×3'	
66	3627		Gx	Leo	11 20.2	+12 59	9.0	9'×4'	
	3628		Gx	Leo	11 20.3	+13 36	9.5	15'×4'	
	2419		GC	Lyn	07 38.1	+38 53	10.4	4'	
	2683		Gx	Lyn	08 52.7	+33 25	9.7	9'×3'	
	2818		OC	Pyx	09 16.0	–36 37	8.2	9'	
	2818A		PN	Pyx	09 16.0	–36 38	13.0p	38"	
	3115		Gx	Sex	10 05.2	–07 43	9.1	8'×3'	Spindle Galaxy
	2841		Gx	UMa	09 22.0	+50 58	9.3	8'×4'	
	2976		Gx	UMa	09 47.3	+67 55	10.2	5'×3'	
81	3031		Gx	UMa	09 55.6	+69 04	7.0	26'×14'	
82	3034		Gx	UMa	09 55.8	+69 41	8.4	11'×5'	
	3073		Gx	UMa	10 00.9	+55 37	13.4p	1'	
	3079		Gx	UMa	10 02.2	+55 41	10.6	8'×2'	
	3077		Gx	UMa	10 03.3	+68 44	9.9	5'×4'	
		STAR 19	As	UMa	10 51.0	+56 09		20'	Broken Engagement Ring
108	3556		Gx	UMa	11 11.5	+55 40	10.1	8'×3'	
97	3587		PN	UMa	11 14.8	+55 01	11.2	3'	Owl Nebula
		Zeta	**	UMa	13 23.9	+54 56	2.3, 4.0	14"	Mizar (ADS 8891)
101	5457		Gx	UMa	14 03.2	+54 21	7.7	27'×26'	Pinwheel Galaxy
		Gamma	**	Vel	08 09.5	–47 20	1.9, 4.2	41"	quintuple star
							8.2	62"	
							9.1	94"	
							12.5	2"	
	2547		OC	Vel	08 10.7	–49 16	4.7	20'	
		IC 2391	OC	Vel	08 40.2	–53 04	2.5	50'	
	2669		OC	Vel	08 44.9	–52 58	6.1	12'	
	3132		PN	Vel	10 07.7	–40 26	8.2	84"×53"	Eight-Burst Nebula
	3201		GC	Vel	10 17.6	–46 25	6.8	18'	
84	4374		Gx	Vir	12 25.1	+12 53	9.3	5'×4'	
	4387		Gx	Vir	12 25.7	+12 49	12.0	2'×1'	
	4388		Gx	Vir	12 25.8	+12 40	11.0	5'×1'	
	4402		Gx	Vir	12 26.1	+13 07	11.7	4'×1'	
86	4406		Gx	Vir	12 26.2	+12 57	9.2	7'×6'	
		3C 273	Q	Vir	12 29.1	+02 03	12.8		
	4476		Gx	Vir	12 30.0	+12 21	12.3	2'×1'	
	4478		Gx	Vir	12 30.3	+12 20	11.2	2'	
87	4486		Gx	Vir	12 30.8	+12 24	8.6	7'	
		STAR 21	As	Vir	12 38.5	–11 30		15'	Jaws
104	4594		Gx	Vir	12 40.0	–11 37	8.3	9'×4'	Sombrero Galaxy

Object M	NGC	Other	Type	Const.	R.A. h m	Dec. ° '	Mag.	Size, Sep., or Per.	Remarks
SUMMER									
	6188		BN	Ara	16 40.5	–48 47		20'×12'	
	6193		OC	Ara	16 41.3	–48 46	5.2	15'	
		h4876	**	Ara	16 41.3	–48 46	5.6, 8.9	2"	sextuple star
							6.8	10"	in NGC 6193
							10.4	13"	
							11.3	14"	
							12.4	21"	
		IC 4651	OC	Ara	17 24.7	–49 57	6.9	12'	
	6397		GC	Ara	17 40.7	–53 40	5.7	26'	
	5823		OC	Cir	15 05.7	–55 36	7.9	10'	chapter 7 under Lupus
	6541		GC	CrA	18 08.0	–43 42	6.6	13'	
		Beta	**	Cyg	19 30.7	+27 58	3.2, 5.4	34"	Albireo (ADS 12540)
	6819		OC	Cyg	19 41.3	+40 11	7.3	5'	
	6826		PN	Cyg	19 44.8	+50 31	9.8p	30"	Blinking Planetary
		Chi	Vr	Cyg	19 50.6	+32 55	3.3–14.2	407d	long-period variable
	6888		BN	Cyg	20 12.0	+38 21		20'×10'	Crescent Nebula
		STAR 26	As	Cyg	20 13.8	+36 30		45'	Red-Necked Emu
	6960		BN	Cyg	20 45.7	+30 43		70'×6'	Filamentary Nebula
			BN	Cyg	20 48.5	+31 09		45'×30'	Pickering's Wedge
	6974		BN	Cyg	20 50.8	+31 52			
		IC 5070	BN	Cyg	20 50.8	+44 21		80'×70'	Pelican Nebula
	6979		BN	Cyg	20 51.0	+32 09			
	6992		BN	Cyg	20 56.4	+31 43		60'×8'	Veil Nebula
	6997		OC	Cyg	20 56.5	+44 38	10.0p	7'	Great Lakes Cluster
		Lynds 935	DN	Cyg	20 56.8	+43 52		150'×40'	
		B352	DN	Cyg	20 57.1	+45 54		20'×10'	Hudson Bay Nebula
	7000		BN	Cyg	20 58.8	+44 20		120'×100'	North America Neb.
		STAR 28	As	Cyg	21 08.3	+47 14		25'	Horseshoe
39	7092		OC	Cyg	21 32.2	+48 26	4.6	32'	
		B168	DN	Cyg	21 53.2	+47 12		100'×10'	
		IC 5146	BN	Cyg	21 53.5	+47 16		12"	Cocoon Nebula
		STAR 9	As	Del	20 38	+13 10		60'×30'	Theta Delphini
		Gamma	**	Del	20 46.7	+16 07	4.3, 5.2	10"	ADS 14279
		STAR 27	As	Del	21 07.3	+16 20		15'	Dolphin's Diamonds
	6543		PN	Dra	17 58.6	+66 38	8.8p	18"	
		STAR 25	As	Dra	18 35.0	+72 25		10'×20'	Little Queen
		STAR 23	As	Her	16 37.8	+31 05		20'	Backwards S
13	6205		GC	Her	16 41.7	+36 28	5.9	16'	Great Hercules Cluster
	6207		Gx	Her	16 43.1	+36 50	11.6	3'×1'	
	6210		PN	Her	16 44.5	+23 49	9.3p	>14"	
		Alpha	**	Her	17 14.6	+14 23	3.5, 5.4	5"	Rasalgethi (ADS 10418)
92	6341		GC	Her	17 17.1	+43 08	6.5	11'	
		95	**	Her	18 01.5	+21 36	5.0, 5.1	6"	ADS 10993
		STAR 24	As	Her	18 02.5	+26 18		15'	
	5822		OC	Lup	15 05.2	–54 21	6.5	40'	
		Epsilon	**	Lyr	18 44.3	+39 40		208"	Double Double (ADS 11653)

Object			Type	Const.	R.A.	Dec.	Mag.	Size, Sep.,	Remarks
M	NGC	Other			h m	° '		or Per.	
							5.0, 6.1	3"	Epsilon-1
							5.2, 5.5	2"	Epsilon-2
57	6720		PN	Lyr	18 53.6	+33 02	9.7	70" × 150"	Ring Nebula
56	6779		GC	Lyr	19 16.6	+30 11	8.2	7'	
		606		OC	Nor	16 13.2	–54 13	5.6	13'
	6087		OC	Nor	16 18.9	–57 54	5.4	12'	
	6152		OC	Nor	16 32.7	–52 37	8.1p	30'	
12	6218		GC	Oph	16 47.2	–01 57	6.6	15'	
10	6254		GC	Oph	16 57.1	–04 06	6.6	15'	
62	6266		GC	Oph	17 01.2	–30 07	6.6	14'	
		B59, 65–7	DN	Oph	17 21	–27		300' × 60'	stem of Pipe Nebula
		B72	DN	Oph	17 23.5	–23 28		4'	Snake Nebula
	6369		PN	Oph	17 29.3	–23 46	12.9p	30"	
		B78	DN	Oph	17 33	–26		200' × 140'	bowl of Pipe Nebula
		IC 4665	OC	Oph	17 46.3	+05 43	4.2	41'	
		70	**	Oph	18 05.5	+02 30	4.2, 6.0	4"	
	6572		PN	Oph	18 12.1	+06 51	9.0p	8"	
	6752		GC	Pav	19 10.9	–59 59	5.4	20'	
	6440		GC	Sgr	17 48.9	–20 22	9.7p	5'	
	6445		PN	Sgr	17 49.2	–20 01	13.2p	›34"	
23	6494		OC	Sgr	17 56.8	–19 01	5.5	27'	
20	6514		BN	Sgr	18 02.6	–23 02	9	29' × 27'	Trifid Nebula
	6520		OC	Sgr	18 03.4	–27 54	7.6p	6'	
		B88–9, 296	DN	Sgr	18 03.8	–24 23			regions in M8
8	6523		BN	Sgr	18 03.8	–24 23	6	90' × 40'	Lagoon Nebula
		B87	DN	Sgr	18 04.3	–32 30		12'	Parrot's Head Nebula
21	6531		OC	Sgr	18 04.6	–22 30	5.9	13'	
		B92	DN	Sgr	18 15.5	–18 11		12' × 6'	in M24
		B93	DN	Sgr	18 16.9	–18 04		12' × 2'	in M24
24			OC	Sgr	18 16.9	–18 29	4.5	90'	Small Sgr Star Cloud
	6603		OC	Sgr	18 18.4	–18 25	11.1p	5'	in M24
17	6618		BN	Sgr	18 20.8	–16 11	7	46' × 37'	Omega Nebula
25		IC 4725	OC	Sgr	18 31.6	–19 15	4.6	32'	
	6645		OC	Sgr	18 32.6	–16 54	8.5p	10'	
22	6656		GC	Sgr	18 36.4	–23 54	5.1	24'	
55	6809		GC	Sgr	19 40.0	–30 58	7.0	19'	
	6818		PN	Sgr	19 44.0	–14 09	9.9p	›17"	
	6822		Gx	Sgr	19 44.9	–14 48	9.4p	10'	Barnard's Galaxy
75	6864		GC	Sgr	20 06.1	–21 55	8.6	6'	
		Beta	**	Sco	16 05.4	–19 48	2.6, 4.9	14"	Graffias (ADS 9913)
80	6093		GC	Sco	16 17.0	–22 59	7.2	9'	
4	6121		GC	Sco	16 23.6	–26 32	6.0	26'	
	6124		OC	Sco	16 25.6	–40 40	5.8	29'	
	6231		OC	Sco	16 54.0	–41 48	2.6	15'	
	6281		OC	Sco	17 04.8	–37 54	5.4	8'	
	6322		OC	Sco	17 18.5	–42 57	6.0	10'	
	6388		GC	Sco	17 36.3	–44 44	6.9	9'	
6	6405		OC	Sco	17 40.1	–32 13	4.2	15'	Butterfly Cluster
	6453		GC	Sco	17 50.9	–34 36	9.9	4'	
7	6475		OC	Sco	17 53.9	–34 49	3.3	80'	

Object M	NGC	Other	Type	Const.	R.A. h m	Dec. ° '	Mag.	Size, Sep., or Per.	Remarks
	6496		GC	Sco	17 59.0	–44 16	9.2	7'	
11	6705		OC	Sct	18 51.1	–06 16	5.8	14'	Wild Duck Cluster
5	5904		GC	Ser	15 18.6	+02 05	5.8	17'	
16	6611		OC/BN	Ser	18 18.8	–13 47	6	35' × 28'	Eagle Nebula
		IC 4756	OC	Ser	18 39.0	+05 27	5.4p	52'	
		Alpha	**	UMi	02 31.8	+89 16	2.0, 8.9	18"	Polaris (ADS 1477)
		STAR 22	As	UMi	16 29.0	+80 13		15'	Mini-Coathanger
		Cr 399	OC	Vul	19 25.4	+20 11	3.6	60'	Coathanger Cluster
27	6853		PN	Vul	19 59.6	+22 43	8.0	8' × 4'	Dumbbell Nebula
	6940		OC	Vul	20 34.6	+28 18	6.3	31'	
AUTUMN									
	7662		PN	And	23 25.9	+42 33	9.2	20"	
110	205		Gx	And	00 40.4	+41 41	8.0	17' × 10'	
	206		OC	And	00 40.6	+40 44			star cloud in M31
32	221		Gx	And	00 42.7	+40 52	8.2	8' × 6'	
31	224		Gx	And	00 42.7	+41 16	3.5	160' × 40'	Andromeda Galaxy
	404		Gx	And	01 09.4	+35 43	10.1	4'	
		STAR 14	As	And	01 52.5	+37 30		20' × 95'	Golf Putter
	752		OC	And	01 57.8	+37 41	5.7	50'	
		Gamma	**	And	02 03.9	+42 20	2.3, 5.5	10"	ADS 1630
							6.3	0".5	
	891		Gx	And	02 22.6	+42 21	10.0	13' × 3'	
	7009		PN	Aqr	21 04.2	–11 22	8.3	25"	Saturn Nebula
2	7089		GC	Aqr	21 33.5	–00 49	6.5	13'	
	7293		PN	Aqr	22 29.6	–20 48	6.5	15' × 12'	Helix Nebula
		Gamma	**	Ari	01 53.5	+19 18	4.8, 4.8	8"	ADS 1507
		STAR 29	As	Cas	23 03	+59 30		70' × 125'	Lucky 7
		STAR 12	As	Cas	23 20	+62 20		60'	Airplane
52	7654		OC	Cas	23 24.2	+61 35	6.9	13'	
	7789		OC	Cas	23 57.0	+56 44	6.7	16'	
	147		Gx	Cas	00 33.2	+48 30	9.3	13' × 8'	
	185		Gx	Cas	00 39.0	+48 20	9.2	12' × 10'	
		Eta	**	Cas	00 49.1	+57 49	3.5, 7.5	12"	ADS 671
	281		BN	Cas	00 52.8	+56 36		35' × 30'	
	457		OC	Cas	01 19.1	+58 20	6.4	13'	
103	581		OC	Cas	01 33.2	+60 42	7.4	6'	
		STAR 13	As	Cas	01 38	+58 30		220' × 160'	Queen's Kite
	663		OC	Cas	01 46.0	+61 15	7.1	16'	
		Stock 2	OC	Cas	02 15.0	+5916	4.4	60'	
		Iota	**	Cas	02 29.1	+67 24	4.6, 6.9	3"	ADS 1860
						8.4	7"		
		STAR 15	As	Cas	03 28	+72 00		90' × 30'	Kemble's Kite
	6939		OC	Cep	20 31.4	+60 38	7.8	8'	
	6946		Gx	Cep	20 34.8	+60 09	8.8	11' × 10'	
		Delta	**/Vr	Cep	22 29.2	+58 25	4, 7.5	41"	Cepheid prototype
	40		PN	Cep	00 13.0	+72 32	10.7p	›37"	
	246		PN	Cet	00 47.0	–11 53	8.5p	225"	
		Omicron	Vr	Cet	02 19.3	–02 59	2.0–10.1	332d	Mira
	1055		Gx	Cet	02 41.8	+00 26	10.6	8' × 3'	
77	1068		Gx	Cet	02 42.7	–00 01	8.9	6' × 5'	

Object M	NGC	Other	Type	Const.	R.A. h m	Dec. ° '	Mag.	Size, Sep., or Per.	Remarks
	7209		OC	Lac	22 05.2	+46 30	6.7	25'	
15	7078		GC	Peg	21 30.0	+12 10	6.4	12'	
	7217		Gx	Peg	22 07.9	+31 22	10.2	4'×3'	
	7317		Gx	Peg	22 35.9	+33 57	13.6	1'	
	7318A		Gx	Peg	22 35.9	+33 58	13.3	1'	
	7318B		Gx	Peg	22 36.0	+33 58	13.1	2'×1'	
	7320		Gx	Peg	22 36.1	+33 57	12.7	2'×1'	
	7319		Gx	Peg	22 36.1	+33 59	13.1	2'×1'	
	7331		Gx	Peg	22 37.1	+34 25	9.5	10'×4'	
76	650–1		PN	Per	01 42.4	+51 34	11.5	2'×1'	Little Dumbbell
	869		OC	Per	02 19.0	+57 09	4.3	30'	} Double Cluster
	884		OC	Per	02 22.4	+57 07	4.4	30'	
34	1039		OC	Per	02 42.0	+42 47	5.2	35'	
		Beta	Vr	Per	03 08.2	+40 57	2.1–3.4	2.87d	Algol, eclipsing binary
		Mel 20	OC	Per	03 22	+49	1.2	185'	Alpha Persei Cluster
	1499		BN	Per	04 00.7	+36 37		145'×40'	California Nebula
	55		Gx	Scl	00 14.9	–39 11	8.2	32'×7'	
	253		Gx	Scl	00 47.6	–25 17	7.1	25'×7'	
	288		GC	Scl	00 52.8	–26 35	8.1	14'	
33	598		Gx	Tri	01 33.9	+30 39	5.7	62'×35'	Triangulum Galaxy
	604		BN	Tri	01 34.5	+30 48		2'×1'	knot in M33
WINTER									
		STAR 4	As	Aur	05 19	+33 40		75'	Flying Minnow
38	1912		OC	Aur	05 28.7	+35 50	6.4	21'	
	1931		BN	Aur	05 31.4	+34 15		3'	
36	1960		OC	Aur	05 36.1	+34 08	6.0	12'	
37	2099		OC	Aur	05 52.4	+32 33	5.6	24'	
		STAR 3	As	Cam	04 00	+63		150'	Kemble's Cascade
	1501		PN	Cam	04 07.0	+60 55	13.3p	52"	
	1502		OC	Cam	04 07.7	+62 20	5.7	8'	
	2403		Gx	Cam	07 36.9	+65 36	8.4	18'×10'	
41	2287		OC	CMa	06 46.0	–20 44	4.5	38'	
		h3945	**	CMa	07 16.6	–23 18	4.8, 6.8	27"	ADS 5951
	2359		BN	CMa	07 18.6	–13 12		8'×6'	
		Cr 140	OC	CMa	07 23.9	–32 12	3.5	42'	
	1851		GC	Col	05 14.1	–40 03	7.3	11'	
		Theta	**	Eri	02 58.3	–40 18	3.4, 4.5	8"	
	1297		Gx	Eri	03 19.2	–19 06	12.7p	2'	
	1300		Gx	Eri	03 19.7	–19 25	10.4	7'×4'	
		32	**	Eri	03 54.3	–02 58	4.7, 6.2	7"	ADS 2850
	1535		PN	Eri	04 14.2	–12 44	9.3p	18"	
	1316		Gx	For	03 22.7	–37 12	8.9	7'×6'	
	1317		Gx	For	03 22.8	–37 06	11.0	3'	
	1360		PN	For	03 33.3	–25 51		390"	
	1365		Gx	For	03 33.6	–36 08	9.5	10'×6'	
	2158		OC	Gem	06 07.5	+24 06	8.6	5'	SW of M35
35	2168		OC	Gem	06 08.9	+24 20	5.1	28'	
	2392		PN	Gem	07 29.2	+20 55	9.9p	13"	Eskimo Nebula
		R	Vr	Lep	04 59.6	–14 48	5.5–11.7	432d	Hind's Crimson Star
79	1904		GC	Lep	05 24.5	–24 33	8.0	9'	

Object M	NGC	Other	Type	Const.	R.A. h m	Dec. ° '	Mag.	Size, Sep., or Per.	Remarks
	2170		BN	Mon	06 07.5	–06 24		2'	
	2182		BN	Mon	06 09.5	–06 20		3'	
	2183		BN	Mon	06 10.8	–06 13		1'	
	2185		BN	Mon	06 11.1	–06 13		3'	
		Beta	**	Mon	06 28.8	–07 01	4.7, 5.2	7"	quadruple star
							6.1	10"	(ADS 5107)
							12.2	26"	
	2237		BN	Mon	06 32.3	+05 03		80' × 60'	Rosette Nebula
	2244		OC	Mon	06 32.4	+04 52	4.8	24'	in Rosette Nebula
	2261		BN	Mon	06 39.2	+08 44	10	2'	Hubble's Variable Neb.
		STAR 17	As	Mon	06 40.5	–09 00		10'	Unicorn's Horn
	2264		OC	Mon	06 41.1	+09 53	3.9	20'	Christmas Tree
	2301		OC	Mon	06 51.8	+00 28	6.0	12'	Great Bird of Galaxy
		STAR 18	As	Mon	06 52.5	–10 10			Pakan's 3
	2316		BN	Mon	06 59.7	–07 46		4' × 3'	
50	2323		OC	Mon	07 02.8	–08 23	5.9	16'	
	1788		BN	Ori	05 06.9	–03 21		8' × 5'	
		Theta-1	**	Ori	05 35.2	–05 23	6.7, 7.9	9"	Trapezium (ADS 4186)
							5.1	13"	
							6.7	22"	
42	1976		BN	Ori	05 35.4	–05 27	4	66' × 60'	Orion Nebula
	1973–5–7		BN	Ori	05 35.5	–04 52	5	20' × 10'	
43	1982		BN	Ori	05 35.6	–05 16	7	20' × 15'	NW part of M42
	2022		PN	Ori	05 42.1	+09 05	12.4p	›18"	
	2024		BN	Ori	05 40.7	–02 27		30' × 30'	
		B33	DN	Ori	05 40.9	–02 28		6' × 4'	Horsehead Nebula
		IC 434	BN	Ori	05 41.0	–02 24		60' × 10'	behind B33
	2023		BN	Ori	05 41.6	–02 14		10'	
	2064		BN	Ori	05 46.3	00 00		12' × 2'	
	2067		BN	Ori	05 46.5	+00 06		8' × 3'	
78	2068		BN	Ori	05 46.7	+00 03	8	8' × 6'	
	2071		BN	Ori	05 47.2	+00 18		4' × 3'	
47	2422		OC	Pup	07 36.6	–14 30	4.5	30'	
	2438		PN	Pup	07 41.8	–14 44	10.1	›66"	in M46
46	2437		OC	Pup	07 41.8	–14 49	6.1	27'	
	2440		PN	Pup	07 41.9	–18 13	10.8p	14"	
93	2447		OC	Pup	07 44.6	–23 52	6.2	22'	
	2451		OC	Pup	07 45.4	–37 58	2.8	45'	
	2477		OC	Pup	07 52.3	–38 33	5.8	27'	
	2467		OC/BN	Pup	07 52.6	–26 23		8' × 7'	
	2546		OC	Pup	08 12.4	–37 38	6.3	41'	
	1432		BN	Tau	03 45.8	+24 22		30'	nebulosity in M45
	1435		BN	Tau	03 46.1	+23 47		30'	nebulosity in M45
45			OC	Tau	03 47.0	+24 07	1.2	110'	Pleiades
		STAR 16	As	Tau	04 22.5	+21 45		200' × 90'	Little Dog
		Mel 25	OC	Tau	04 27	+16	0.5	330'	Hyades
1	1952		BN	Tau	05 34.5	+22 00	8	6' × 4'	Crab Nebula

Bibliography

Books

Atlas of the Andromeda Galaxy
Paul W. Hodge
University of Washington Press, 1981

Bedford Catalogue, The
Admiral William H. Smyth
Willmann-Bell, Inc., 1986
(reprint of 1844 edition)

Burnham's Celestial Handbook (3 volumes)
Robert Burnham, Jr.
Dover Publications, Inc., 1978
(reprint of 1966 edition)

Cambridge Star Atlas 2000.0, 2nd edition
Wil Tirion
Cambridge University Press, 1996

Celestial Objects for Common Telescopes
(2 volumes)
Rev. T. W. Webb
Dover Publications, Inc. 1962
(reprint of 1917 edition)

Constellations
Lloyd Motz and Carol Nathanson
Doubleday, 1988

Deep-Sky Field Guide to Uranometria
Murray Craigin, James Lucyk, and
Barry Rappaport
Willmann-Bell, Inc., 1993

Edmund Mag 6 Star Atlas
Terence Dickinson, Victor Costanzo
and Glenn F. Chaple, Jr.
Edmund Scientific Co., 1982

Hartung's Astronomical Objects for Southern Telescopes, 2nd edition
Revised and illustrated by David Malin
and David J. Frew
Cambridge University Press, 1995

Messier Album, The
John H. Mallas and Evered Kreimer
Sky Publishing Corp., 1978
P.O. Box 9111, Belmont, MA 02178
(800) 253-0245

Messier's Nebulae and Star Clusters,
2nd edition
Kenneth Glyn Jones
Cambridge University Press, 1991

Monthly Star Charts
George Lovi and Graham Blow
Sky Publishing Corp., 1995
(details above)

Norton's 2000.0 Star Atlas and Reference Handbook, 18th edition
Ian Ridpath, editor
Longman/Addison Wesley, 1989

Observing Handbook and Catalogue of Deep-Sky Objects
Christian Luginbuhl and Brian Skiff
Cambridge University Press, 1989

1000+: The Amateur Astronomer's Field Guide to Deep Sky Observing
Tom Lorenzin
(available from P.O. Box 700,
Davidson, NC 28036, 1987)

Sky Atlas 2000.0, 2nd edition
Wil Tirion and Roger W. Sinnott
Sky Publishing Corp./Cambridge
University Press, 1998

Star Ware
Philip S. Harrington
John Wiley & Sons, Inc., 1994

Star Ware homepage:
http://ourworld.compuserve.com/homepages/pharrington/

Star-Hopping for Backyard Astronomers
Alan M. MacRobert
Sky Publishing Corp., 1993
(details above)

Touring the Universe Through Binoculars
Philip S. Harrington
John Wiley & Sons, 1990

Universe from Your Backyard, The
David J. Eicher
Kalmbach Publishing Co./
Cambridge University Press, 1988

Uranometria 2000.0 (2 volumes)
Wil Tirion, Barry Rappaport and
George Lovi
Willmann-Bell, Inc., 1987

Webb Society Deep-Sky Observer's Handbook (8 volumes)
edited by Kenneth Glyn Jones
Enslow, 1979-1990 (available only from the Webb Society; see p. 226 for address)

oftware

Deep-Sky Planner
Phyllis Lang
Sky Publishing Corp. (details above)
http://www.skypub.com/catalog/dsp20/dsplan.html

Deep Space
David Chandler Co.
P.O. Box 309, La Verne, CA 91750
(800) 516-9756
http://www.csz.com/dschandler/

Earth-Centered Universe
Nova Astronomics
P.O. Box 31013, Halifax, NS
B3K 5T9 Canada
http://www.nova-astro.com/

Guide
Project Pluto
Ridge Rd., Bowdoinham, ME 04008
(800) 777-5886
http://www.projectpluto.com/

Megastar
Willmann-Bell, Inc.
Box 35025, Richmond, VA 23235
(804) 320-7016
http://www.willbell.com/

NGC 2000.0, NGC 2000.0 Optimizer
by Dan Gray, and
Hyper NGC 2000.0 by Jonathan Ausubel
Sky Publishing Corp. (details as above)
http://www.skypub.com/catalog/atlascat.html

New General Program of Nonstellar Astronomical Objects
Dean Williams
http://www.cei.net/~deanw/software.html

PCSky
CapellaSoft
P.O. Box 3964, La Mesa, CA 91944
(619) 460-8265
http://www.psnw.com/~crink/cs.html

Redshift 2
Maris Multimedia, Inc.
726 Alia St., Novato, CA 94945
(415) 492-2819
http://www.maris.com/

Starry Night Deluxe
Sienna Software
538-366 Adelaide St. E., Toronto, ON
M5A 3X9, Canada
(416) 410-0259
http://www.siennasoft.com/

TheSky
Software Bisque
912 Twelfth Street, Suite A
Golden, CO 80401
(800) 843-7599
http://www.bisque.com/thesky/

Voyager II
Carina Software
12919 Alcosta Blvd., #7
San Ramon, CA 94543
(800) 493-8555
http://www.carinasoft.com/

Useful Addresses

Manufacturers of telescopes (R refractors, F reflectors, C catadioptrics)

Astro-Physics, Inc. (R)
11250 Forest Hills Rd.
Rockford, IL 61115
(815) 282-1513
http://www.astro-physics.com/

Celestron International (R,F,C)
2835 Columbia St.
Torrance, CA 90503
(310) 328-9560
http://www.celestron.com/

Edmund Scientific Co. (R,F,C,)
N964 Edscorp Building
101 E. Gloucester Pike
Barrington, NJ 08007
(609) 573-6250
E-mail: scientifics@edsci.com/

Jim's Mobile, Inc. (JMI) (F)
810 Quail St., Unit E
Lakewood, CO 80215
(303) 233-5353

Meade Instruments Corp. (R,F,C)
6001 Oak Canyon
Irvine, CA 92620
(714) 451-1450
http://www.meade.com/

Obsession Telescopes (F)
P.O. Box 804s
Lake Mills, WI 53551
(920) 648-2328
E-mail:obsessiontscp@globaldialog.com

Telescope & Binocular Center (R,F)
P.O. Box 1815-S
Santa Cruz, CA 95061
(800) 447-1001
E-mail: sales@oriontel.com
http://www.oriontel.com

StarMaster Telescopes (R)
Rt. 1, Box 780
Arcadia, KS 66711
(316) 638-4743
E-mail: starmaster@ckt.net
http://www.icstars.com/starmaster/

Starsplitter Telescopes (F)
3228 Rikkard Dr.
Thousand Oaks, CA 91362
(805) 492-0489
E-mail: strspltr@aol.com
http://www.ez2.net/starsplitter/

Tectron Telescopes (F)
3544 Oak Grove Dr.
Sarasota, FL 34243
(941) 758-9890
E-mail: aatclark@aol.com
http://www.icstars.com/tectron

Tele Vue Optics, Inc. (R)
100 Route 59
Suffern, NY 10901
(914) 357-9522
http://www.televue.com/

Manufacturers of eyepieces

Celestron International (details as above

Meade Instruments Corp.
(details as above)

Telescope & Binocular Center
(details as above)

Tele Vue Optics, Inc. (details as above)

VERNONscope & Company
5 Ithaca Road, Candor, NY 13743
(607) 659-7000

Manufacturers of filters

DayStar Filter Corp.
P.O. Box 5110
Diamond Bar, CA 91765
(909) 591-4673

Lumicon
2111 Research Dr., Suites 4 & 5
Livermore, CA 94550
(800) 767-9576
E-mail: marling@pacbell.net
http://www.astronomy-mall.com/

Telescope & Binocular Center
(details as above)

Thousand Oaks Optical
Box 4813
Thousand Oaks, CA 91359
(805) 491-3642

Roger W. Tuthill, Inc.
11 Tanglewood Lane
Mountainside, NJ 07092
(908) 232-1786

Astronomical equipment dealers outside the United States

Canada

CosmicConnection
32 Ashgrove Blvd.
Brandon, Manitoba R7B 1C2
(204) 727-3111

EfstonScience Inc.
3350 Dufferin St.
Toronto, Ontario M6A 3A4
(416) 787-4581

Khan Scope Centre
3243 Dufferin St.,
Toronto, Ontario M6A 2T2
(416) 783-4140
E-mail: khan@globalserve.net
http://www.khanscope.com

United Kingdom

Beacon Hill Telescopes
112 Mill Rd., Cleethorpes,
South Humberside DN35 8JD
+44 (0) 1472 692959

Broadhurst, Clarkson & Fuller, Ltd.,
Telescope House
63 Farringdon Rd., London EC1M 3JB
+44 (0) 171 405 2156

Orion Optics
Unit 12, Quakers Coppice
Crewe Gates Industrial Estate
Crewe, Cheshire CW1 1XU
+44 (0) 1270 500089

Venturescope
Wren Centre
Westbourne Road, Emsworth
Hampshire PO10 7RN
+44 (0) 1243-379322
E-mail: orders@venturescope.co.uk

Periodicals

Sky & Telescope, SkyWatch
Sky Publishing Corp.
P.O. Box 9111
Belmont, MA 02178-9111
(800) 253-0245
http://www.skypub.com/

Astronomy, Explore the Universe
Kalmbach Publishing Co.
21027 Crossroads Circle
Waukesha, WI 53187
(414) 796-8776
http://www.astronomy.com/

Griffith Observer
Griffith Observatory
2800 E. Observatory Rd.
Los Angeles, CA 90027
(213) 664-1181
http://www.GriffithObs.org/

Mercury
Astronomical Society of the Pacific
390 Ashton Avenue
San Francisco, CA 94112
(415) 337-1100
http://www.aspsky.org/

Starry Messenger
Starry Messenger Press
P. O. Box 6422, Ithaca, NY 14851
(607) 277-6836
http://www.starrymessenger.com/

Further information on deep-sky observing

American Association of Variable Star Observers (AAVSO)
25 Birch Street
Cambridge, MA 02138-1205
(617) 354-0484
http://www.aavso.org/

Astronomical League
1719 N Street, NW
Washington, DC 20036
(847) 398-0562
http://www.mcs.net/bstevens/al/

Royal Astronomical Society of Canada
136 Dupont St.
Toronto, Ontario M5R 1V2 Canada
(416) 924-7973
http://www.rasc.ca/

National Deep Sky Observers Society
Alan Goldstein
1607 Washington Blvd.
Louisville, KY 40242
(502) 426-4399
http://www.cismall.com/deepsky/index.html

Webb Society of North America
John E. Isles, Secretary
11105 Tremont Lane
Plymouth, MI 48170
(313) 451-7069
http://www.webbsociety.org/

Internet sites that list astronomy clubs

SKY Online
http://www.skypub.com/

AstroNet
http://www.rahul.resource/regular/clubs-etc/clubsetc.html

Astronomical Society of the Pacific
http://maxwell.sfsu.edu/asp/amateur.html

The Messier Catalogue

1. Object type:

**	=	Double Star
As	=	Asterism
BN	=	Bright Nebula
GC	=	Globular Cluster
Gx	=	Galaxy
OC	=	Open Cluster
PN	=	Planetary Nebula

2. Size: Apparent size of object in either minutes of arc, seconds of arc, or degrees. Most measurements were made from photographs; visual appearance may be smaller. For double stars, this number is a measure of the stars' separation from one another.

M	NGC	Const.	Type[1]	R.A.	Dec.	Mag.	Size[2]	Remarks
1	1952	Tau	BN	05 34.5	+22 00	8	6' × 4'	Crab Nebula
2	7089	Aqr	GC	21 33.5	–00 49	6.5	13'	
3	5272	CVn	GC	13 42.2	+28 23	6.4	16'	
4	6121	Sco	GC	16 23.6	–26 32	6.0	26'	
5	5904	Ser	GC	15 18.6	+02 05	5.8	17'	
6	6405	Sco	OC	17 40.1	–32 13	4.2	15'	Butterfly Cluster
7	6475	Sco	OC	17 53.9	–34 49	3.3	80'	
8	6523	Sgr	BN	18 03.8	–24 23	6	90' × 40'	Lagoon Nebula
9	6333	Oph	GC	17 19.2	–18 31	7.9	9'	
10	6254	Oph	GC	16 57.1	–04 06	6.6	15'	
11	6705	Sct	OC	18 51.1	–06 16	5.8	14'	Wild Duck Cluster
12	6218	Oph	GC	16 47.2	–01 57	6.6	15'	
13	6205	Her	GC	16 41.7	+36 28	5.9	16'	Great Hercules Cluster
14	6402	Oph	GC	17 37.6	–03 15	7.6	12'	
15	7078	Peg	GC	21 30.0	+12 10	6.4	12'	
16	6611	Ser	OC/BN	18 18.8	–13 47	6	35'	Eagle Nebula
17	6618	Sgr	BN	18 20.8	–16 11	7	46' × 37'	Omega Nebula
18	6613	Sgr	OC	18 19.9	–17 08	6.9	9'	
19	6273	Oph	GC	17 02.6	–26 16	7.1	14'	
20	6514	Sgr	BN	18 02.6	–23 02	9	29' × 27'	Trifid Nebula
21	6531	Sgr	OC	18 04.6	–22 30	5.9	13'	
22	6656	Sgr	GC	18 36.4	–23 54	5.1	24'	
23	6494	Sgr	OC	17 56.8	–19 01	5.5	27'	
24		Sgr		18 16.9	–18 29	4.5	90'	Small Sgr Star Cloud
25	IC 4725	Sgr	OC	18 31.6	–19 15	4.6	32'	
26	6694	Sct	OC	18 45.2	–09 24	8.0	15'	
27	6853	Vul	PN	19 59.6	+22 43	8.0	8' × 4'	Dumbbell Nebula
28	6626	Sgr	GC	18 24.5	–24 52	6.9	11'	
29	6913	Cyg	OC	20 23.9	+38 22	6.6	7'	
30	7099	Cap	GC	21 40.4	–23 11	7.5	11'	
31	224	And	Gx	00 42.7	+41 16	3.5	160' × 40'	Andromeda Galaxy
32	221	And	Gx	00 42.7	+40 52	8.2	8' × 6'	
33	598	Tri	Gx	01 33.9	+30 39	5.7	60' × 35'	Triangulum Galaxy
34	1039	Per	OC	02 42.0	+42 47	5.2	35'	
35	2168	Gem	OC	06 08.9	+24 20	5.1	28'	
36	1960	Aur	OC	05 36.1	+34 08	6.0	12'	
37	2099	Aur	OC	05 52.4	+32 33	5.6	24'	
38	1912	Aur	OC	05 28.7	+35 50	6.4	21'	
39	7092	Cyg	OC	21 32.2	+48 26	4.6	32'	

M	NGC	Const.	Type	R.A.	Dec.	Mag.	Size	Remarks
40		UMa	**	12 22.4	+58 05	9.0, 9.6	50"	Winnecke 4
41	2287	CMa	OC	06 46.0	–20 44	4.5	38'	
42	1976	Ori	BN	05 35.4	–05 27	4	66' × 60'	Orion Nebula
43	1982	Ori	BN	05 35.6	–05 16	7	20' × 15'	
44	2632	Cnc	OC	08 40.1	+19 46	3.1	95'	Beehive or Praesepe
45		Tau	OC	03 47.0	+24 07	1.2	110'	Pleiades
46	2437	Pup	OC	07 41.8	–14 49	6.1	27'	
47	2422	Pup	OC	07 36.6	–14 30	4.5	30'	
48	2548	Hya	OC	08 13.8	–05 48	5.8	55'	
49	4472	Vir	Gx	12 29.8	+08 00	8.4	9' × 7'	
50	2323	Mon	OC	07 02.8	–08 23	5.9	16'	
51	5194-5	CVn	Gx	13 29.9	+47 12	8.4	11' × 8'	Whirlpool Galaxy
52	7654	Cas	OC	23 24.2	+61 35	6.9	13'	
53	5024	Com	GC	13 12.9	+18 10	7.7	13'	
54	6715	Sgr	GC	18 55.1	–30 29	7.7	9'	
55	6809	Sgr	GC	19 40.0	–30 58	7.0	19'	
56	6779	Lyr	GC	19 16.6	+30 11	8.2	7'	
57	6720	Lyr	PN	18 53.6	+33 02	9.7	70" × 150"	Ring Nebula
58	4579	Vir	Gx	12 37.7	+11 49	9.8	5' × 4'	
59	4621	Vir	Gx	12 42.0	+11 39	9.8	5' × 3'	
60	4649	Vir	Gx	12 43.7	+11 33	8.8	7' × 6'	
61	4303	Vir	Gx	12 21.9	+04 28	9.7	6' × 5'	
62	6266	Oph	GC	17 01.2	–30 07	6.6	14'	
63	5055	CVn	Gx	13 15.8	+42 02	8.6	12' × 8'	
64	4826	Com	Gx	12 56.7	+21 41	8.5	9' × 5'	Black-Eye Galaxy
65	3623	Leo	Gx	11 18.9	+13 05	9.3	10' × 3'	
66	3627	Leo	Gx	11 20.2	+12 59	9.0	9' × 4'	
67	2682	Cnc	OC	08 51.4	+11 49	6.9	30'	
68	4590	Hya	GC	12 39.5	–26 45	8.2	12'	
69	6637	Sgr	GC	18 31.4	–32 21	7.7	7'	
70	6681	Sgr	GC	18 43.2	–32 18	8.1	8'	
71	6838	Sge	GC	19 53.8	+18 47	8.3	7'	
72	6981	Aqr	GC	20 53.5	–12 32	9.4	6'	
73	6994	Aqr	AS	20 58.9	–12 38			
74	628	Psc	Gx	01 36.7	+15 47	9.2	10' × 9'	
75	6864	Sgr	GC	20 06.1	–21 55	8.6	6'	
76	650–1	Per	PN	01 42.4	+51 34	11.5	2' × 1'	Little Dumbbell Nebula
77	1068	Cet	Gx	02 42.7	–00 01	8.9	6' × 5'	
78	2068	Ori	BN	05 46.7	+00 03	8	8' × 6'	
79	1904	Lep	GC	05 24.5	–24 33	8.0	9'	
80	6093	Sco	GC	16 17.0	–22 59	7.2	9'	
81	3031	UMa	Gx	09 55.6	+69 04	7.0	26' × 14'	
82	3034	UMa	Gx	09 55.8	+69 41	8.4	11' × 5'	
83	5236	Hya	Gx	13 37.0	–29 52	7.6	11' × 10'	
84	4374	Vir	Gx	12 25.1	+12 53	9.3	5' × 4'	
85	4382	Com	Gx	12 25.4	+18 11	9.2	7' × 5'	
86	4406	Vir	Gx	12 26.2	+12 57	9.2	7' × 6'	
87	4486	Vir	Gx	12 30.8	+12 24	8.6	7'	
88	4501	Com	Gx	12 32.0	+14 25	9.5	7' × 4'	
89	4552	Vir	Gx	12 35.7	+12 33	9.8	4'	
90	4569	Vir	Gx	12 36.8	+13 10	9.5	9' × 5'	
91	4548	Com	Gx	12 35.4	+14 30	10.2	5' × 4'	

M	NGC	Const.	Type	R.A.	Dec.	Mag.	Size	Remarks
92	6341	Her	GC	17 17.1	+43 08	6.5	11'	
93	2447	Pup	OC	07 44.6	–23 52	6.2	22'	
94	4736	CVn	Gx	12 50.9	+41 07	8.2	11' × 9'	
95	3351	Leo	Gx	10 44.0	+11 42	9.7	7' × 5'	
96	3368	Leo	Gx	10 46.8	+11 49	9.2	7' × 5'	
97	3587	UMa	PN	11 14.8	+55 01	11.2	3'	Owl Nebula
98	4192	Com	Gx	12 13.8	+14 54	10.1	10' × 3'	
99	4254	Com	Gx	12 18.8	+14 25	9.8	5'	
100	4321	Com	Gx	12 22.9	+15 49	9.4	7' × 6'	
101	5457	UMa	Gx	14 03.2	+54 21	7.7	27' × 26'	Pinwheel Galaxy
102			duplicate observation of M101					
103	581	Cas	OC	01 33.2	+60 42	7.4	6'	
104	4594	Vir	Gx	12 40.0	–11 37	8.3	9' × 4'	Sombrero Galaxy
105	3379	Leo	Gx	10 47.8	+12 35	9.3	5' × 4'	
106	4258	CVn	Gx	12 19.0	+47 18	8.3	18' × 8'	
107	6171	Oph	GC	16 32.5	–13 03	8.1	10'	
108	3556	UMa	Gx	11 11.5	+55 40	10.1	8' × 3'	
109	3992	UMa	Gx	11 57.6	+53 23	9.8	8' × 5'	
110	205	And	Gx	00 40.4	+41 41	8.0	17' × 10'	

Star Atlas

The following set of thirty-six star-atlas charts has been prepared especially for use with this book. All stars down to magnitude 6.5 are plotted for epoch 2000.0 coordinates. As is customary, brighter stars are shown as large black circles, while fainter stars are represented by increasingly smaller circles.

Many of the brighter stars on the charts are connected by lines to form the familiar constellations. These patterns agree with those shown on the four seasonal charts found in the beginning of chapters 6 through 9, and are designed to help the reader compare the naked-eye seasonal sky with this detailed star atlas.

Scattered throughout the charts are all of the double stars, variable stars, star clusters, asterisms, nebulae, and galaxies discussed in this book. All are identified by their catalog numbers. Members of the NGC are referred to only by their number (e.g., "2903" for NGC 2903, etc.), the "NGC" being omitted to reduce clutter. Each entry is also identified by a symbol corresponding to the object type (open cluster, bright nebula, etc.). Symbols are defined in the legend adjacent to each chart.

The atlas divides the sky into four "tiers:" two charts show the north circumpolar region, from declination +90° to approximately +60° ; 11 north-equatorial charts show the sky from +70° to +20° ; 12 equatorial charts range from +25° to –25° ; and 11 south-equatorial charts cover from –20° to –70°. All individual charts within a tier are ordered according to decreasing right ascension (east to west). (Note that the south-circumpolar region is ignored, since this book is limited to deep-sky objects north of –60° declination.)

A grid of celestial coordinate lines is shown on each star-atlas page. On the two polar charts, right ascension (R.A.) coordinates are indicated around the edge, with values for declination (Dec.) increasing toward Polaris and the North Celestial Pole. On the tiered charts, right ascension values are found along the top edge, while declination positions are denoted along the outer edge of each page.

How many of the objects plotted on these charts will you be able to see from your observing site? The objects farthest to the south that are described in this book lie at –60° declination. This means that an observer could be no farther north than +30° latitude, at least theoretically, to see them. In practice, however, we must subtract at least another 10°, since haze in the Earth's atmosphere usually makes it difficult to spot objects less than 10° above the horizon. So, an observer near, say, New York City (+40° latitude) will probably be able to see objects no farther south than –40° declination. As a reminder, you might want to draw a horizontal line on each south-equatorial chart to show your usable horizon.

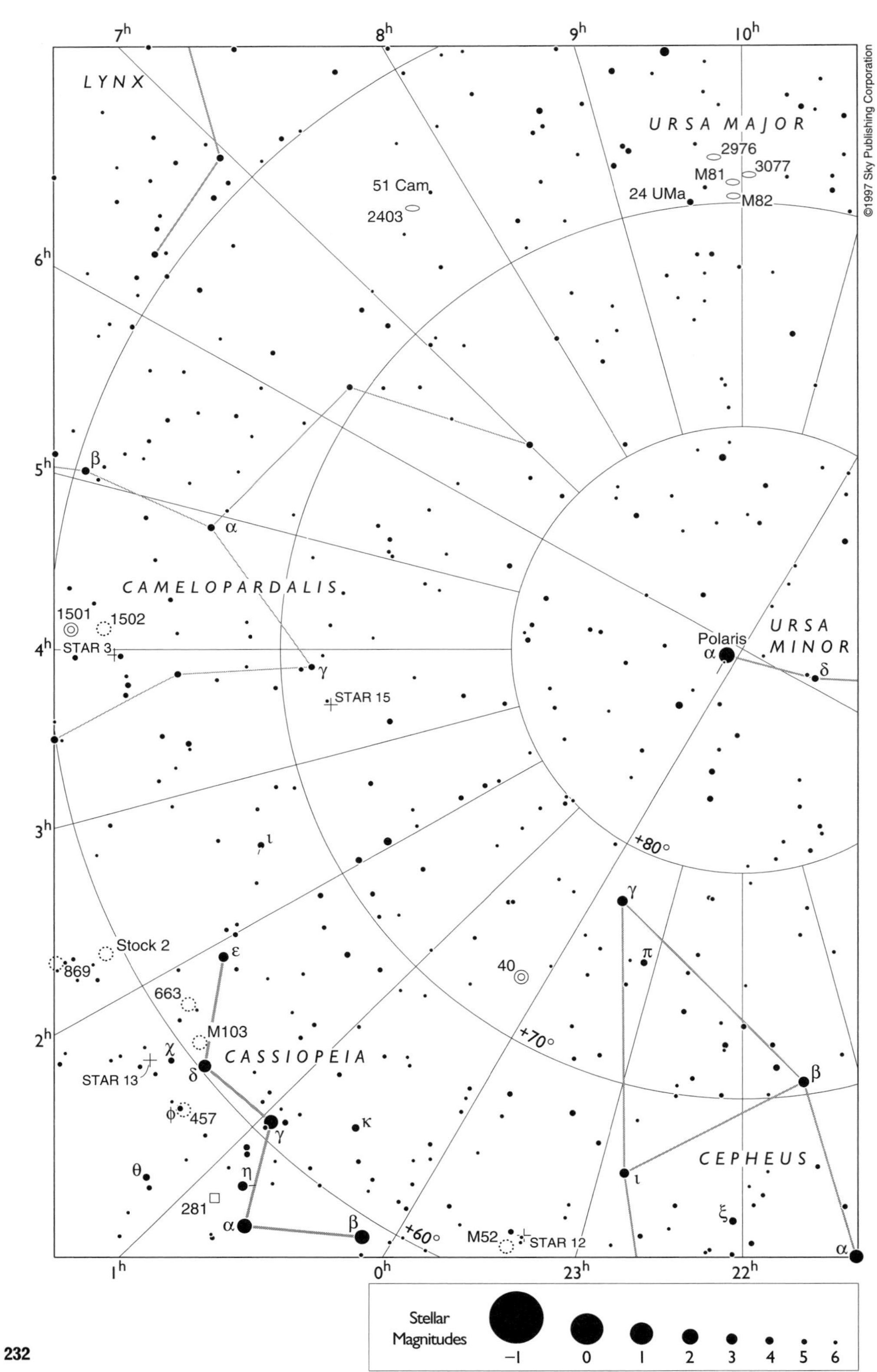

LYNX
URSA MAJOR
2976
3077
M81
M82
24 UMa
51 Cam
2403
CAMELOPARDALIS
1501
1502
STAR 3
STAR 15
Polaris
URSA MINOR
+80°
+70°
+60°
40
Stock 2
869
663
M103
CASSIOPEIA
STAR 13
457
281
M52
STAR 12
CEPHEUS
7h
8h
9h
10h
6h
5h
4h
3h
2h
1h
0h
23h
22h
Stellar Magnitudes
−1
0
1
2
3
4
5
6

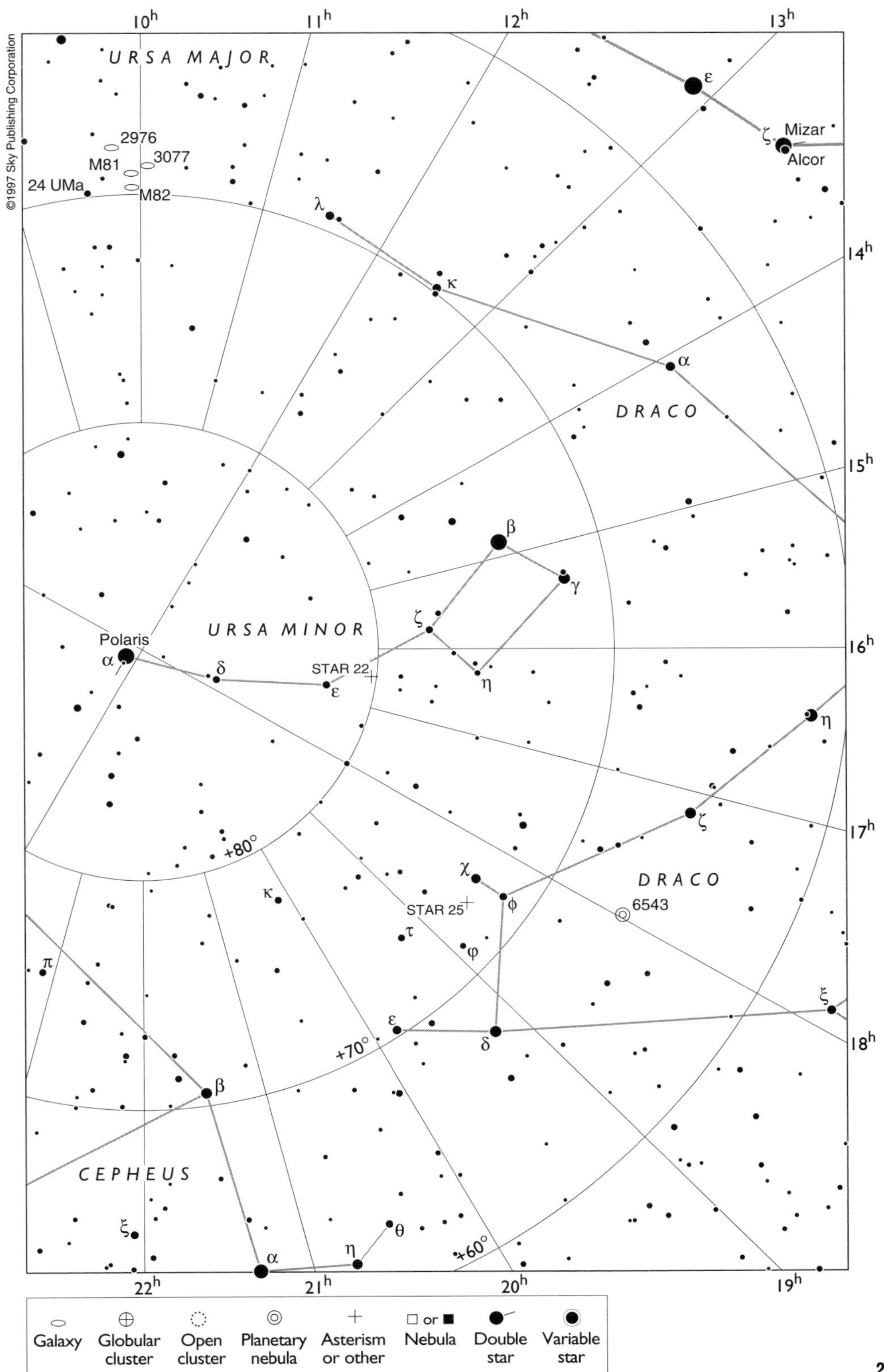

©1997 Sky Publishing Corporation
10h
11h
12h
13h
14h
15h
16h
17h
18h
19h
20h
21h
22h
URSA MAJOR
2976
M81
3077
24 UMa
M82
Mizar
Alcor
DRACO
URSA MINOR
Polaris
STAR 22
+80°
+70°
+60°
STAR 25
6543
DRACO
CEPHEUS
Galaxy
Globular cluster
Open cluster
Planetary nebula
Asterism or other
□ or ■ Nebula
Double star
Variable star

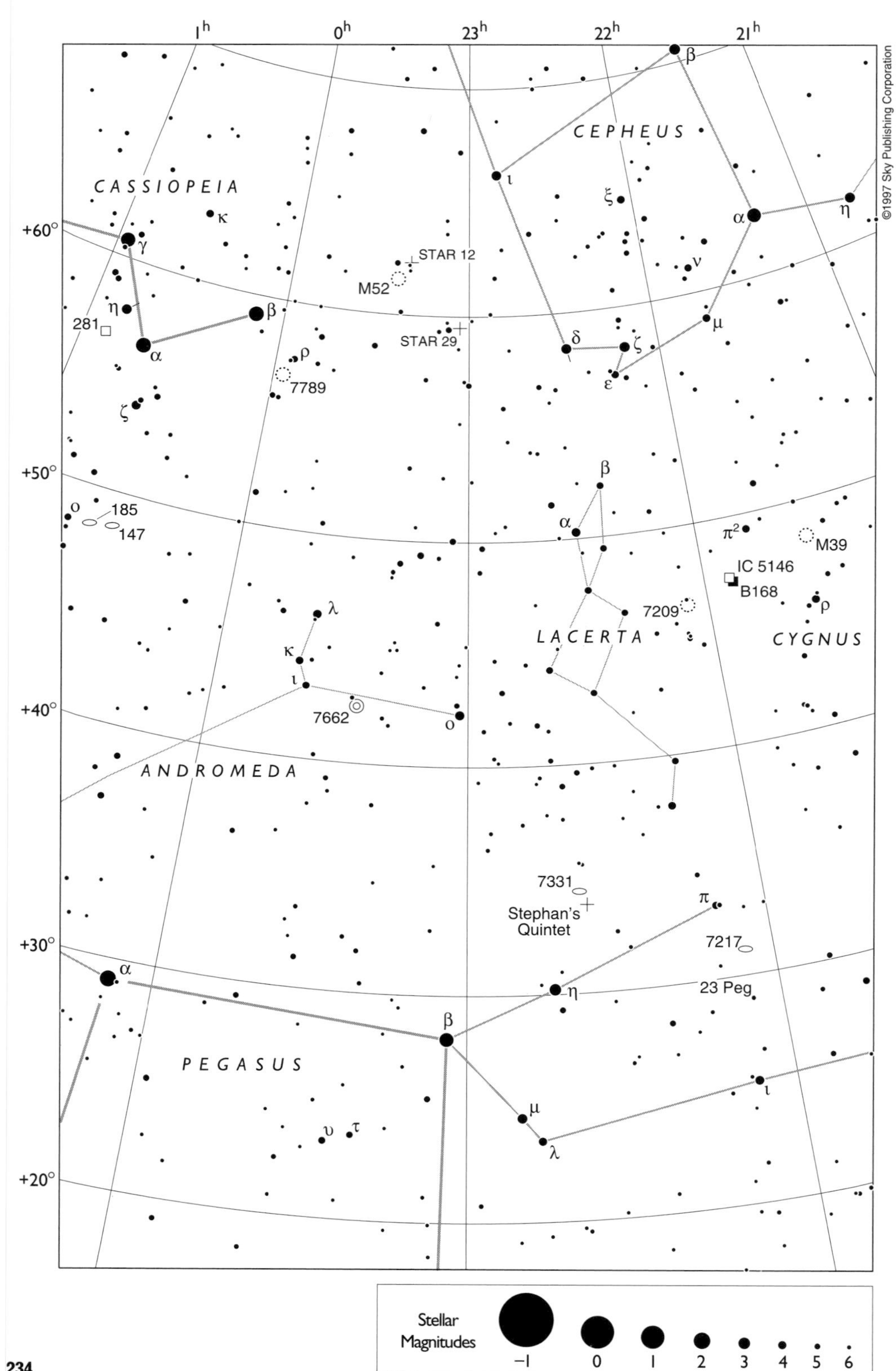

1h
0h
23h
22h
21h
©1997 Sky Publishing Corporation
CEPHEUS
CASSIOPEIA
+60°
STAR 12
M52
281
STAR 29
7789
+50°
185
147
M39
IC 5146
B168
7209
LACERTA
CYGNUS
+40°
7662
ANDROMEDA
7331
Stephan's Quintet
+30°
7217
23 Peg
PEGASUS
+20°
Stellar Magnitudes
–1
0
1
2
3
4
5
6

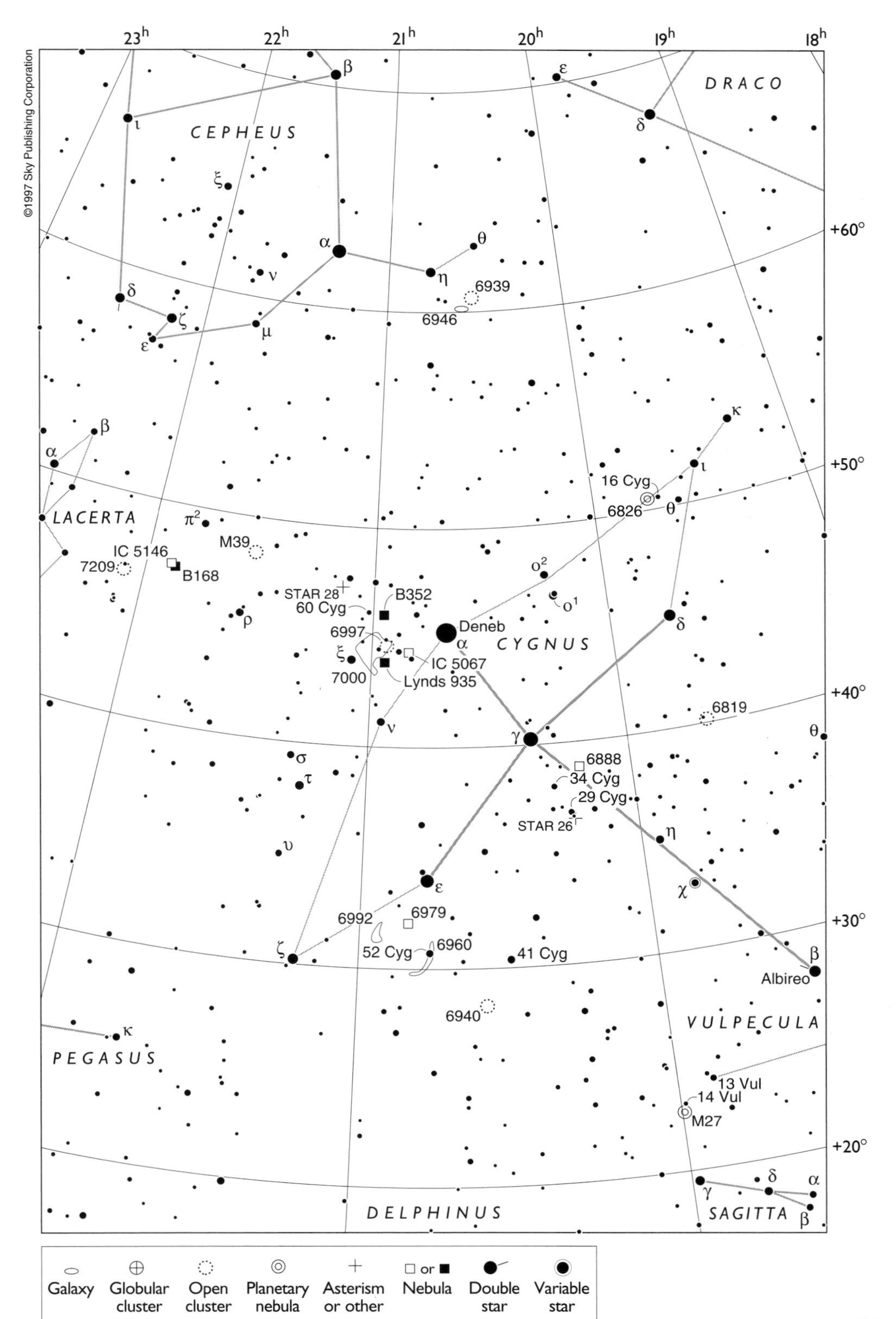

23h
22h
21h
20h
19h
18h
+60°
+50°
+40°
+30°
+20°
CEPHEUS
DRACO
LACERTA
CYGNUS
PEGASUS
VULPECULA
DELPHINUS
SAGITTA
6939
6946
16 Cyg
6826
IC 5146
7209
B168
M39
STAR 28
B352
60 Cyg
Deneb
6997
IC 5067
7000
Lynds 935
6819
6888
34 Cyg
29 Cyg
STAR 26
6992
6979
52 Cyg
6960
41 Cyg
Albireo
6940
13 Vul
14 Vul
M27
Galaxy
Globular cluster
Open cluster
Planetary nebula
Asterism or other
□ or ■ Nebula
Double star
Variable star

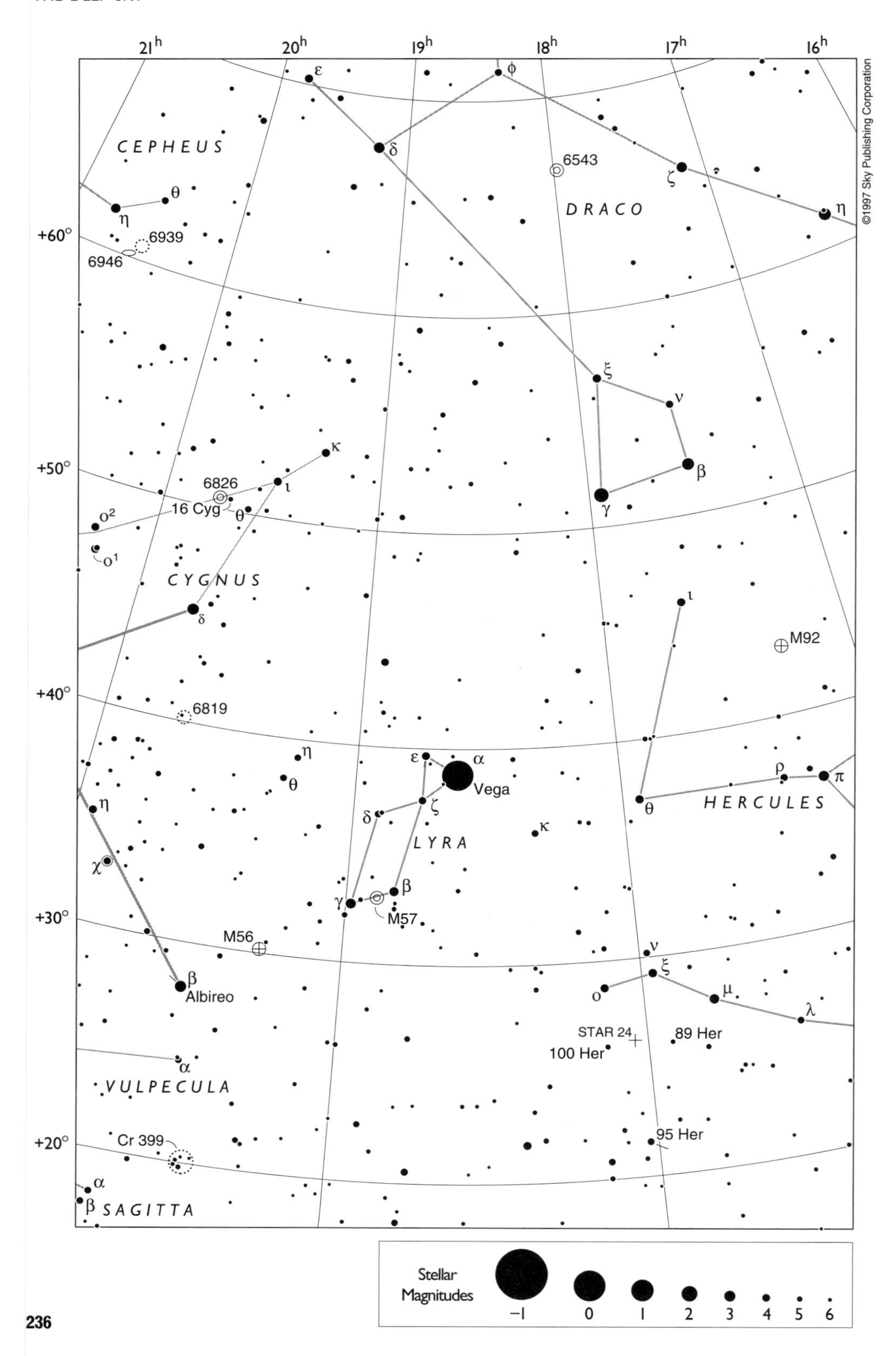

21h
20h
19h
18h
17h
16h
+60°
+50°
+40°
+30°
+20°
CEPHEUS
DRACO
CYGNUS
LYRA
HERCULES
VULPECULA
SAGITTA
6543
6939
6946
6826
16 Cyg
6819
M92
M56
M57
Vega
Albireo
STAR 24
89 Her
100 Her
95 Her
Cr 399
Stellar Magnitudes
−1
0
1
2
3
4
5
6
©1997 Sky Publishing Corporation

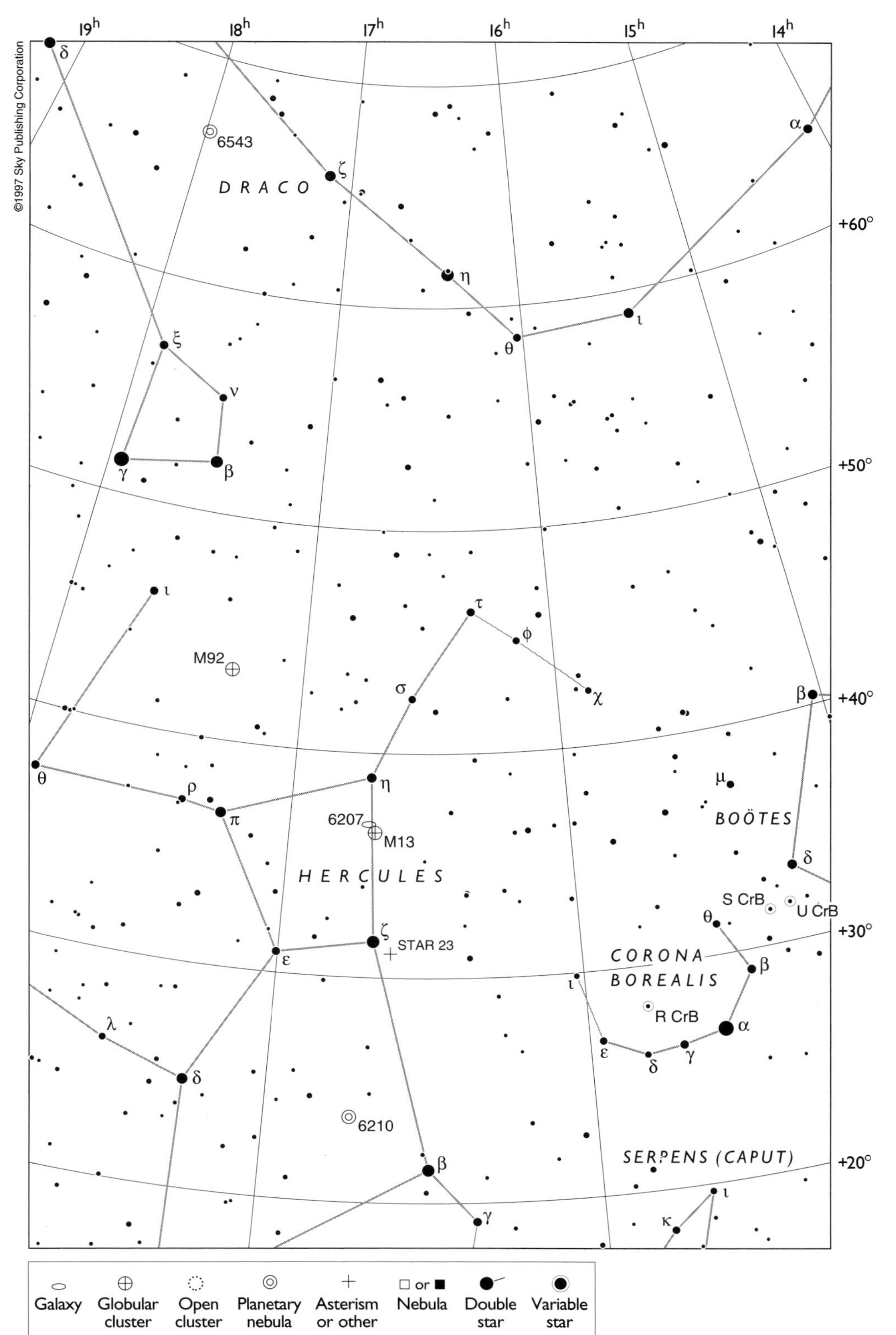
19h
18h
17h
16h
15h
14h
+60°
+50°
+40°
+30°
+20°
6543
DRACO
M92
6207
M13
HERCULES
STAR 23
6210
BOÖTES
S CrB
U CrB
CORONA
BOREALIS
R CrB
SERPENS (CAPUT)
Galaxy
Globular cluster
Open cluster
Planetary nebula
Asterism or other
Nebula
Double star
Variable star

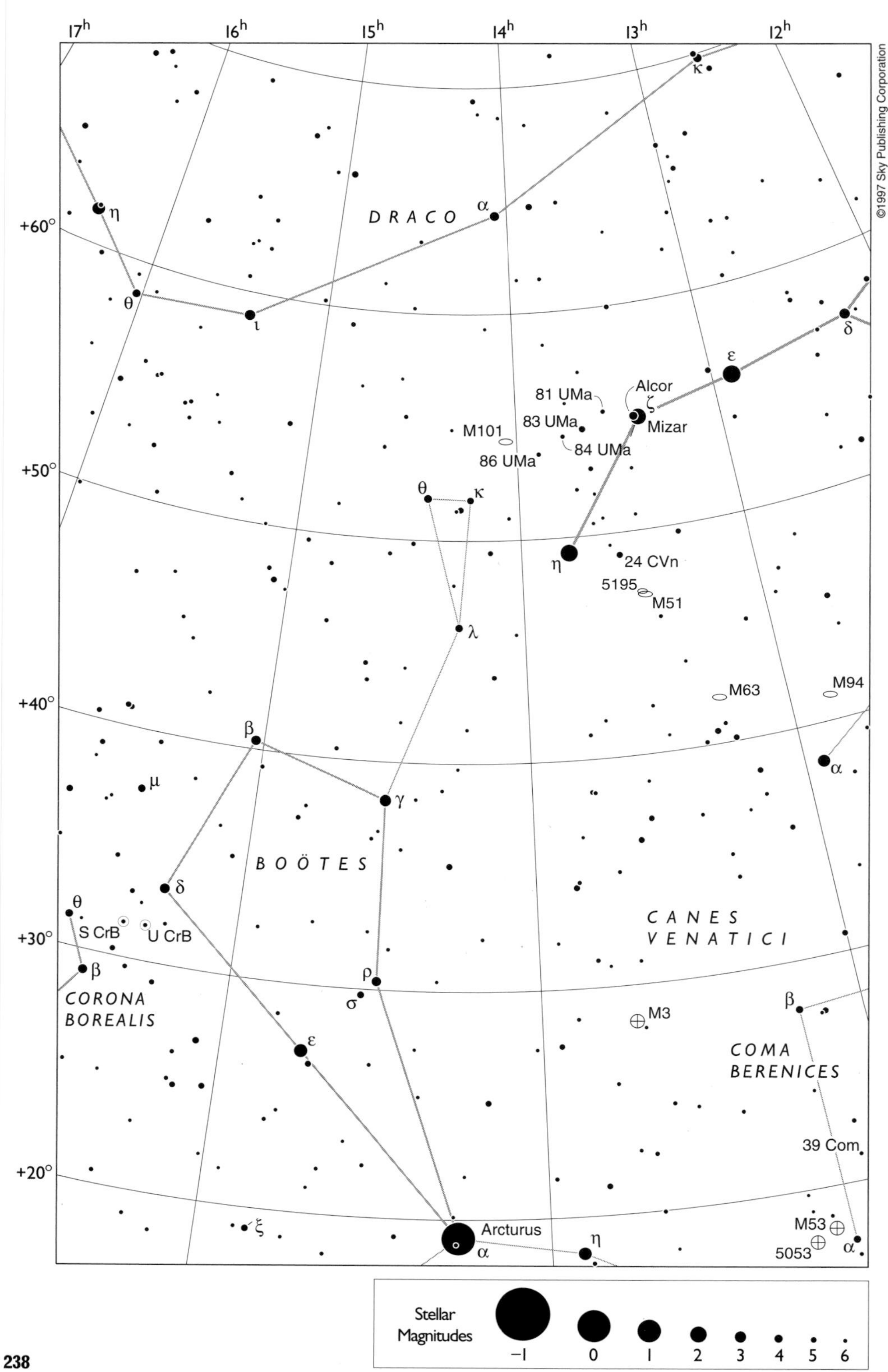
17h
16h
15h
14h
13h
12h
+60°
+50°
+40°
+30°
+20°
DRACO
BOÖTES
CANES VENATICI
CORONA BOREALIS
COMA BERENICES
Alcor
Mizar
81 UMa
83 UMa
84 UMa
86 UMa
M101
24 CVn
5195
M51
M63
M94
S CrB
U CrB
M3
39 Com
M53
5053
Arcturus
Stellar Magnitudes
−1
0
1
2
3
4
5
6

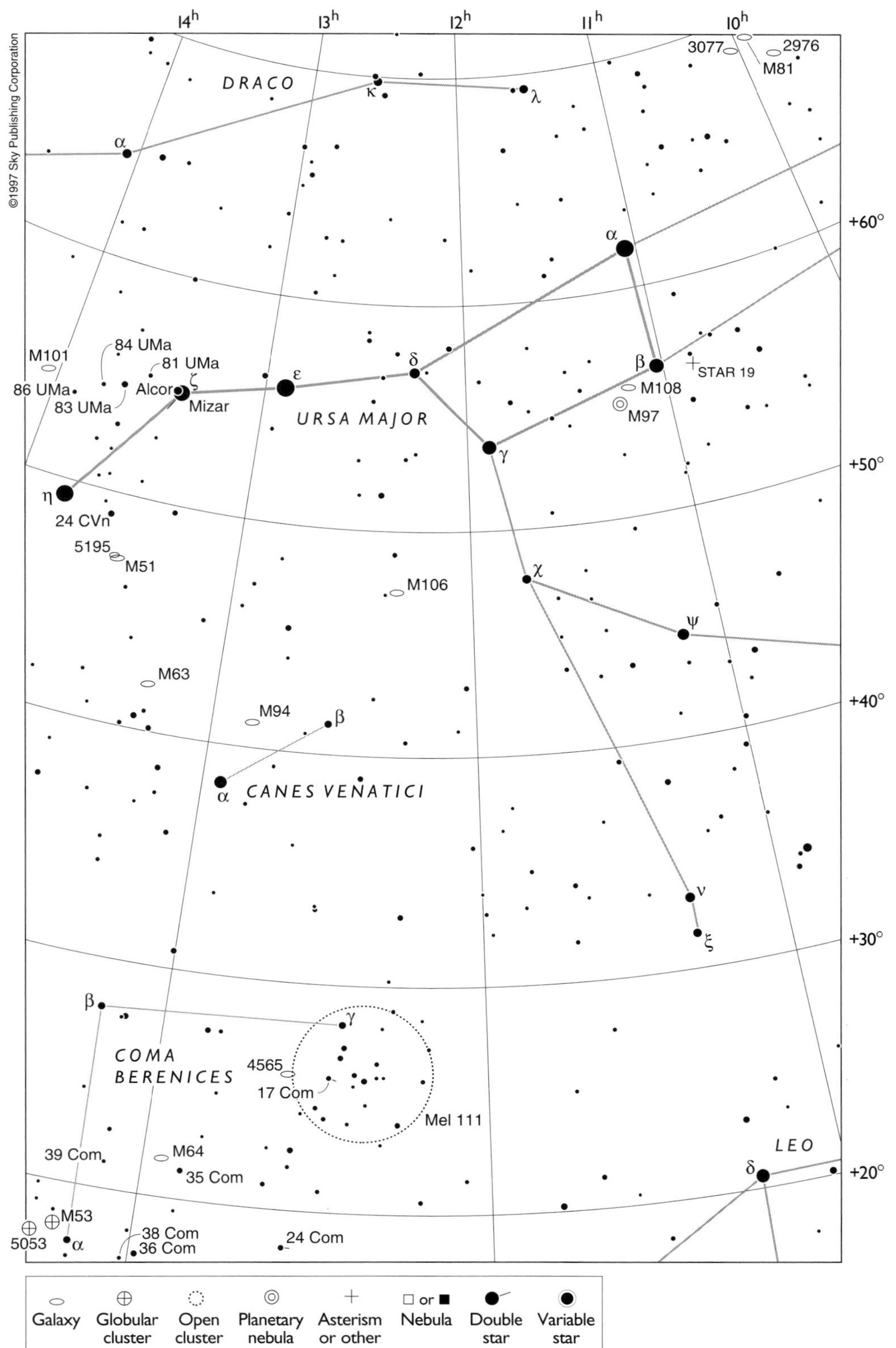

©1997 Sky Publishing Corporation
14h
13h
12h
11h
10h
+60°
+50°
+40°
+30°
+20°
DRACO
URSA MAJOR
CANES VENATICI
COMA BERENICES
LEO
3077
2976
M81
M101
84 UMa
81 UMa
86 UMa
83 UMa
Alcor
Mizar
STAR 19
M108
M97
24 CVn
5195
M51
M106
M63
M94
4565
17 Com
Mel 111
39 Com
M64
35 Com
M53
5053
38 Com
36 Com
24 Com
Galaxy
Globular cluster
Open cluster
Planetary nebula
Asterism or other
or Nebula
Double star
Variable star

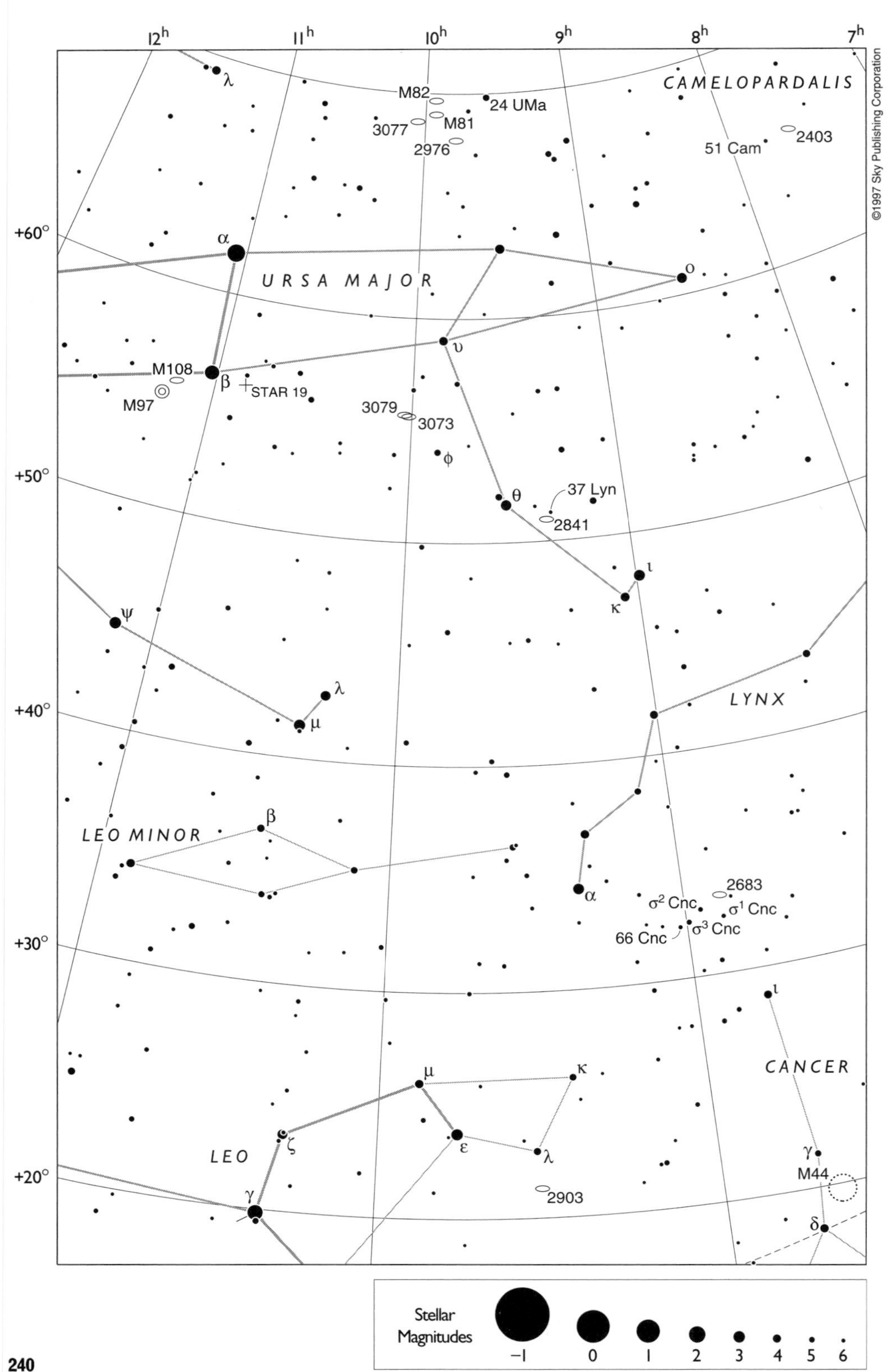

12h
11h
10h
9h
8h
7h
+60°
+50°
+40°
+30°
+20°
CAMELOPARDALIS
URSA MAJOR
LYNX
LEO MINOR
CANCER
LEO
M82
24 UMa
3077
M81
2976
2403
51 Cam
M108
M97
STAR 19
3079
3073
37 Lyn
2841
2683
σ2 Cnc
σ1 Cnc
σ3 Cnc
66 Cnc
M44
2903
©1997 Sky Publishing Corporation
Stellar Magnitudes
–1
0
1
2
3
4
5
6

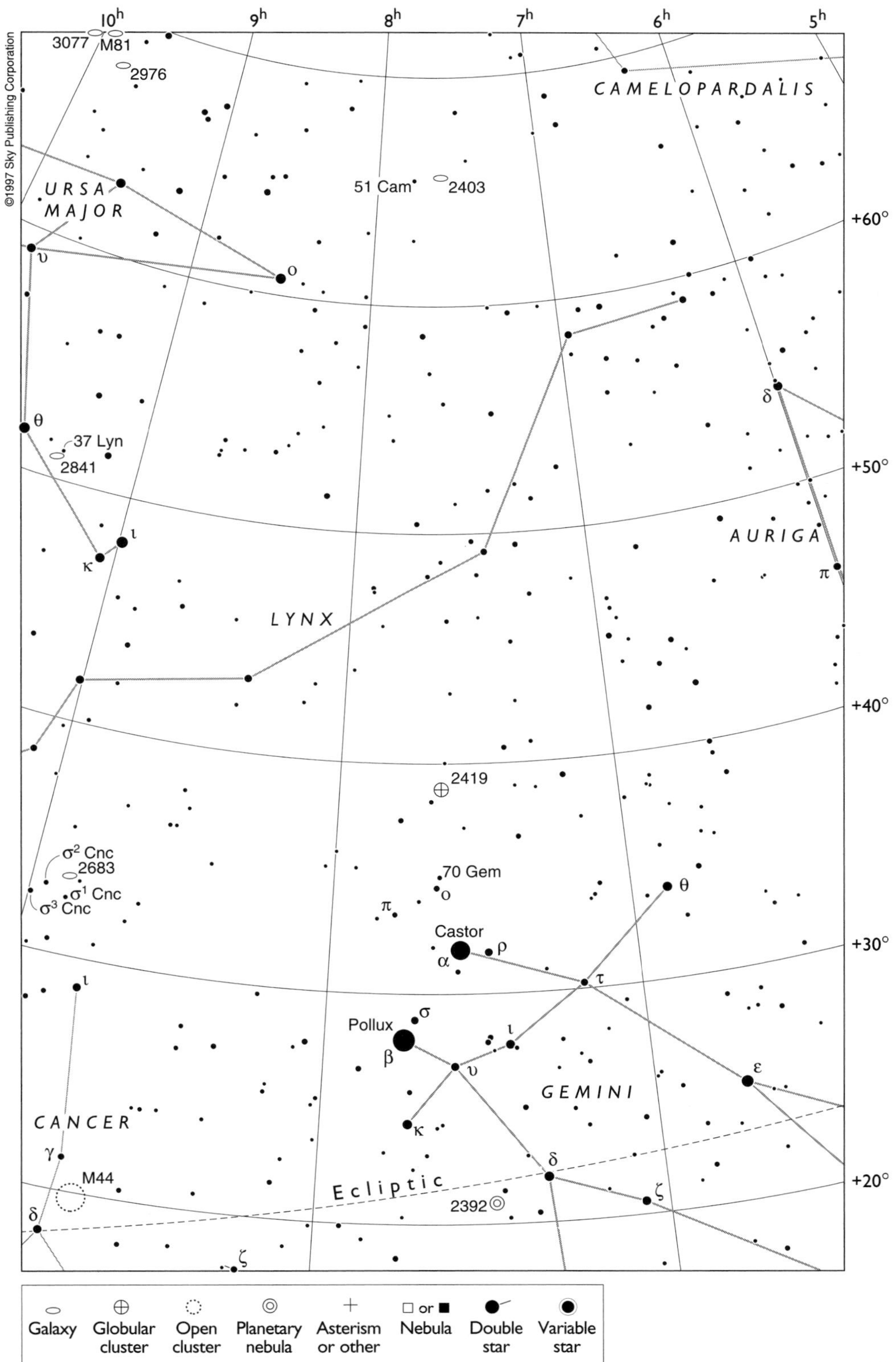
10h
9h
8h
7h
6h
5h
3077
M81
2976
CAMELOPARDALIS
URSA MAJOR
51 Cam
2403
+60°
37 Lyn
2841
+50°
AURIGA
LYNX
+40°
2419
σ2 Cnc
2683
σ1 Cnc
σ3 Cnc
70 Gem
Castor
+30°
Pollux
GEMINI
CANCER
M44
Ecliptic
2392
+20°
Galaxy
Globular cluster
Open cluster
Planetary nebula
Asterism or other
Nebula
Double star
Variable star
©1997 Sky Publishing Corporation

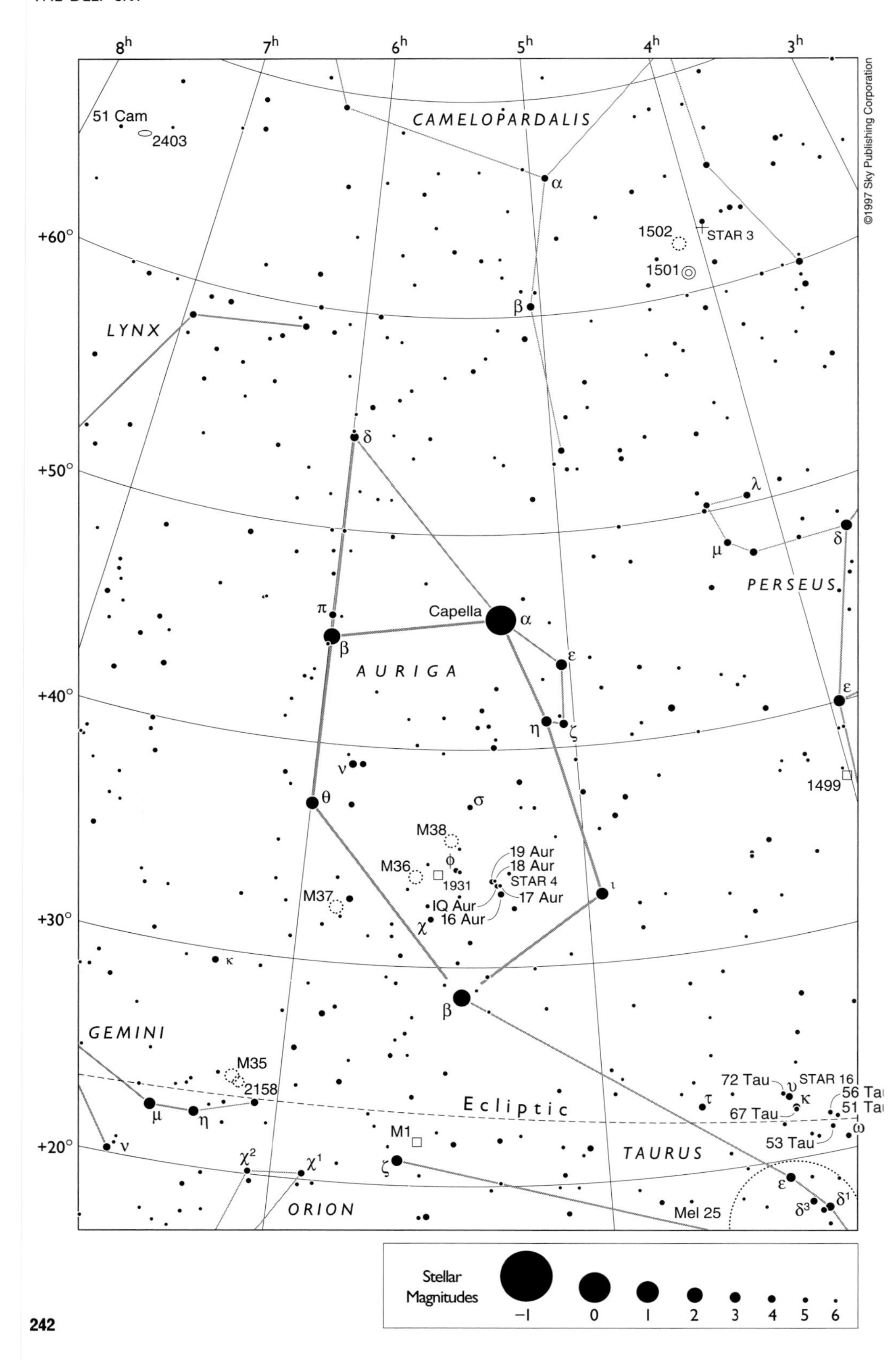
8h
7h
6h
5h
4h
3h
©1997 Sky Publishing Corporation
51 Cam
2403
CAMELOPARDALIS
α
+60°
1502
STAR 3
1501
β
LYNX
δ
+50°
λ
δ
μ
PERSEUS
π
Capella
α
β
ε
AURIGA
+40°
η
ζ
ε
ν
1499
θ
σ
M38
19 Aur
φ
18 Aur
M36
1931
STAR 4
ι
17 Aur
M37
IQ Aur
χ
16 Aur
+30°
κ
β
GEMINI
M35
72 Tau
STAR 16
υ
2158
κ
56 Tau
τ
51 Tau
Ecliptic
67 Tau
μ
η
M1
53 Tau
ω
+20°
ν
χ2
χ1
TAURUS
ζ
ε
ORION
Mel 25
δ3
δ1
Stellar Magnitudes
–1
0
1
2
3
4
5
6

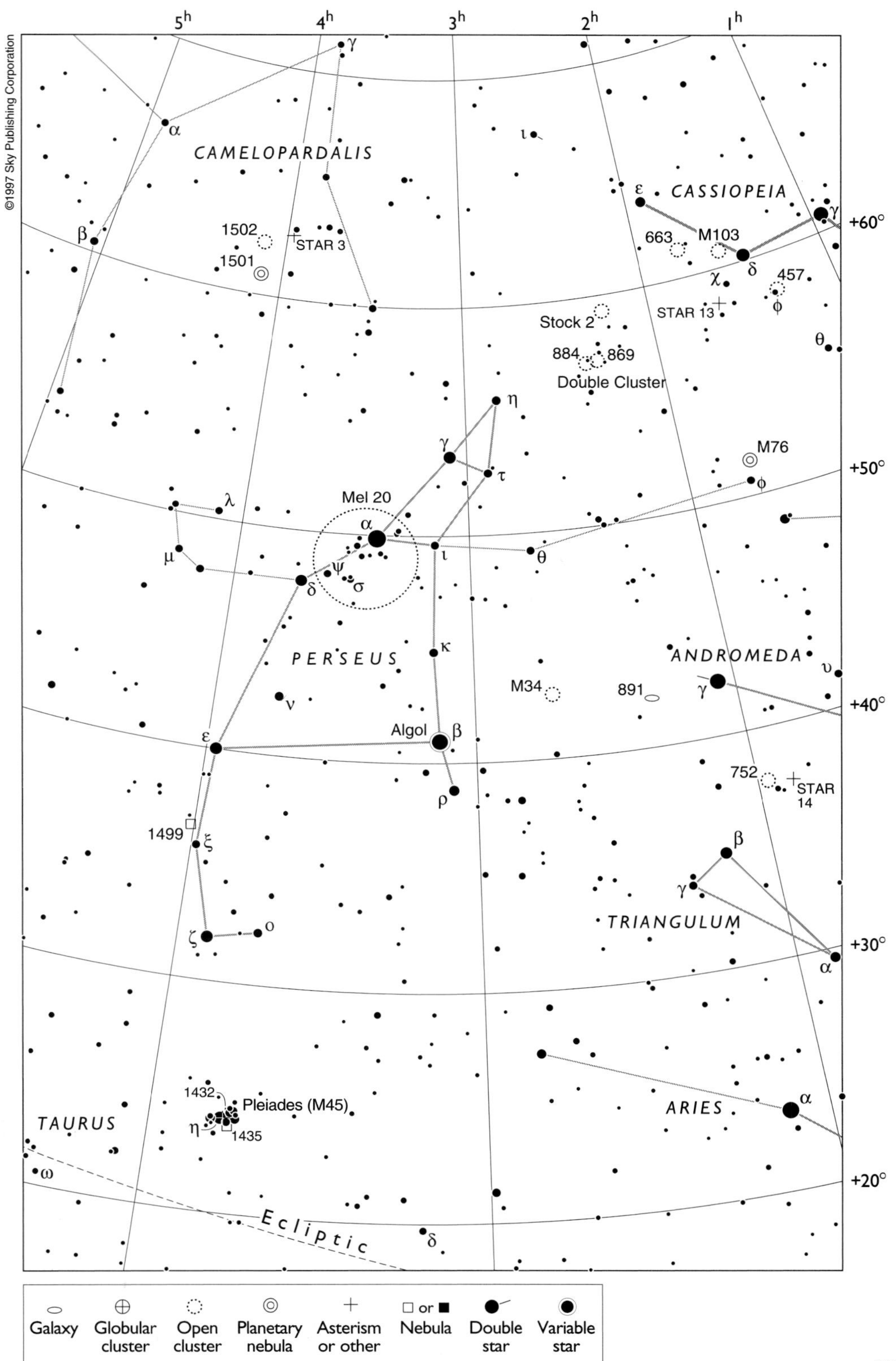

5h
4h
3h
2h
1h
+60°
+50°
+40°
+30°
+20°
CAMELOPARDALIS
CASSIOPEIA
PERSEUS
ANDROMEDA
TRIANGULUM
ARIES
TAURUS
Ecliptic
1502
STAR 3
1501
663
M103
457
STAR 13
Stock 2
884
869
Double Cluster
M76
Mel 20
M34
891
Algol
752
STAR 14
1499
1432
Pleiades (M45)
1435
Galaxy
Globular cluster
Open cluster
Planetary nebula
Asterism or other
□ or ■ Nebula
Double star
Variable star

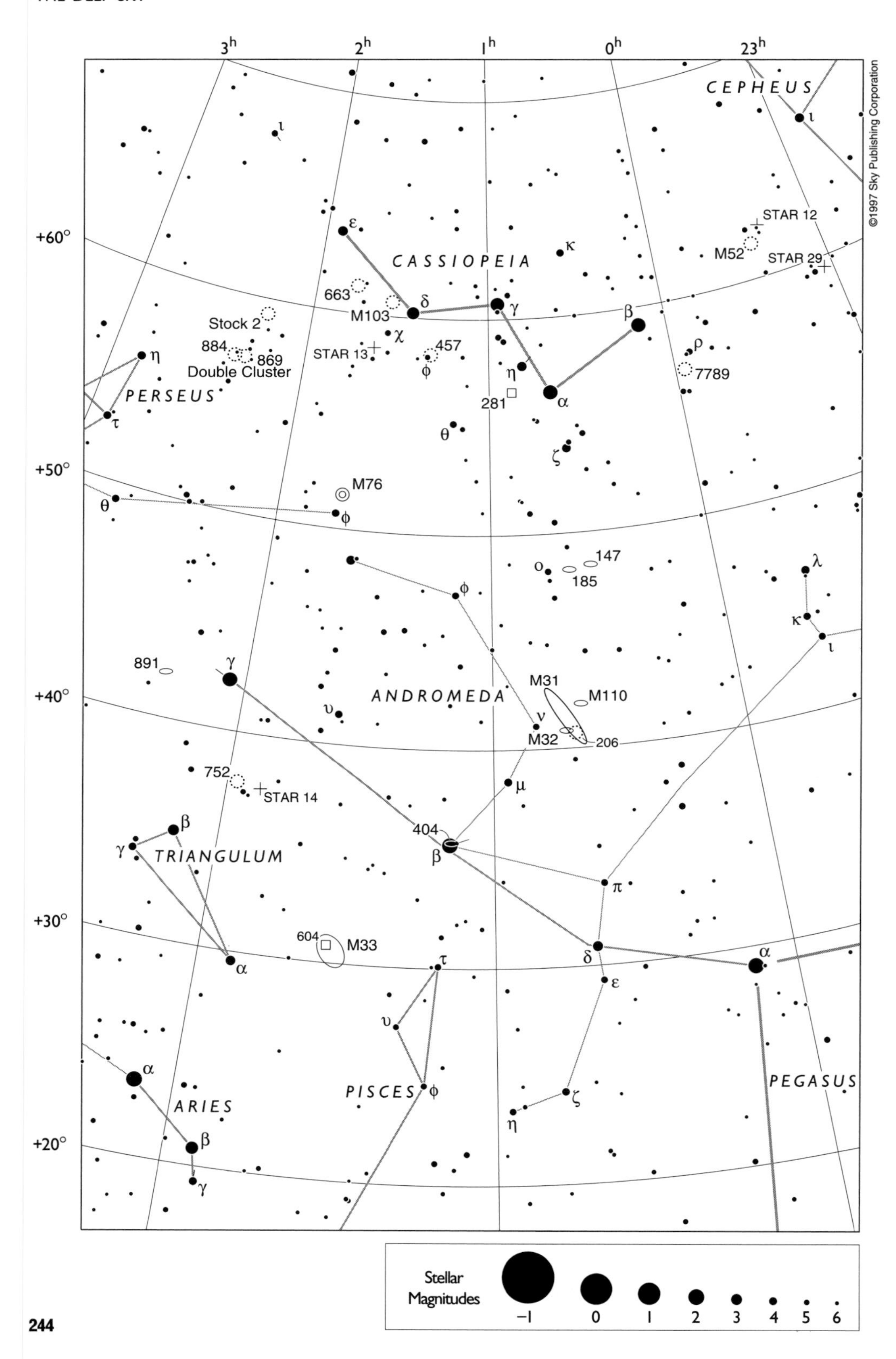
3h
2h
1h
0h
23h
+60°
+50°
+40°
+30°
+20°
CEPHEUS
CASSIOPEIA
PERSEUS
ANDROMEDA
TRIANGULUM
PISCES
ARIES
PEGASUS
STAR 12
M52
STAR 29
663
M103
Stock 2
884
869
Double Cluster
STAR 13
457
281
7789
M76
147
185
891
M31
M110
M32
206
752
STAR 14
404
604
M33
Stellar Magnitudes
–1
0
1
2
3
4
5
6
©1997 Sky Publishing Corporation

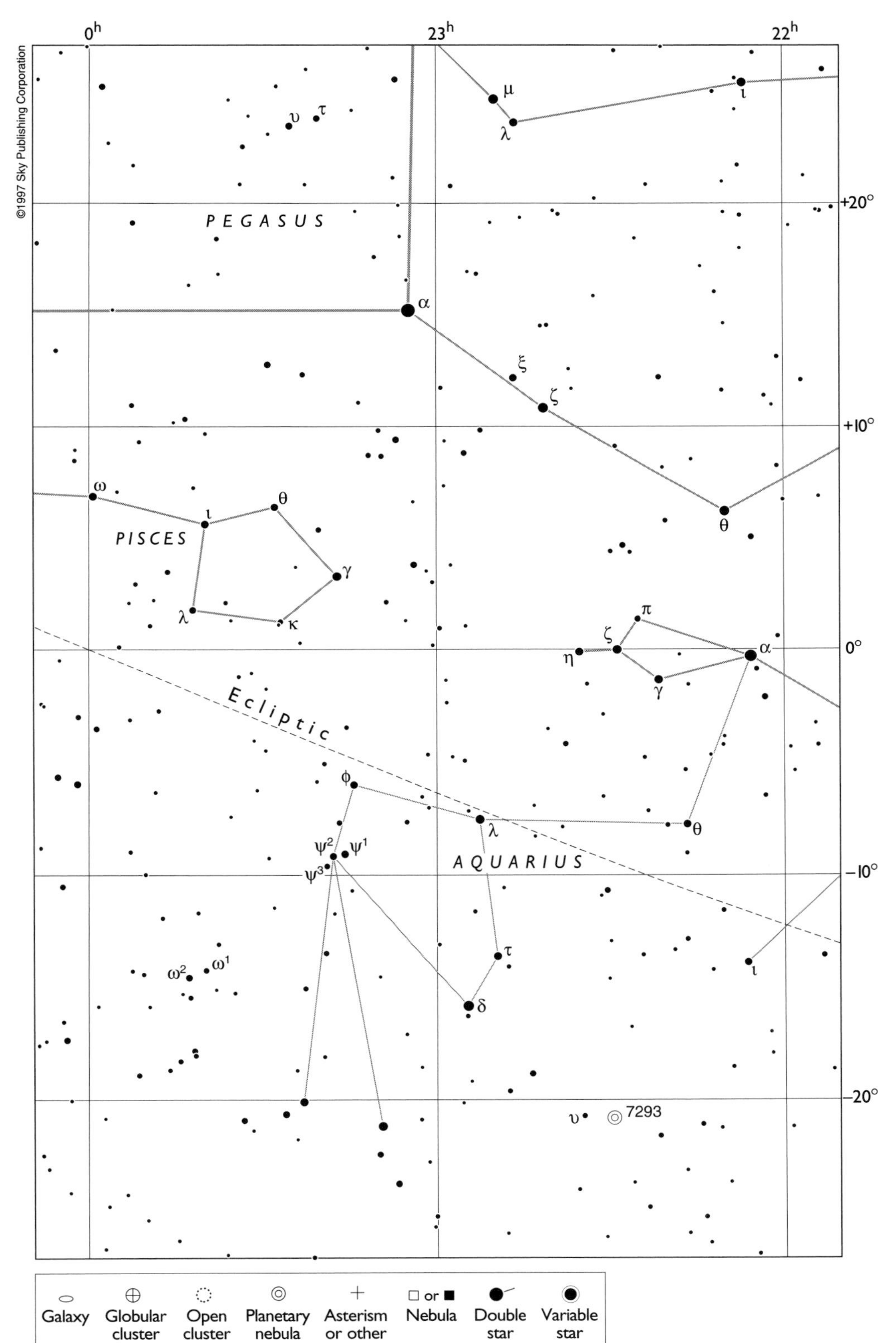

©1997 Sky Publishing Corporation
0h
23h
22h
+20°
+10°
0°
−10°
−20°
PEGASUS
PISCES
AQUARIUS
Ecliptic
7293
Galaxy
Globular cluster
Open cluster
Planetary nebula
Asterism or other
□ or ■ Nebula
Double star
Variable star

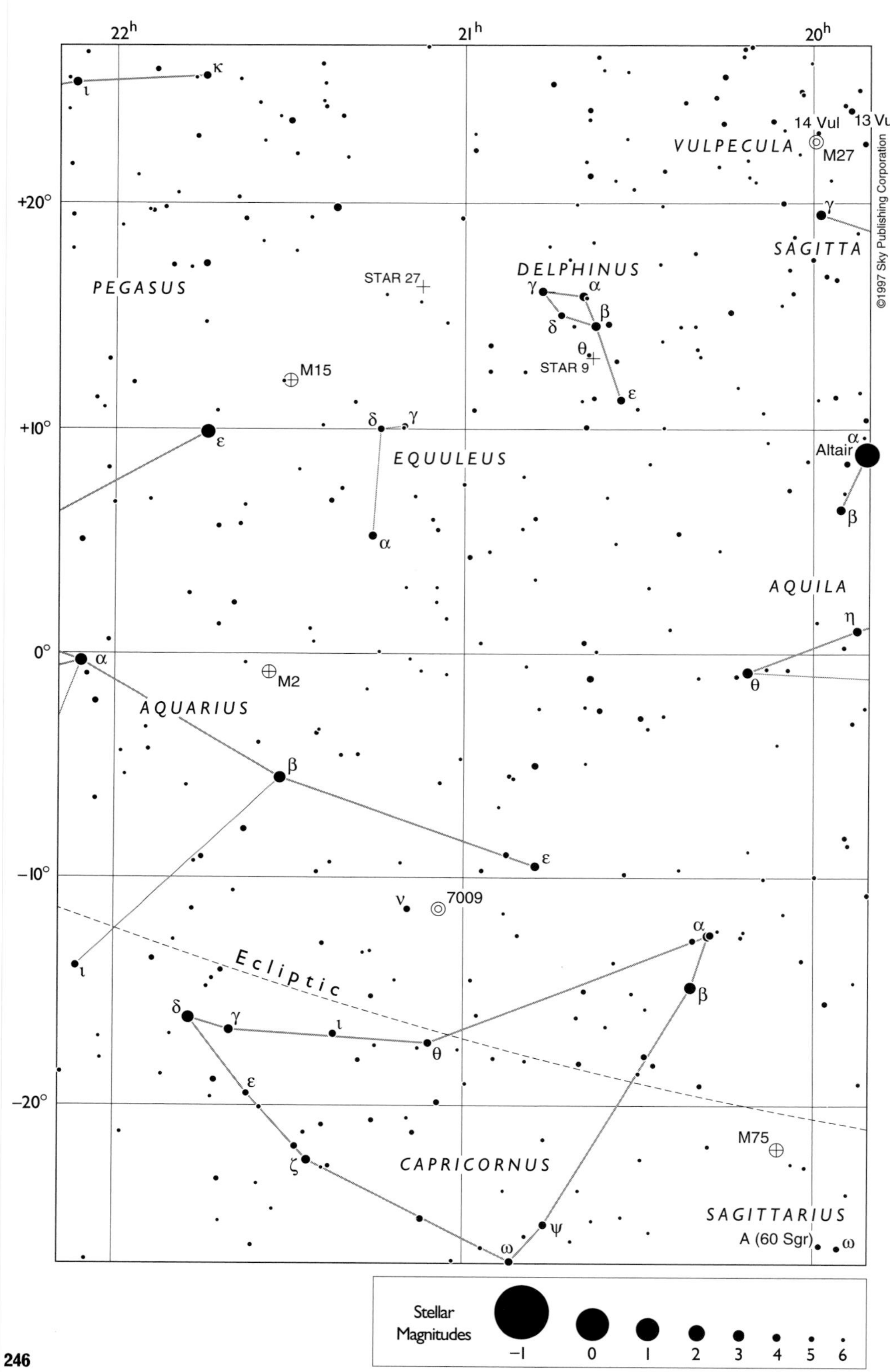
22h
21h
20h
+20°
+10°
0°
−10°
−20°
PEGASUS
VULPECULA
14 Vul
13 Vul
M27
SAGITTA
DELPHINUS
STAR 27
STAR 9
M15
EQUULEUS
Altair
AQUILA
M2
AQUARIUS
7009
Ecliptic
CAPRICORNUS
M75
SAGITTARIUS
A (60 Sgr)
Stellar Magnitudes
−1
0
1
2
3
4
5
6
©1997 Sky Publishing Corporation

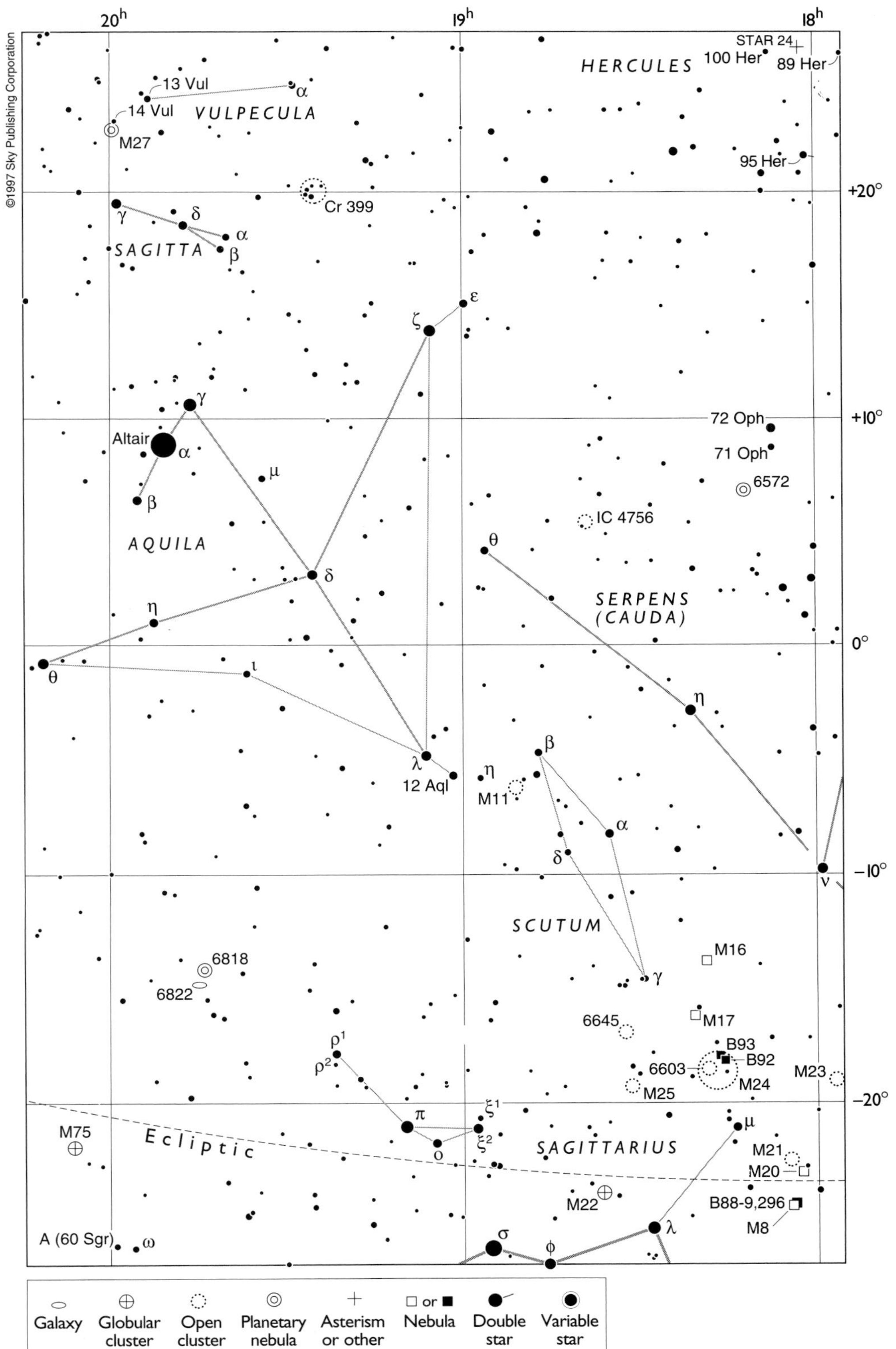
20h
19h
18h
+20°
+10°
0°
−10°
−20°
HERCULES
STAR 24
100 Her
89 Her
95 Her
13 Vul
14 Vul
VULPECULA
M27
Cr 399
SAGITTA
Altair
AQUILA
72 Oph
71 Oph
6572
IC 4756
SERPENS (CAUDA)
12 Aql
M11
SCUTUM
M16
M17
6645
B93
B92
6603
M24
M25
M23
6818
6822
M75
Ecliptic
SAGITTARIUS
M21
M20
M22
B88-9,296
M8
A (60 Sgr)
Galaxy
Globular cluster
Open cluster
Planetary nebula
Asterism or other
□ or ■ Nebula
Double star
Variable star

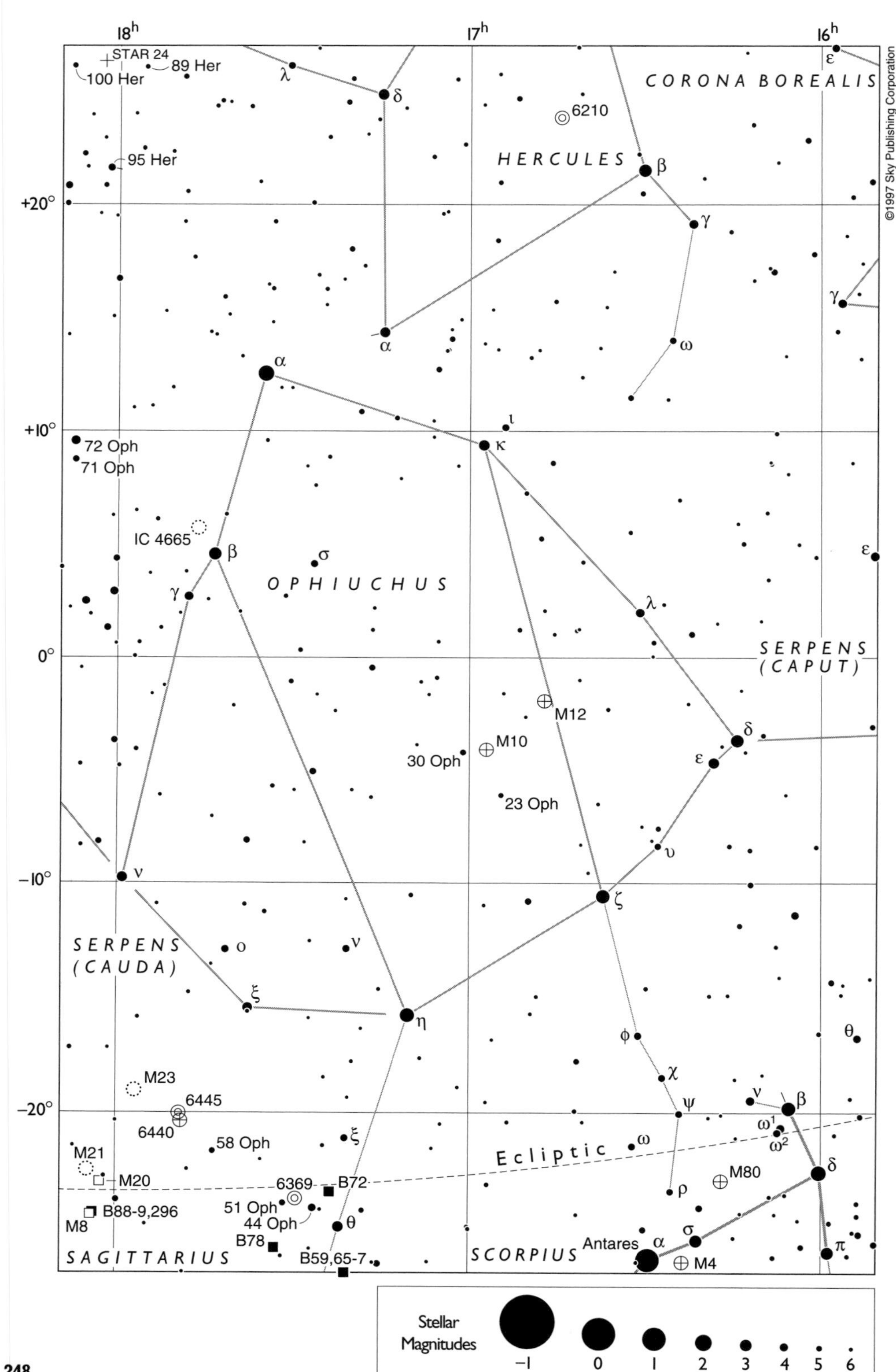
18h
17h
16h
+20°
+10°
0°
–10°
–20°
STAR 24
89 Her
100 Her
95 Her
CORONA BOREALIS
HERCULES
6210
72 Oph
71 Oph
IC 4665
OPHIUCHUS
SERPENS (CAPUT)
M12
M10
30 Oph
23 Oph
SERPENS (CAUDA)
M23
6445
6440
58 Oph
M21
M20
M8
B88-9,296
6369
B72
51 Oph
44 Oph
B78
B59,65-7
SAGITTARIUS
SCORPIUS
Antares
M80
M4
Ecliptic
Stellar Magnitudes
–1
0
1
2
3
4
5
6
©1997 Sky Publishing Corporation

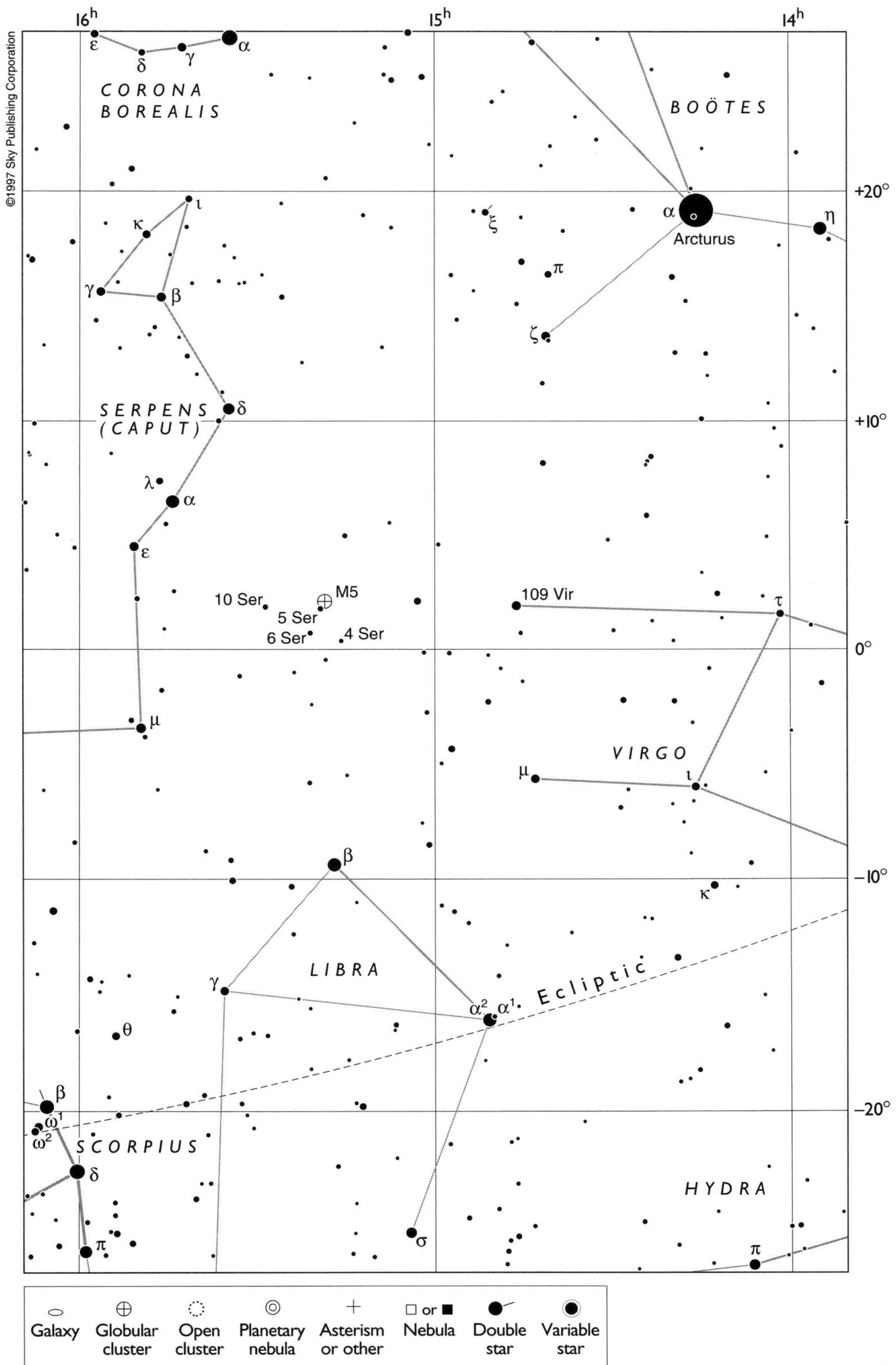
16h
15h
14h
+20°
+10°
0°
–10°
–20°
CORONA BOREALIS
BOÖTES
Arcturus
SERPENS (CAPUT)
10 Ser
M5
5 Ser
6 Ser
4 Ser
109 Vir
VIRGO
LIBRA
Ecliptic
SCORPIUS
HYDRA
Galaxy
Globular cluster
Open cluster
Planetary nebula
Asterism or other
□ or ■ Nebula
Double star
Variable star

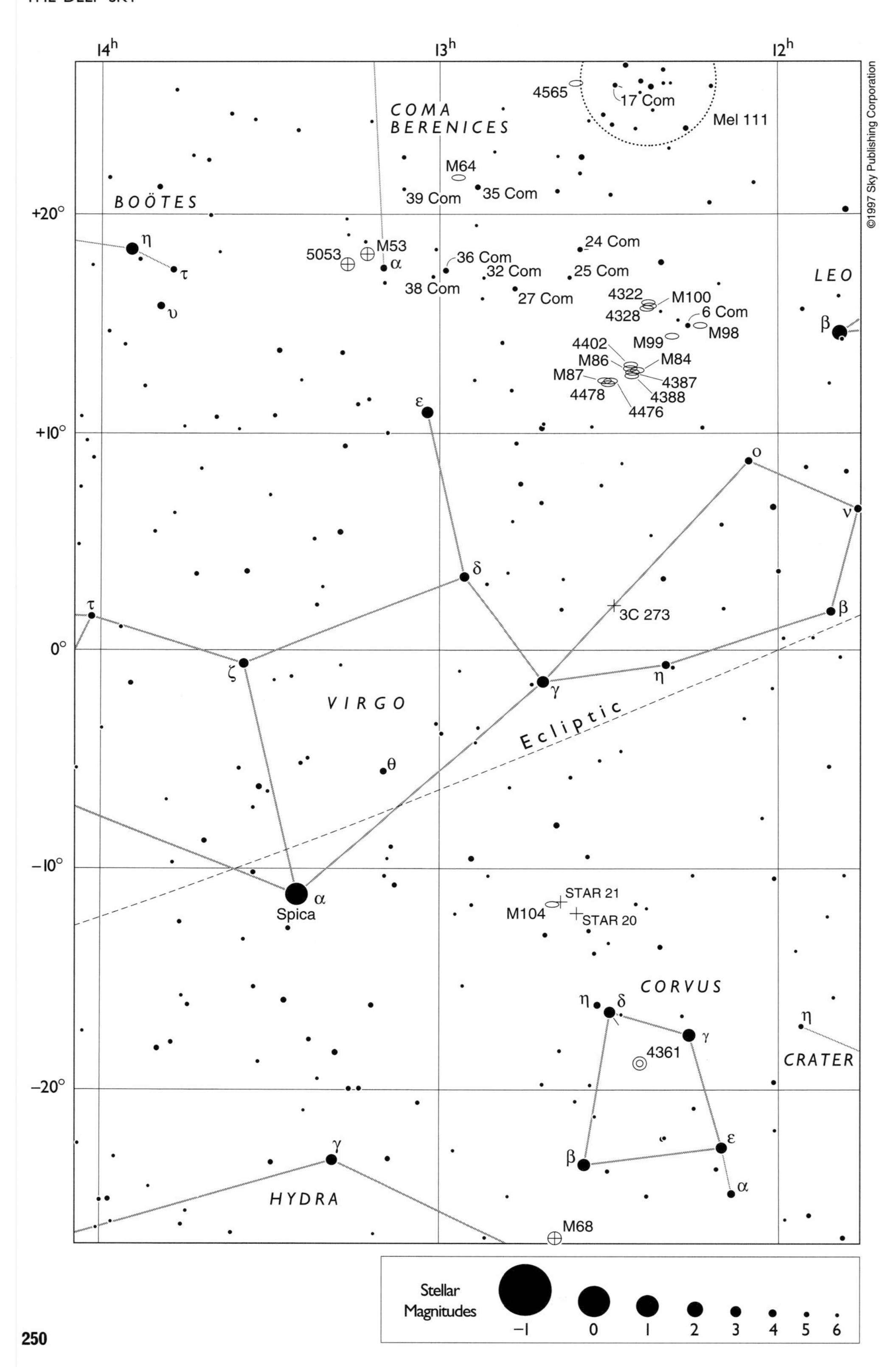

14h
13h
12h
+20°
+10°
0°
−10°
−20°
COMA BERENICES
BOÖTES
LEO
VIRGO
CORVUS
CRATER
HYDRA
Ecliptic
4565
17 Com
Mel 111
M64
39 Com
35 Com
5053
M53
24 Com
36 Com
32 Com
25 Com
38 Com
27 Com
4322
M100
4328
6 Com
M98
M99
4402
M86
M84
M87
4387
4478
4388
4476
3C 273
Spica
STAR 21
M104
STAR 20
4361
M68
Stellar Magnitudes
−1
0
1
2
3
4
5
6

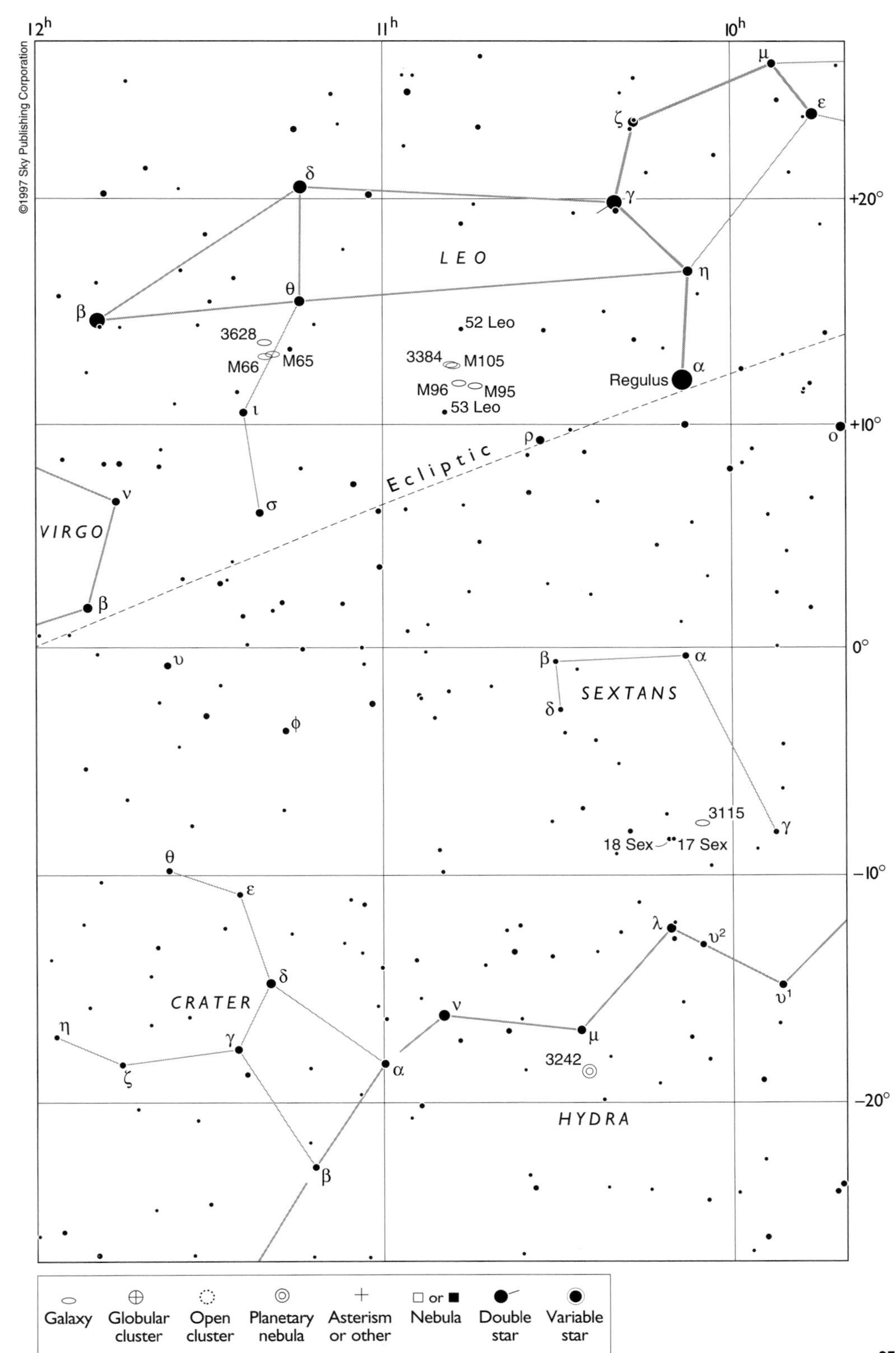
©1997 Sky Publishing Corporation
12h
11h
10h
+20°
+10°
0°
–10°
–20°
LEO
VIRGO
SEXTANS
CRATER
HYDRA
Ecliptic
Regulus
52 Leo
53 Leo
3628
M66
M65
3384
M105
M96
M95
3115
18 Sex
17 Sex
3242
Galaxy
Globular cluster
Open cluster
Planetary nebula
Asterism or other
Nebula
Double star
Variable star

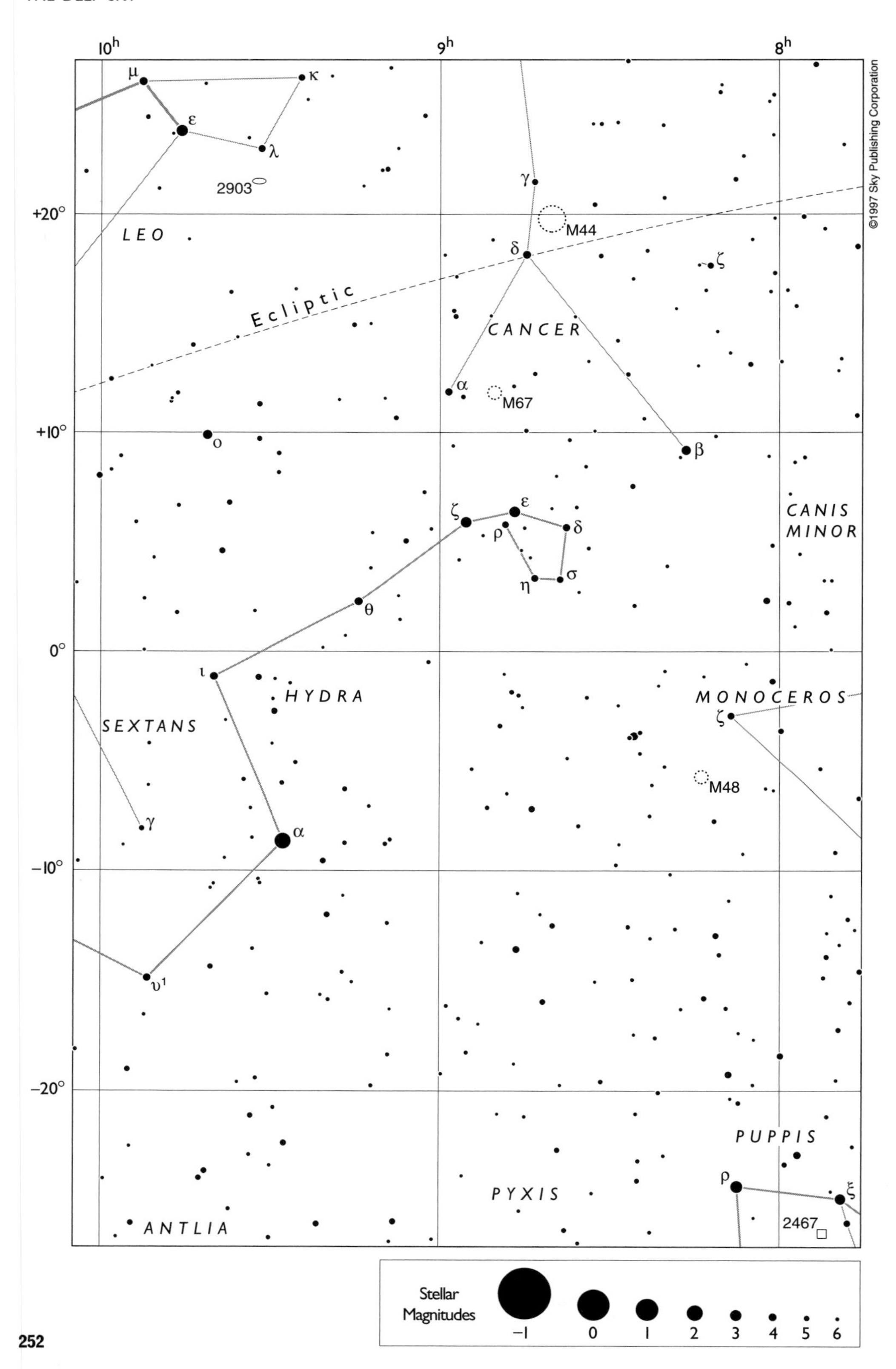

10h
9h
8h
+20°
+10°
0°
−10°
−20°
μ
κ
ε
λ
2903
LEO
γ
M44
δ
ζ
Ecliptic
CANCER
α
M67
o
β
ζ
ε
δ
ρ
η
σ
θ
CANIS
MINOR
ι
HYDRA
SEXTANS
MONOCEROS
ζ
M48
γ
α
υ1
PUPPIS
PYXIS
ρ
ξ
2467
ANTLIA
Stellar
Magnitudes
−1
0
1
2
3
4
5
6

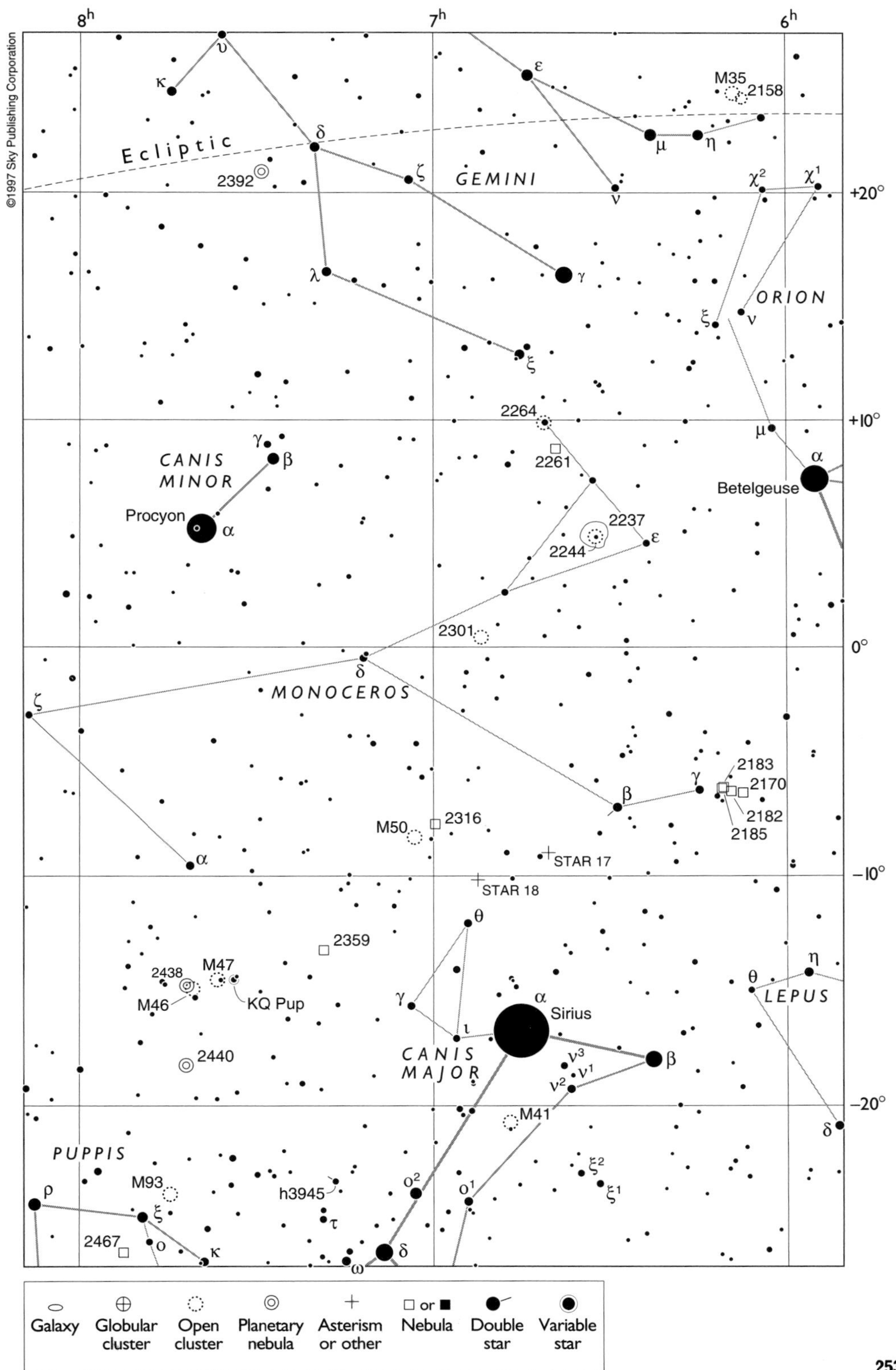

8h
7h
6h
+20°
+10°
0°
−10°
−20°
Ecliptic
GEMINI
ORION
CANIS MINOR
Procyon
Betelgeuse
MONOCEROS
CANIS MAJOR
Sirius
LEPUS
PUPPIS
M35
2158
2392
2264
2261
2237
2244
2301
2183
2170
2182
2185
M50
2316
STAR 17
STAR 18
2359
2438
M47
M46
KQ Pup
2440
M41
M93
h3945
2467
Galaxy
Globular cluster
Open cluster
Planetary nebula
Asterism or other
□ or ■ Nebula
Double star
Variable star

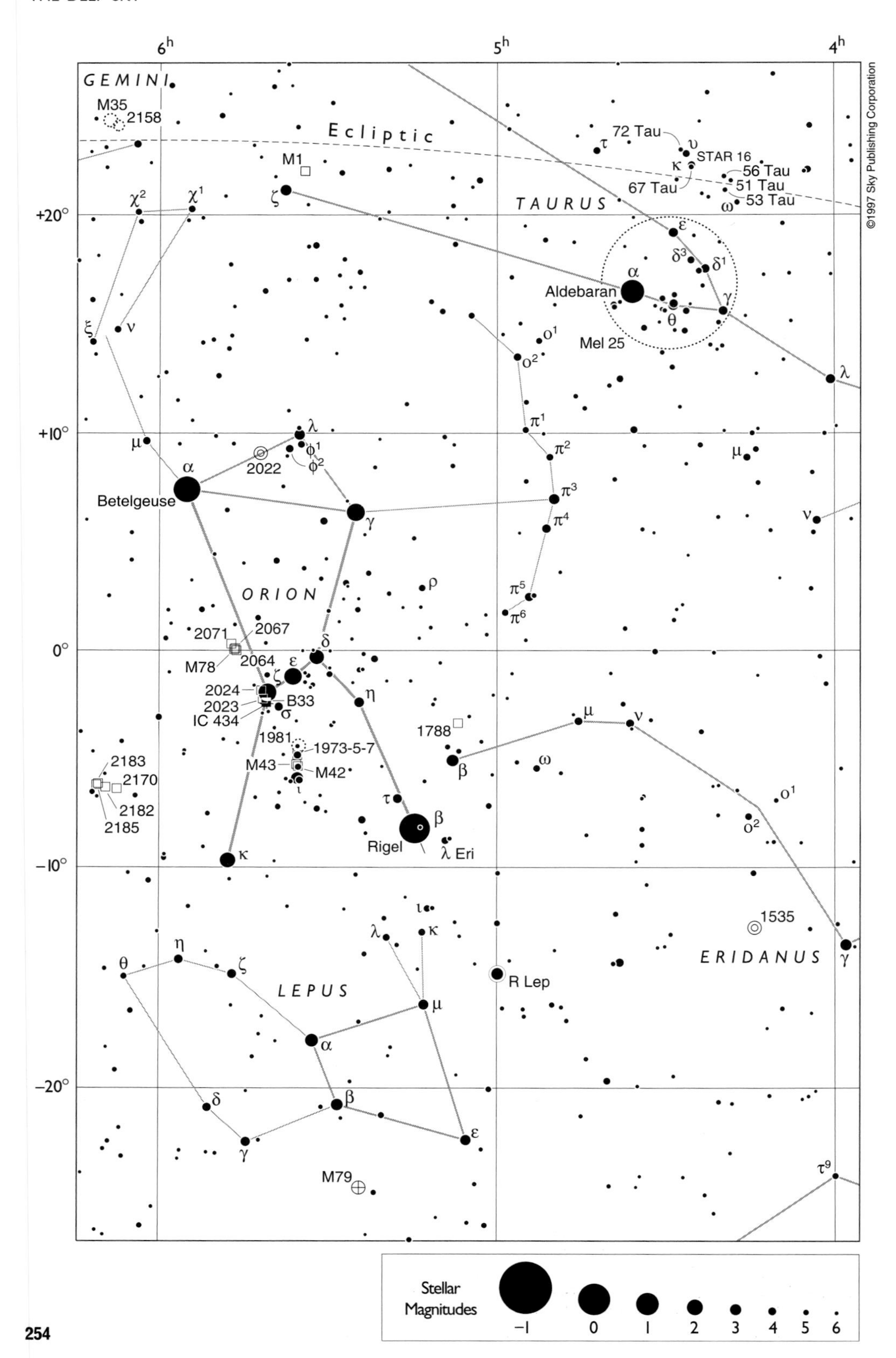

6h
5h
4h
+20°
+10°
0°
−10°
−20°
GEMINI
M35
2158
Ecliptic
M1
TAURUS
72 Tau
STAR 16
56 Tau
51 Tau
53 Tau
67 Tau
Aldebaran
Mel 25
Betelgeuse
2022
ORION
2071
2067
M78
2064
2024
2023
B33
IC 434
1981
1973-5-7
M43
M42
1788
2183
2170
2182
2185
Rigel
λ Eri
1535
ERIDANUS
LEPUS
R Lep
M79
Stellar Magnitudes
−1
0
1
2
3
4
5
6
©1997 Sky Publishing Corporation

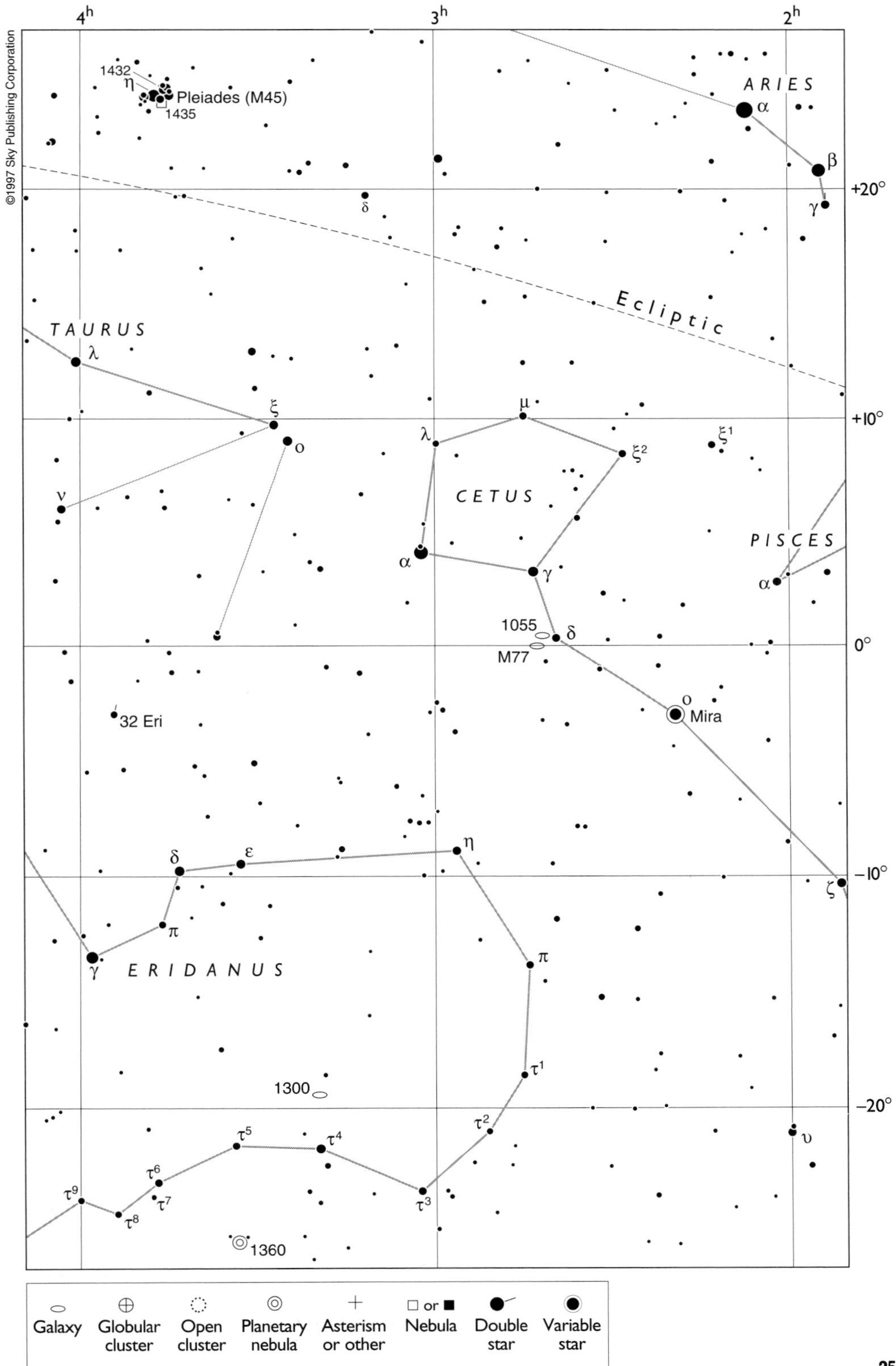

Galaxy	Globular cluster	Open cluster	Planetary nebula	Asterism or other	□ or ■ Nebula	Double star	Variable star

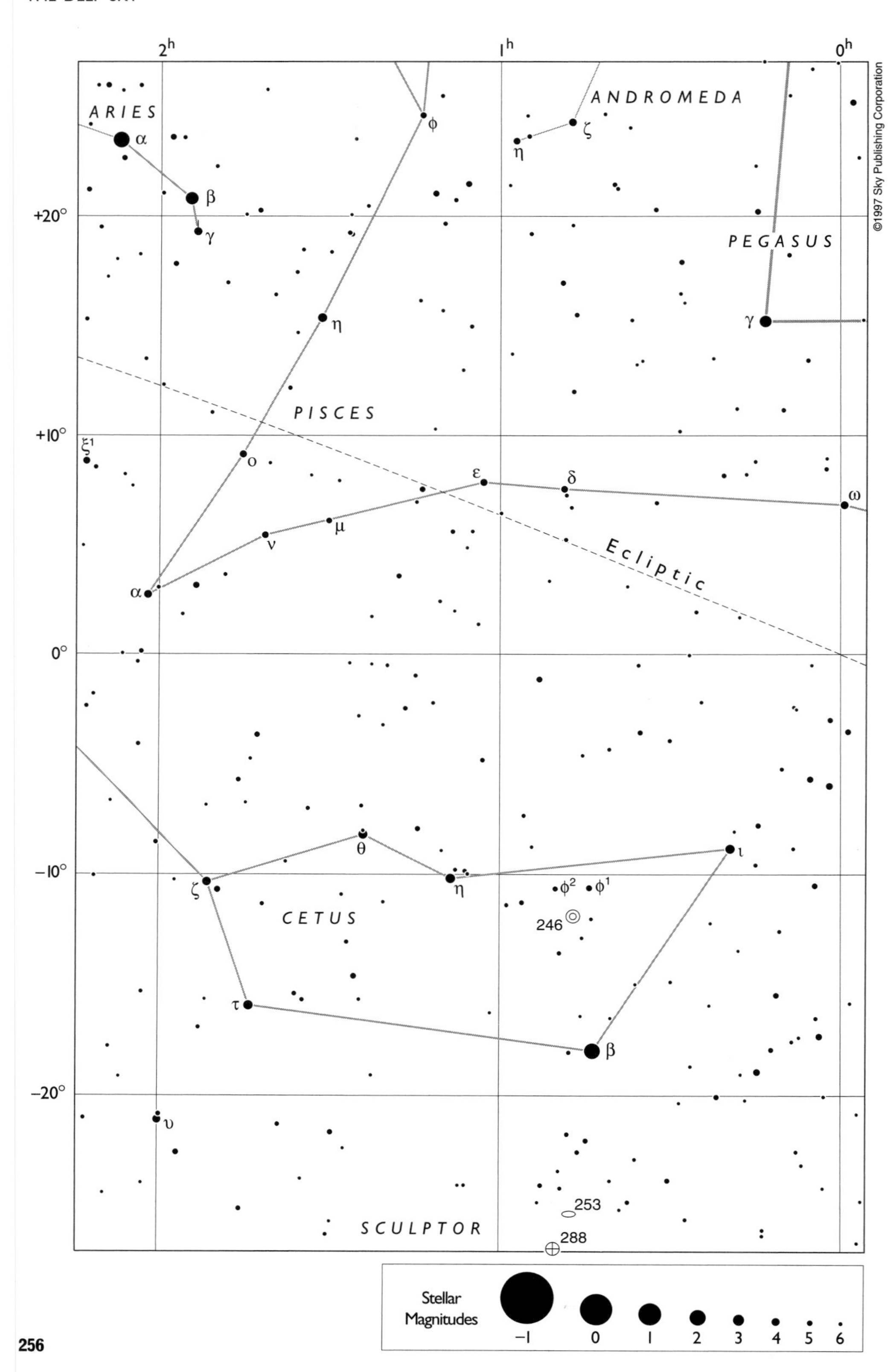
2h
1h
0h
ARIES
ANDROMEDA
PEGASUS
PISCES
Ecliptic
CETUS
SCULPTOR
+20°
+10°
0°
−10°
−20°
246
253
288
Stellar Magnitudes
−1
0
1
2
3
4
5
6
©1997 Sky Publishing Corporation

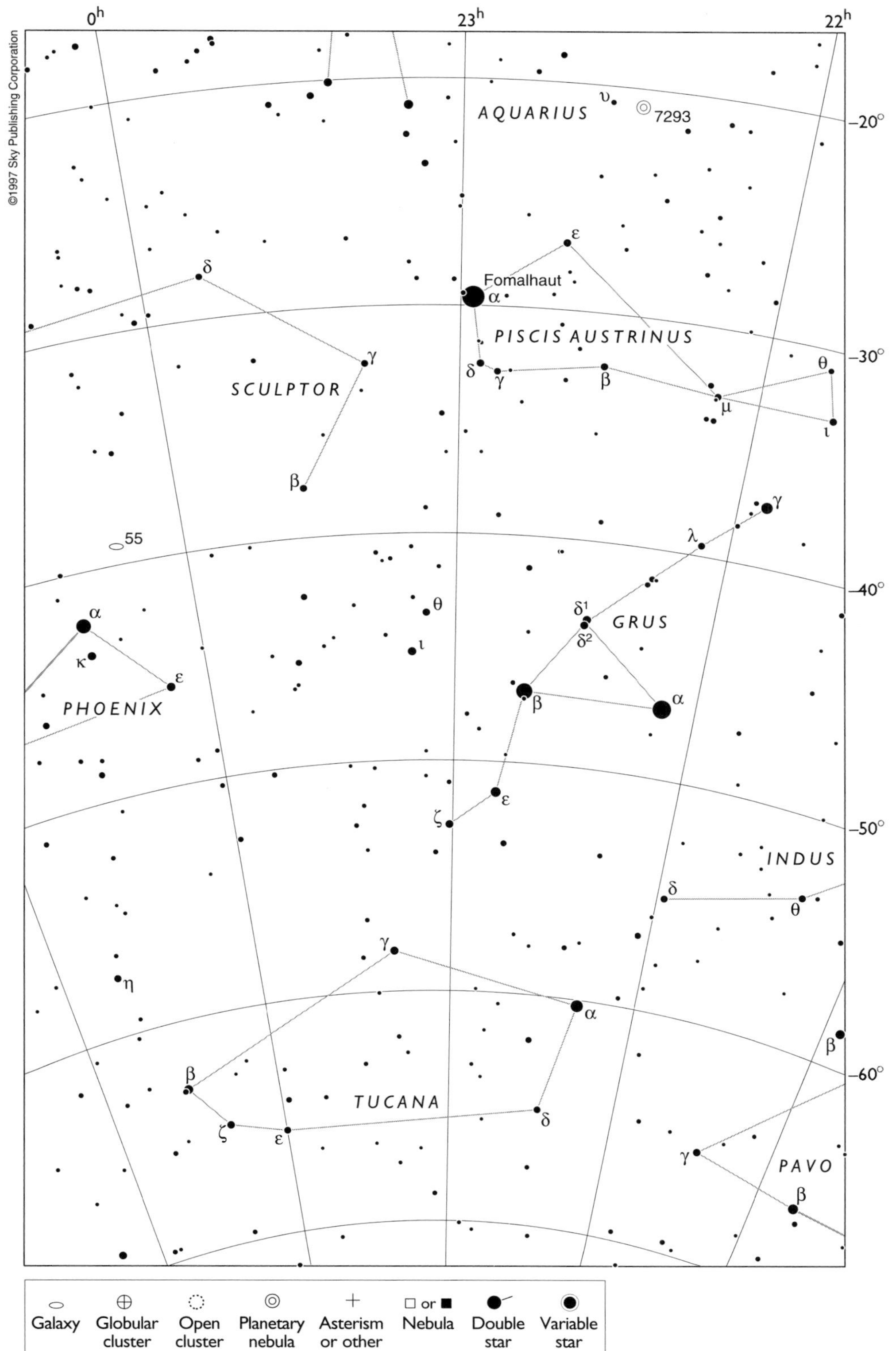

0h
23h
22h
AQUARIUS
υ
7293
−20°
ε
δ
Fomalhaut
α
PISCIS AUSTRINUS
−30°
γ
SCULPTOR
δ
γ
β
θ
μ
ι
β
γ
55
λ
−40°
α
θ
δ1
GRUS
δ2
ι
κ
ε
PHOENIX
β
α
ε
ζ
−50°
INDUS
δ
θ
γ
η
α
β
−60°
β
TUCANA
ζ
ε
δ
γ
PAVO
β
Galaxy
Globular cluster
Open cluster
Planetary nebula
Asterism or other
□ or ■ Nebula
Double star
Variable star

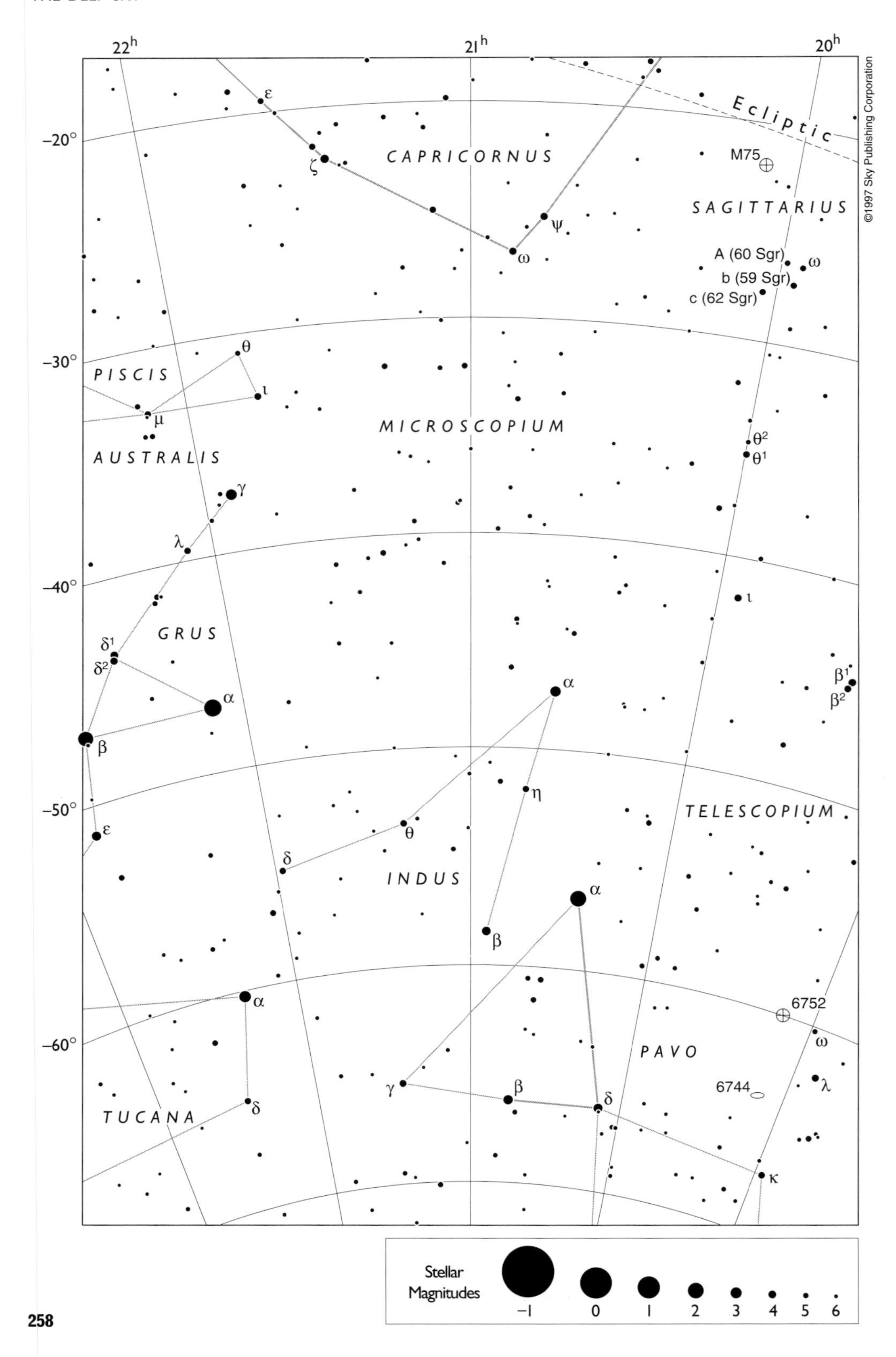
22h
21h
20h
−20°
−30°
−40°
−50°
−60°
Ecliptic
CAPRICORNUS
M75
SAGITTARIUS
A (60 Sgr)
b (59 Sgr)
c (62 Sgr)
PISCIS
AUSTRALIS
MICROSCOPIUM
GRUS
TELESCOPIUM
INDUS
PAVO
6752
6744
TUCANA
©1997 Sky Publishing Corporation
Stellar Magnitudes
−1
0
1
2
3
4
5
6

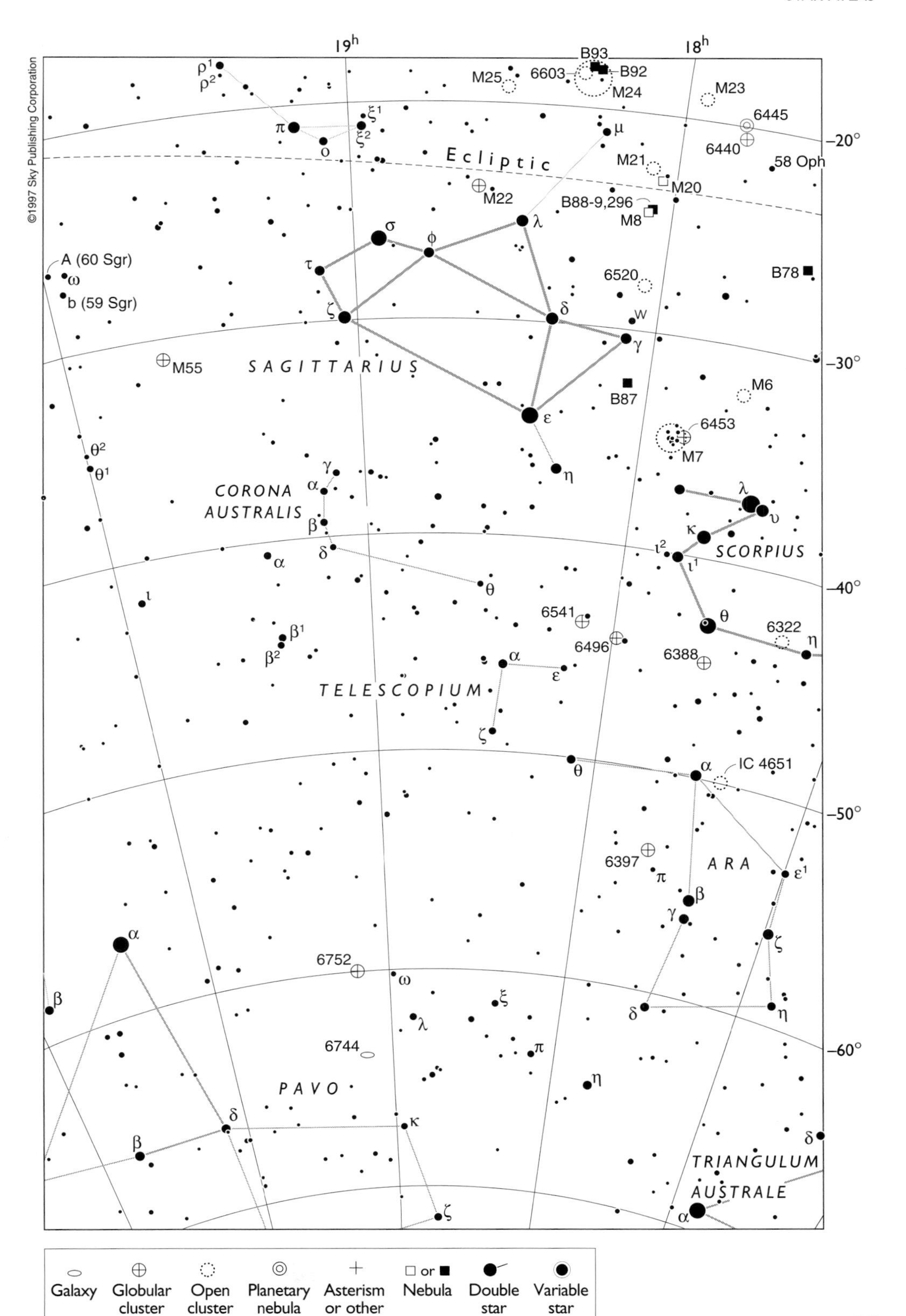

19h
18h
B93
B92
6603
M25
M24
M23
6445
6440
58 Oph
M21
M20
M22
B88-9,296
M8
Ecliptic
A (60 Sgr)
b (59 Sgr)
6520
B78
SAGITTARIUS
M55
B87
M6
6453
M7
CORONA
AUSTRALIS
SCORPIUS
6541
6496
6322
6388
TELESCOPIUM
IC 4651
6397
ARA
6752
6744
PAVO
TRIANGULUM
AUSTRALE
–20°
–30°
–40°
–50°
–60°
Galaxy
Globular cluster
Open cluster
Planetary nebula
Asterism or other
□ or ■ Nebula
Double star
Variable star

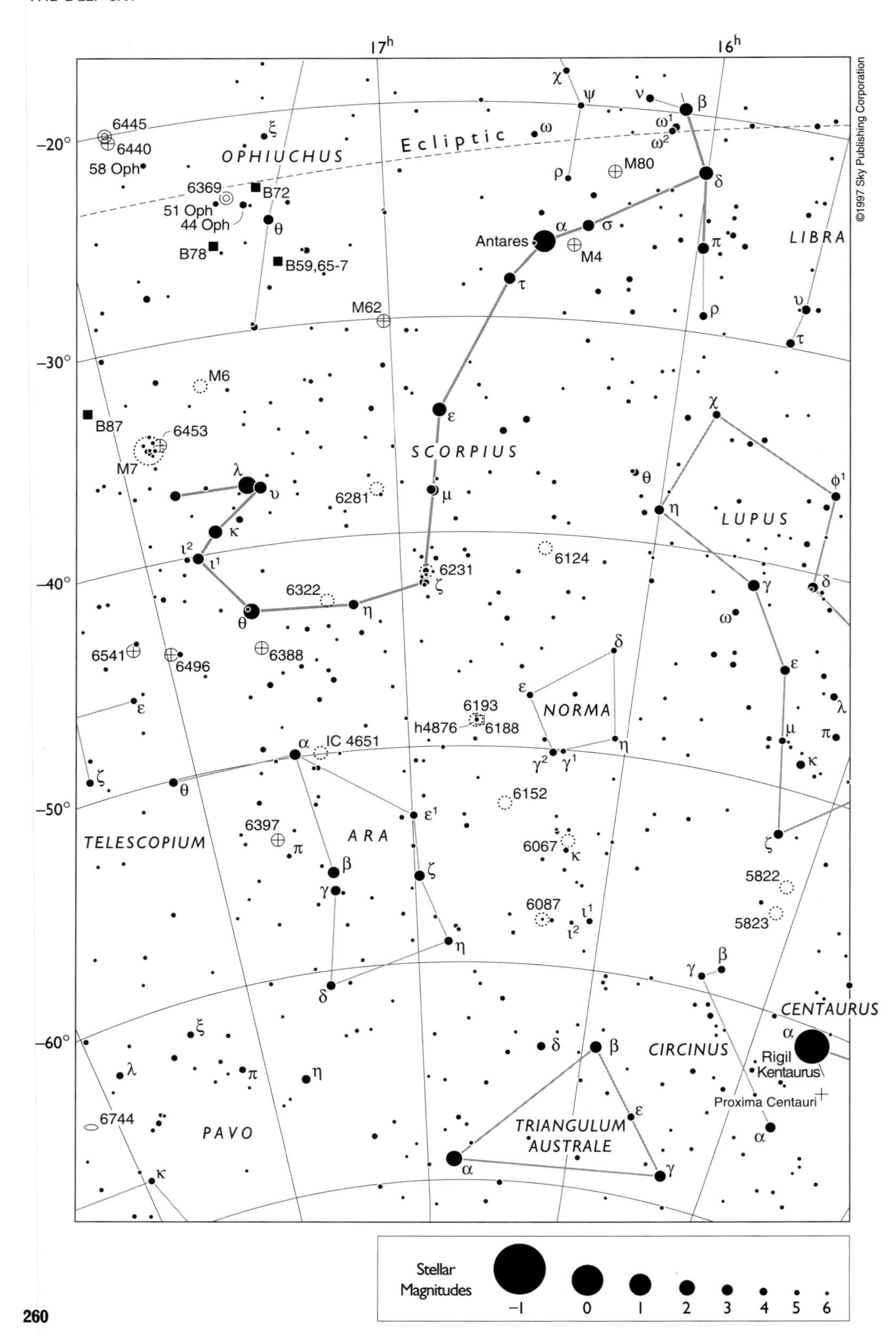

17h
16h
OPHIUCHUS
Ecliptic
SCORPIUS
LIBRA
LUPUS
NORMA
ARA
TELESCOPIUM
CENTAURUS
CIRCINUS
TRIANGULUM AUSTRALE
PAVO
Antares
Rigil Kentaurus
Proxima Centauri
M80
M4
M62
M6
M7
6445
6440
58 Oph
6369
51 Oph
44 Oph
B72
B78
B59,65-7
B87
6453
6281
6124
6231
6322
6541
6496
6388
6193
h4876
6188
IC 4651
6152
6397
6067
6087
5822
5823
6744
−20°
−30°
−40°
−50°
−60°
Stellar Magnitudes
−1
0
1
2
3
4
5
6

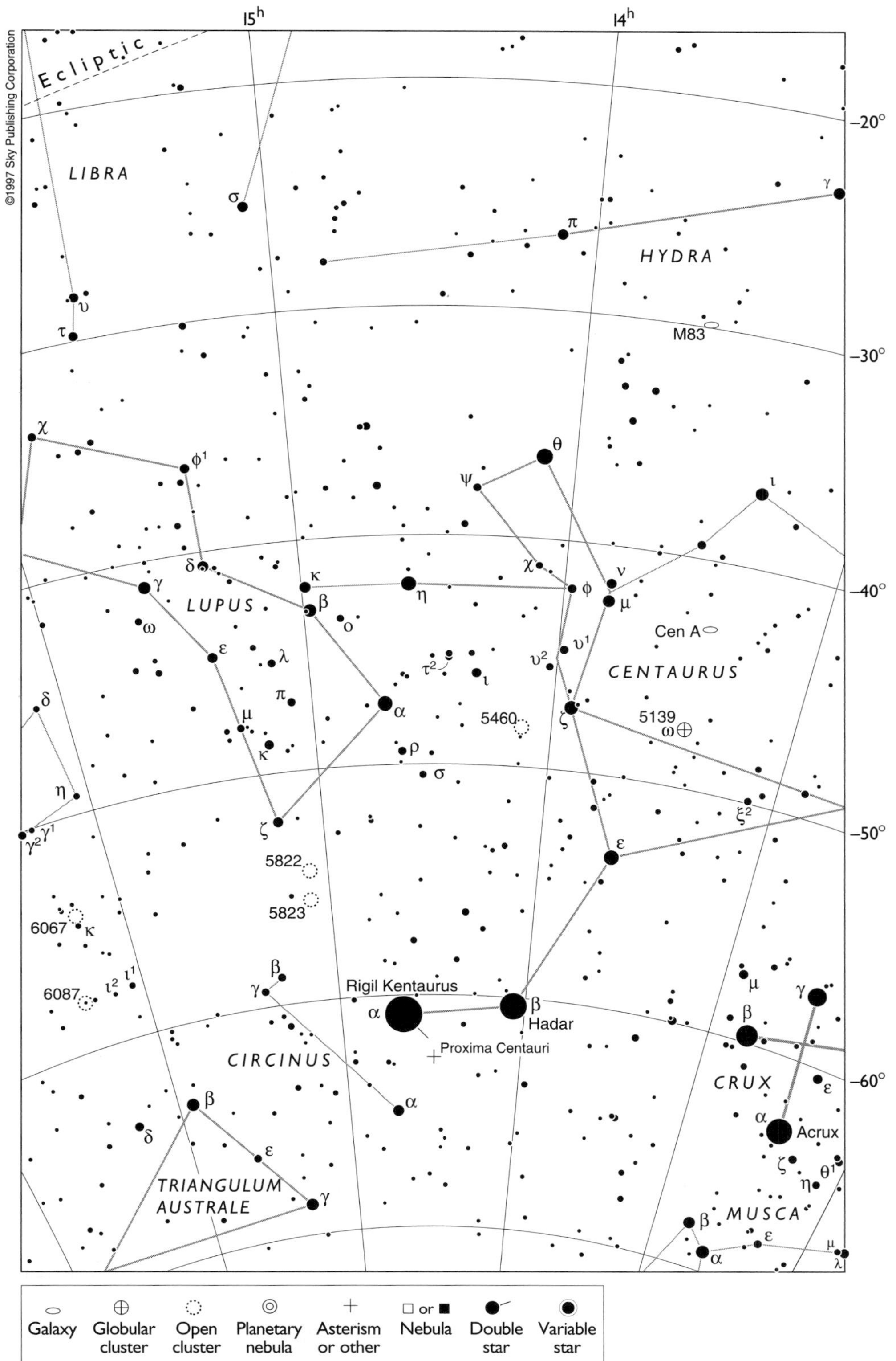

15h
14h
Ecliptic
-20°
-30°
-40°
-50°
-60°
LIBRA
HYDRA
M83
LUPUS
CENTAURUS
Cen A
5460
5139
5822
5823
6067
6087
Rigil Kentaurus
Hadar
Proxima Centauri
CIRCINUS
CRUX
Acrux
TRIANGULUM
AUSTRALE
MUSCA
Galaxy
Globular cluster
Open cluster
Planetary nebula
Asterism or other
□ or ■ Nebula
Double star
Variable star

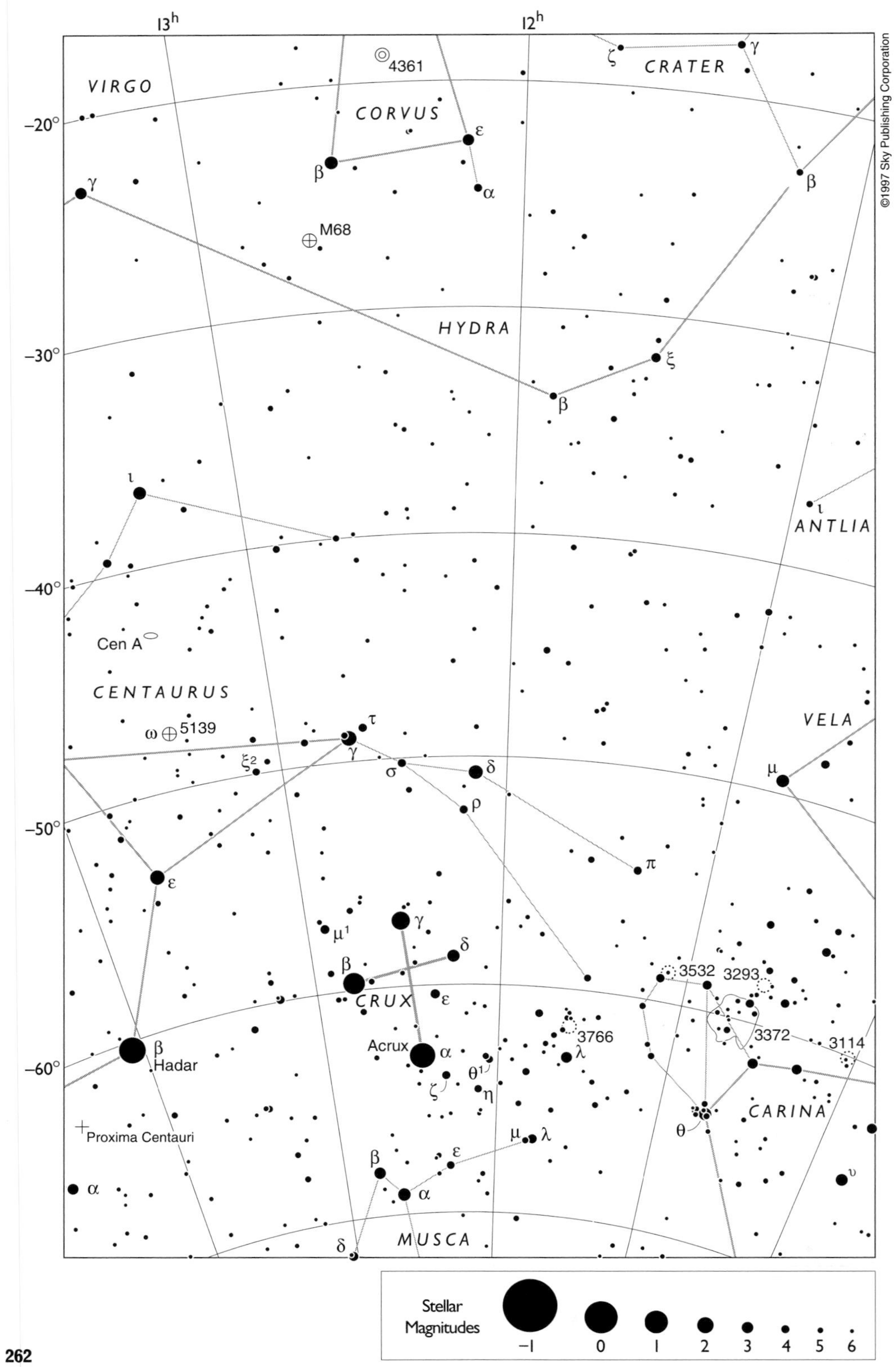

13h
12h
VIRGO
CORVUS
CRATER
HYDRA
ANTLIA
CENTAURUS
VELA
CRUX
CARINA
MUSCA
4361
M68
Cen A
ω 5139
Acrux
Hadar
Proxima Centauri
3532
3293
3766
3372
3114
-20°
-30°
-40°
-50°
-60°
©1997 Sky Publishing Corporation
Stellar Magnitudes
-1
0
1
2
3
4
5
6

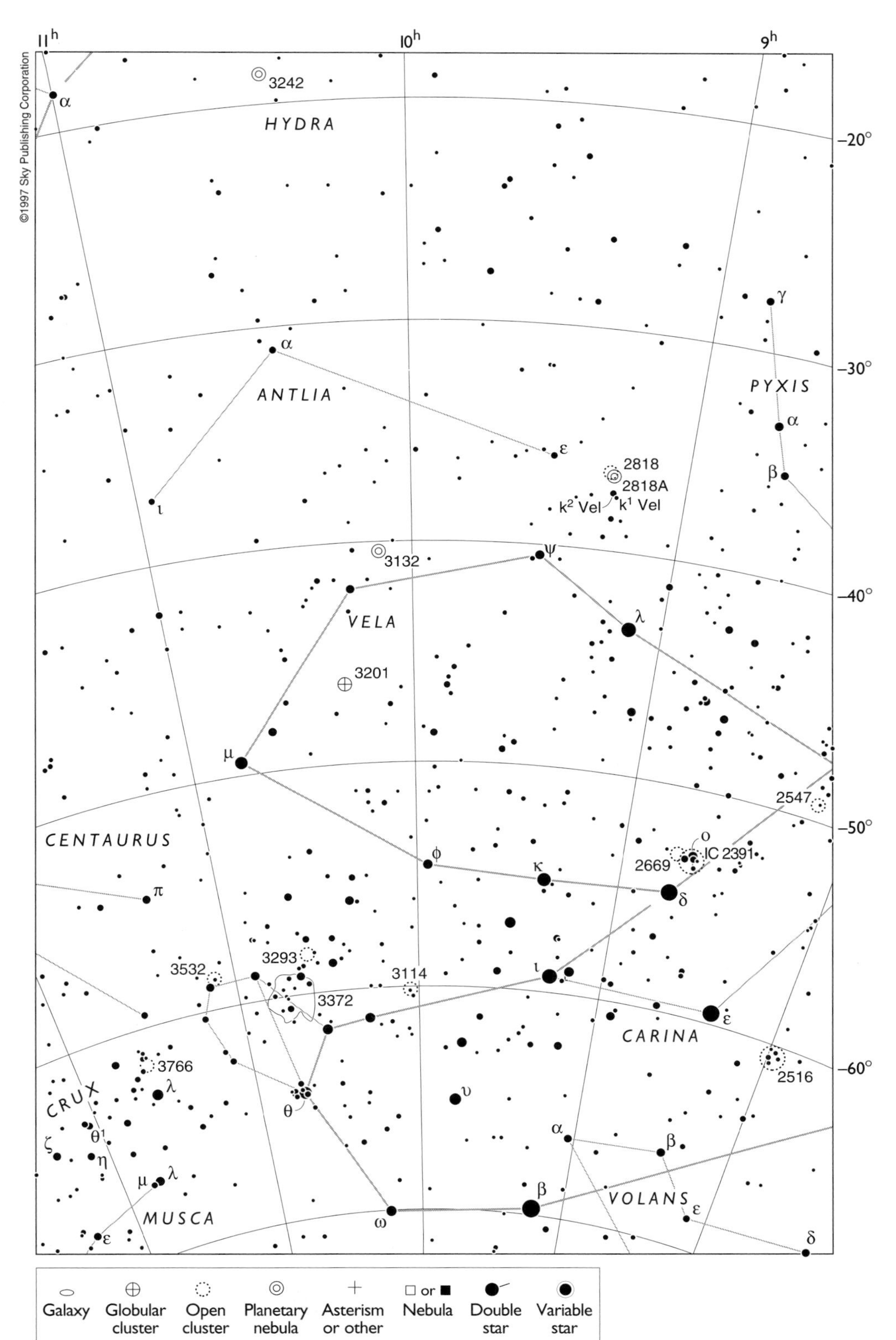

11h
10h
9h
-20°
-30°
-40°
-50°
-60°
HYDRA
ANTLIA
PYXIS
VELA
CENTAURUS
CARINA
CRUX
MUSCA
VOLANS
3242
2818
2818A
k² Vel
k¹ Vel
3132
3201
2547
IC 2391
2669
3293
3532
3114
3372
3766
2516
Galaxy
Globular cluster
Open cluster
Planetary nebula
Asterism or other
□ or ■ Nebula
Double star
Variable star

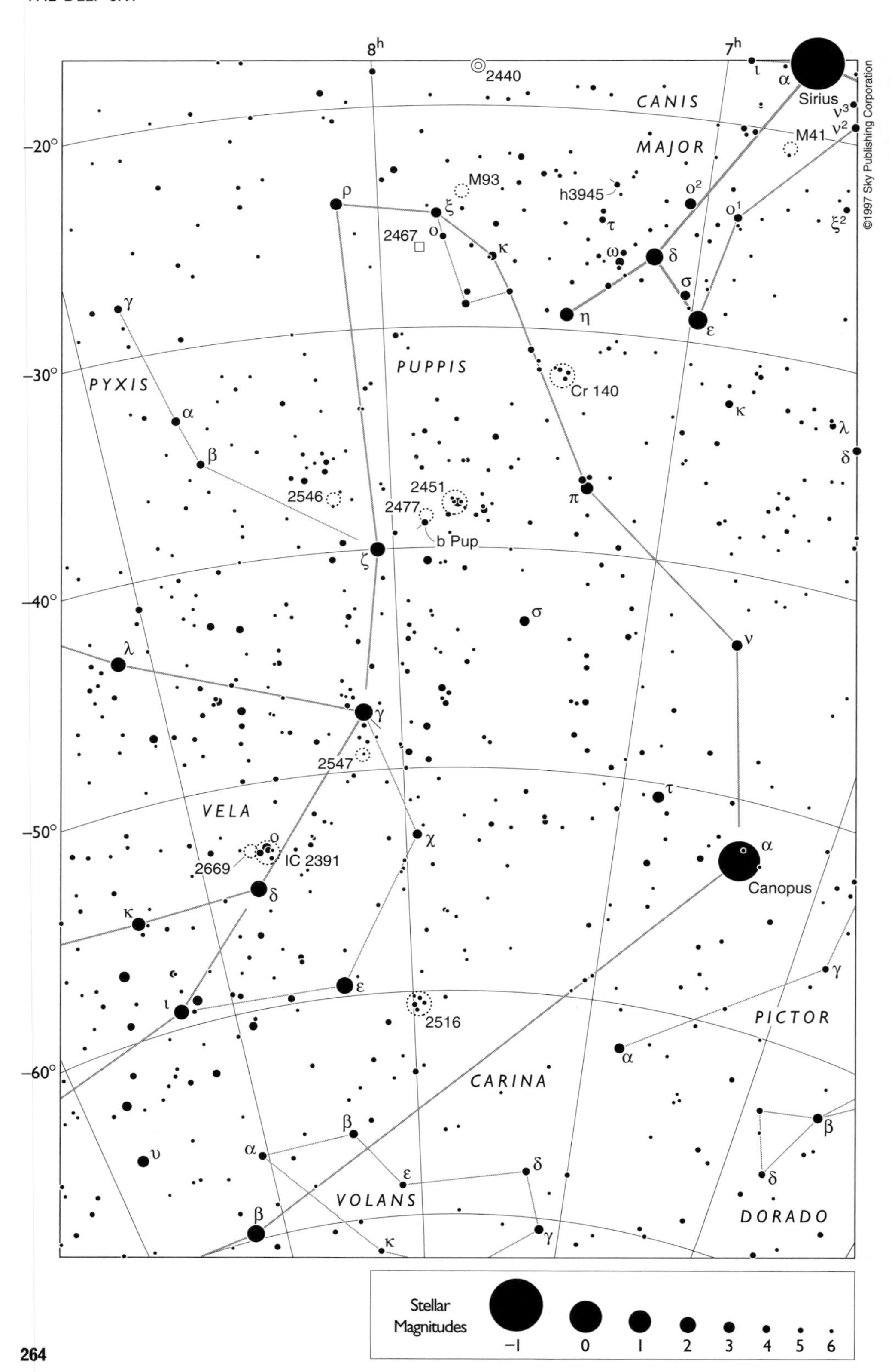
8h
7h
2440
CANIS
MAJOR
Sirius
M41
M93
h3945
2467
PUPPIS
PYXIS
Cr 140
2546
2451
2477
b Pup
2547
VELA
2669
IC 2391
Canopus
2516
PICTOR
CARINA
VOLANS
DORADO
-20°
-30°
-40°
-50°
-60°
Stellar Magnitudes
-1
0
1
2
3
4
5
6
©1997 Sky Publishing Corporation

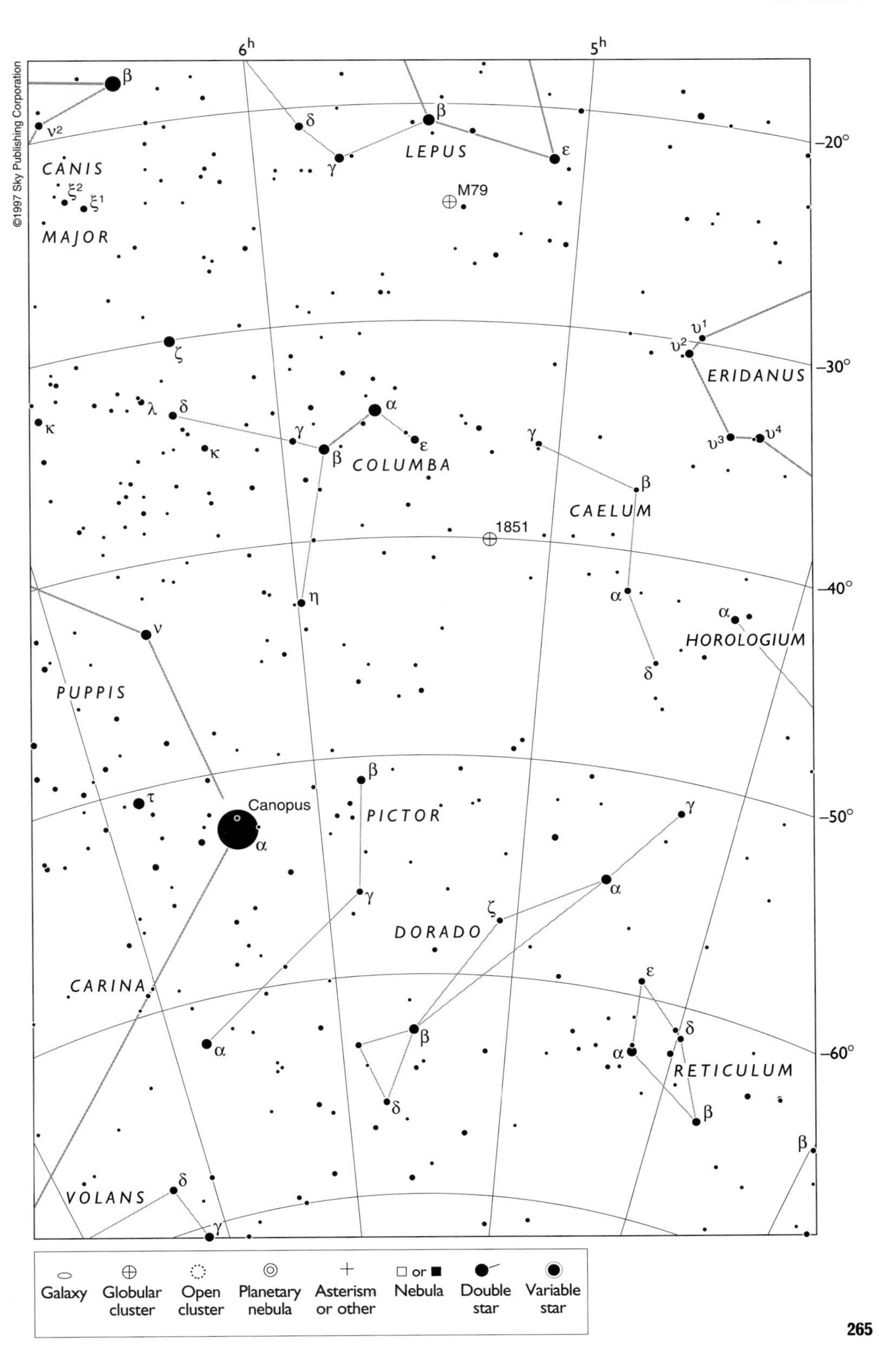

6h
5h
-20°
-30°
-40°
-50°
-60°
CANIS MAJOR
LEPUS
M79
ERIDANUS
COLUMBA
CAELUM
1851
HOROLOGIUM
PUPPIS
Canopus
PICTOR
DORADO
CARINA
RETICULUM
VOLANS
Galaxy
Globular cluster
Open cluster
Planetary nebula
Asterism or other
□ or ■ Nebula
Double star
Variable star

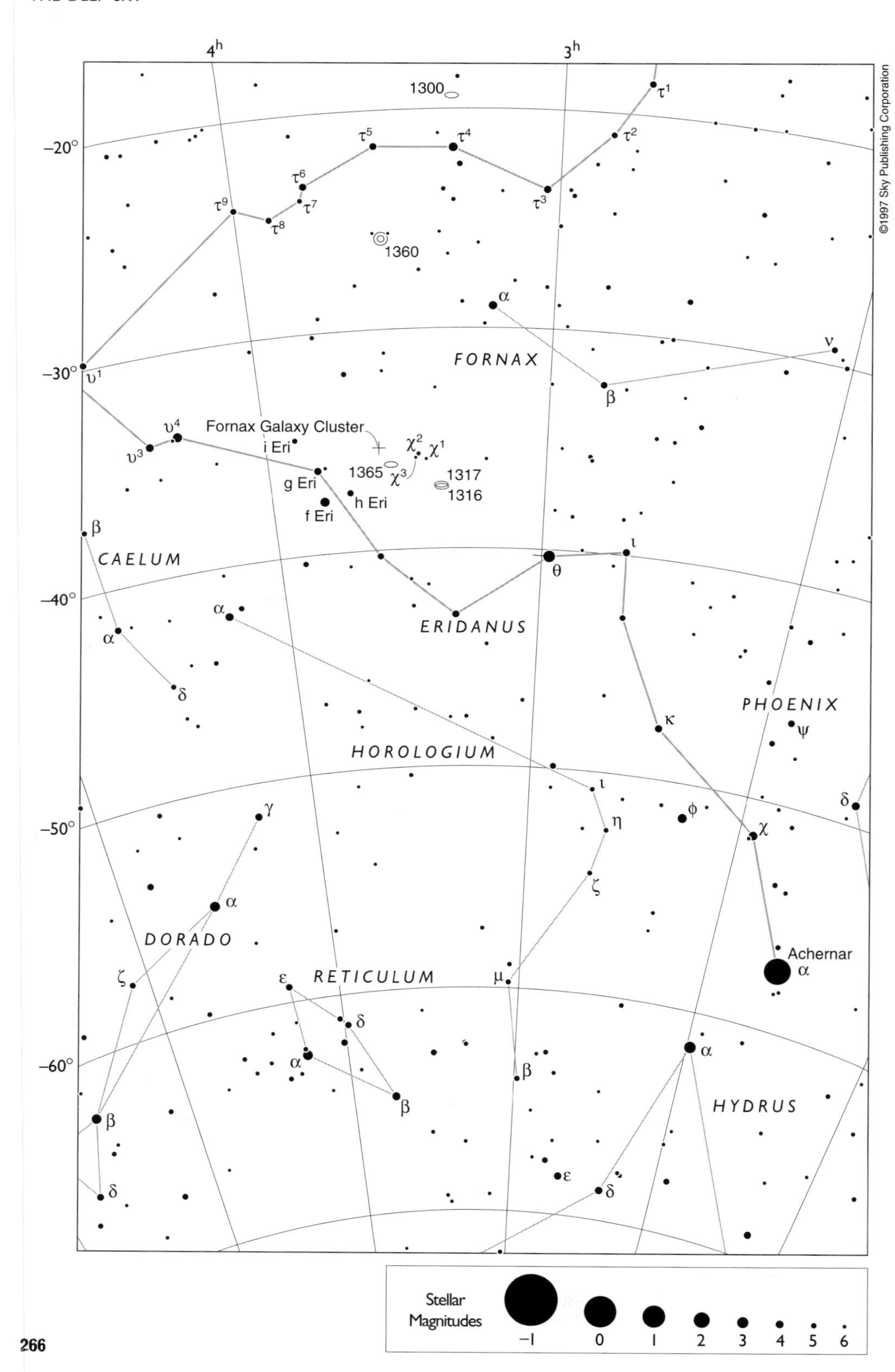

4h
3h
−20°
−30°
−40°
−50°
−60°
1300
1360
FORNAX
Fornax Galaxy Cluster
i Eri
g Eri
h Eri
f Eri
1365
1317
1316
CAELUM
ERIDANUS
HOROLOGIUM
PHOENIX
DORADO
RETICULUM
HYDRUS
Achernar
Stellar Magnitudes
−1
0
1
2
3
4
5
6

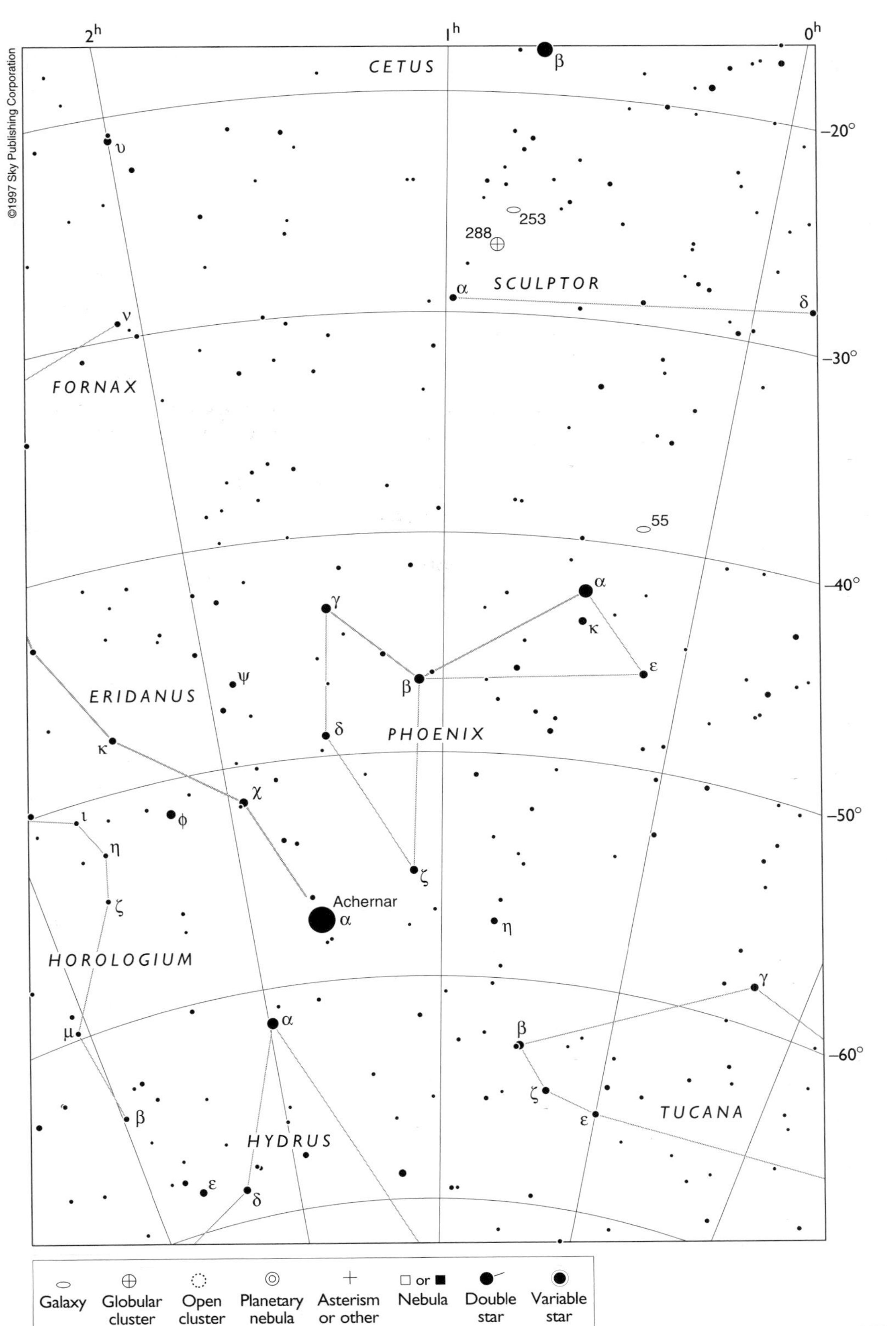
2h
1h
0h
−20°
−30°
−40°
−50°
−60°
CETUS
SCULPTOR
FORNAX
ERIDANUS
PHOENIX
HOROLOGIUM
HYDRUS
TUCANA
Achernar
253
288
55
Galaxy
Globular cluster
Open cluster
Planetary nebula
Asterism or other
□ or ■ Nebula
Double star
Variable star

Index

Page numbers in boldface type refer to figures.